The Derby philosophers

MANCHESTER
1824

Manchester University Press

The Derby philosophers

Science and culture in British urban society, 1700–1850

PAUL A. ELLIOTT

Manchester University Press

Manchester and New York

distributed exclusively in the USA by Palgrave Macmillan

Published by Manchester University Press
Oxford Road, Manchester M13 9NR, UK
and Room 400, 175 Fifth Avenue, New York, NY 10010, USA
www.manchesteruniversitypress.co.uk

Distributed exclusively in the USA by
Palgrave Macmillan, 175 Fifth Avenue, New York,
NY 10010, USA

Distributed exclusively in Canada by
UBC Press, University of British Columbia, 2029 West Mall,
Vancouver, BC, Canada V6T 1Z2

British Library Cataloguing-in-Publication Data
A catalogue record for this book is available from the British Library

Library of Congress Cataloging-in-Publication Data applied for

ISBN   978 0 7190 7922 1 *hardback*

First published 2009

18 17 16 15 14 13 12 11 10 09   10 9 8 7 6 5 4 3 2 1

Typeset in Minion by
Koinonia, Manchester
Printed in Great Britain
by CPI Antony Rowe, Chippenham, Wiltshire

Contents

List of figures

Figures 1, 2, 4, 5, 10, 11, 12, 13, 14, 17, 27, 31, 32 and 35 are from the collections at Derby Local Studies Library.

Preface

This book began life as a PhD thesis at Leicester University during the 1990s, although the final result is very different from how that appeared. I would like to thank the original supervisor Bill Brock for taking on the project, Alex Keller and Roey Sweet for helping to renew my enthusiasm in it, and Ian Inkster for trenchant criticism as the external examiner. More recently, working as a research fellow in cultural and historical geography at Nottingham University, I have received support and encouragement from Stephen Daniels, Charles Watkins and other members of the School of Geography. I am very grateful to Emma Brennan and the staff at Manchester University Press for publishing the book, to Dr Susan Milligan for copy-editing the typescript so efficiently, and to anonymous readers who offered criticisms and suggestions. Of course, in a work of this kind straddling so many different fields, as the length and content of the bibliography demonstrate, there are bound to be lots of debts to other scholars with greater expertise in each of the individual areas. I ask for forbearance on their part when they encounter any errors or misinterpretations that may well have arisen through this attempt at synthesis.

The book has greatly benefited from assistance provided by the staff of many libraries and record offices, especially Mandy Henchliffe, Linda Owen, Patricia Kenny and Mark Young and the other staff at Derby Local Studies Library (past and present), Margaret O'Sullivan and the staff at Derbyshire Record Office, Matlock and the staff of Derby Museum and Art Gallery. There are perhaps eight academic influences that most inspired and shaped the book: Desmond King-Hele's wonderful work on Erasmus Darwin, Albert Musson and Eric Robinson's analyses of British science and industrialisation, Benedict Nicholson's book on Joseph Wright, Paul Sturges' studies of Georgian Derby, Maxwell Craven's histories of Derby and John Whitehurst, Ian Inkster's work on scientific culture in provincial England, Robert Morris's studies of eighteenth- and nineteenth-century British urban culture and society, and John Brewer's *Pleasures of the Imagination* (1997), which appeared when I was finishing the original thesis.

On a more personal level, I am very grateful to my brother Christian and sisters Magdalene and Bernadette for help and support over the years. Special thanks are also offered to my friends Dan Bohane, Martha Clokie and Neil Brumfield for encouragement and forbearance during my time at Leicester and Nottingham. Unfortunately, my father died suddenly and unexpectedly as the book was nearing completion. However, I am sure that he would have enjoyed it had he lived and I would like to dedicate it to my parents Alan and Kathleen Elliott for their love and support over many years.

Introduction

The spirit and ambiguities of the European enlightenment's love for science are wonderfully captured in Joseph Wright's now iconic paintings 'The Orrery' and 'An Experiment on a Bird in the Air Pump', which have been employed in a host of contexts from book covers to television documentaries. Thanks principally to Desmond King-Hele, Erasmus Darwin, the founder and primary inspiration behind the Derby Philosophical Society (1783), is now widely recognised as one of the most important intellectuals of the European enlightenment. Studies of other members of Derby philosophical societies (at least five existed between 1760 and 1860) have also appeared. Likewise, the importance of Herbert Spencer as evolutionary theorist and educational thinker has long been acknowledged in the United States and Europe if not in Britain. Yet, despite work on Wright and two substantial studies of the Lunar Society, no general history of the Derby philosophers has ever been published.

Like Jenny Uglow's highly successful study of the Lunar Society (2002), this book focuses upon a group of Midland intellectuals who had a national and international impact. Like their Lunar and Manchester counterparts, the activities of the Derby *savants* demonstrate the degree to which much British cultural innovation during the eighteenth and nineteenth centuries occurred in the provinces. Besides Darwin and Spencer, members of Derby philosophical societies included Brooke Boothby, the poet, political writer and friend of Rousseau, the Rev. Thomas Gisborne the evangelical philosopher and poet, Robert Bage the novelist, Charles Sylvester the chemist and engineer, William George and his son Herbert Spencer, the internationally renowned evolutionary philosopher, and members of the Wedgwood and Strutt families, including the brothers William and Joseph Strutt. The scope of their ideas and activities are evident from a network of friends which included Benjamin Franklin, Matthew Boulton, Josiah Wedgwood, Richard Lovell and Maria Edgeworth, the Benthams, Robert Owen, John Claudius Loudon, the poet Thomas Moore, Thomas Stuart Traill and the geologist Robert Bakewell. Much more than a parochial history of one intellectual group or town, this book examines science, politics and culture during one of the most turbulent periods of British history, an age of political and industrial revolutions in which the Derby philosophers were closely involved. In so doing it seeks to answer a question posed by George Holyoake in 1904 when he asked how 'dreamy, stagnant, obfuscated Derby', a town which had burnt to death Joan Waste and 'procured a charter to enable it to expel anyone having a new idea', and which seemed so hostile to religious or political innovation, 'gave birth to Herbert Spencer, the greatest master of innovatory thought England has produced'.[1]

Inspired by science and through educational activities, publications and institutional formations, the Derby philosophers strove to promote social, political and urban improvements, many of which had a national and international impact. Natural philosophy was part of broader progressive enlightenment culture which inspired social, religious and political campaigns such as the activities of the 'friends of peace' during the Napoleonic wars, establishment of the Derby Mechanics' Institute (1825) and Derby Arboretum (1840). Important urban institutions as well as intended contributions to intellectual and political objectives, they served a much wider community in addition to scientific enthusiasts. Reflecting this, the book moves outwards chronologically from philosophical societies to public urban scientific institutions. It combines thematic and narrative structures, seeking to balance institutional and associational analysis with studies of 'significant' individuals. It was inspired by the conviction that in contrast to the cultural history of France, Germany, Scotland or the USA, there has been a tendency to drain English history of its intellectual content, even to deny that the English, from shopkeepers and gardeners to physicians and aristocrats, could be learned and motivated by intellectual concerns. The kernel of a 'cultural' approach to the history of science is the need to explore the relationships between science, scientific culture and non-scientific culture rather than 'internal' analyses of changing scientific ideas. Hence, attention is devoted to the economic growth of Derby, manifestations of the urban renaissance such as the physical transformation of the townscape, and the growth of associational and consumer culture.

The proliferation of local history publications and newspaper articles, the popularity of family history and the activities of numerous local study groups attest to the fascination with local history and the strength of the 'heritage industry', as do the recent designation of the Derwent Valley as a World Heritage Site and the restoration of the Derby Arboretum with National Lottery Heritage funding. Every age invents its own past to some degree but the historical journey provides meaning and forms of reconnection and escape in a modern consumption-obsessed world. Recent large-scale business investment into the city, exemplified by the creation of a new giant shopping complex and the major expansion of Derby University, have also encouraged renewed attention to local urban history, raising the question, for instance, of how modern commercial and historical urban environments can be accommodated together. On the whole, these developments help to encourage historical study, not least by generating interest in the past as well as economic prosperity, although heritage-agenda driven local history can degenerate into subjective, distorted promotional activities. The importance of local history to the heritage and tourist industries in British towns such as Derby means that it is all the more important to emphasise the constructedness and contingency of history.

This book tries to 'place' Derby scientific culture within its national and international contexts. Studies of individual towns provide valuable case studies which challenge blander national and international generalisations, enriching our understanding of phenomena such as enlightenment, forms of identity, class consciousness, class formation and industrialisation. However, they can collapse through weight of local detail, which may have little interest or appeal to readers from elsewhere, making comparisons with other urban scientific cultures desirable. It is now much better appreciated that British history, even so-called local history, cannot be studied without constant attention to international as well as national events. As we shall see, this is fully evident in the international networks and concerns of the Derby philosophers and the international status of key individuals such as Wright, Darwin, Herbert Spencer and the Strutts, and the impact of Derby institutions such as the philosophical societies, Derbyshire Infirmary and the Derby Arboretum.

1 Derby Silk Mill in 1794 drawn by Nixon and engraved by Walker and Storer.

Although the importance of science in British late-enlightenment society has been readily acknowledged, different aspects have received attention. It is agreed that the enlightenment aspiration for objectivity and universality and the relationships between natural philosophy and natural theology were important motivations, but the role of science in British industrialisation remains disputed. An older view, articulated between the 1940s and 1970s, took utilitarian pronouncements by Georgian promoters of natural philosophy at face value, accepting that modern science and industry had proceeded together and that aspects of natural philosophy, including practical mathematics, engineering and chemistry, had provided important benefits for British industry.[2] This view was challenged during the 1970s and 1980s by a series of studies that viewed scientific culture as an activity pursued by relatively marginalised social groups, such as dissenters and reformers, and as a vehicle for cultural expression and social advance. This model was supported by detailed prosopographical (collective biographical) work on philosophical communities of 'performers' and audiences at Manchester, Sheffield and other mostly northern and midland towns. The importance of utilitarian rhetoric was not denied, but it was contended that the supposed links between science, industry and technology were tenuous; after all, as was frequently pointed out, French science was second to none in the period yet the country experienced a political rather than industrial revolution. The emphasis upon social position also helped to shift attention from some famous philosophical associations such as those in Manchester, Derby and Birmingham towards the many other sites of scientific culture.[3]

Subsequent studies of British scientific culture have developed this cultural approach. An important example has been the way that natural philosophy and scientific culture have been associated with the economic and cultural 'urban renaissance' of the late seventeenth and eighteenth centuries. This was manifest most visibly in the construction of shops and commercial buildings, taverns, assembly rooms, fashionable neo-classical town houses and new public buildings such as town halls. In a world where the power of gentry and aristocracy had arguably been augmented by the relative peace of the post-Glorious Revolution world, increasingly prosperous traders, merchants and professionals aspired to gain acceptance in fashionable society in order to consolidate their status. Stimulated in various ways by natural theology, practical mechanics, natural history, geography and chorography, natural philosophy became an increasingly important component of practical and polite education. In practical terms this was manifest by the role of science in mechanics, navigation, geography and chorography. In terms of polite culture this was evident in natural theology and the celebrated status accorded to contested presentations of Newtonian natural philosophy. Natural philosophy had tended to embrace all the natural and experimental sciences besides mathematics, yet Newton's work on theoretical physics, gravitational theory and the experimental science of optics was held to signify a new marriage between mathematical knowledge, experimental physics and mechanics. This is readily apparent from textbooks produced by 'Newtonians' such as Jean Desaguliers, whose introduction to experimental philosophy was repeatedly reprinted.[4]

Focusing upon the mid-nineteenth century, a different approach to urban scientific culture emphasised the importance of civic science in studies of the British Association and the role of natural philosophy in fostering local, regional and national identities. This highlighted the special qualities of civic science exemplified by museums, libraries, botanical gardens and mechanics' institutes, which promoted a rational recreational message of co-operation and social inclusiveness rather than fostering separation and antagonism.[5] More recent work on the Georgian 'urban renaissance' and urban sociability by historians such as Peter Borsay, Peter Clark and Robert Morris, and upon enlightenment sociability by Roy Porter and others, has emphasised the role of scientific culture as a polite social, consumer and leisure activity. Although, of course, stimulated by metropolitan emulation, through facilitating the development of a middling sort of professionals, merchants and urban gentry this played an important role in the development of distinctive provincial, regional and urban identities. Renewed attention to the Habermasian concept of the public sphere has highlighted the centrality of British scientific culture in this process, helping to explain the importance and visibility of science in non-industrial and manufacturing centres. There has also been fresh emphasis upon the relationship between natural philosophy and industrialisation and the importance of rhetoric and political and social positioning in the development of audiences for science.[6]

In *Structural Transformation of the Public Sphere*, Habermas famously contended that during the seventeenth and eighteenth centuries, and beginning in England after the Glorious Revolution, a 'bourgeois' public sphere was created out of the relations between capitalism and the state. This was defined as the 'public' of private individuals who joined in debating issues bearing upon state authority, a new sociability based on rational-critical discourse and taking place in centres such as salons and coffee houses. The process led to an idea of society separate from the ruler (or the state) and of a private realm separate from the public. Though Habermas had little to say directly of science in his original politically oriented formulation, the conception of public sphere has been utilised by historians of science to posit the birth of a public science in England in the seventeenth and eighteenth centuries. In these terms, science became a culture of discourse which required social legitimisation through experimentation and public demonstration in addition to the philosophical text – where readers, listeners and observers could prove as important as authors and orators in the advancement of natural philosophy. Science came to be traded in a market economy, part of the commercialisation of leisure associated with the urban renaissance, bought by consumers as a component of polite education. It was also sold by experimental impresarios and popularisers as the key to greater mastery of nature and commercial success through utility, which, it was claimed, allowed practitioners to distinguish the true from the fool's gold. However, while historians continue to utilise the Habermasian concept of the public sphere, there has been criticism of the vagueness and abstract nature of the idea. Political historians, for instance, have contended that Georgian ideas concerning politics and the constitution were shaped by preconceived ideas such as notions of kingship, monarchical authority and precedent rather than on a hypothetical rational *tabula rasa*.[7]

The impact of Georgian scientific culture was evident in the domestic as well as the public sphere and Alice Walters has contended that just as the domestic space of the English home became a stage on which the display of polite consumer goods contributed to social advancement, so 'polite science aligned the acquisition of socially appropriate kinds of scientific knowledge with the acquisition of material goods illustrative and symbolic of that knowledge', such as telescopes, orreries and globes. This polite science was unpractical, 'distinctly non-servile, non-mechanical, and most emphatically non-professional'. Utilitarian and natural theological justifications for scientific education were two of the justifications most commonly made during the Georgian period. We should also be careful about regarding individuals merely as 'bundles of commodities' and over emphasising the role of entrepreneurs and marketeers in promoting natural philosophy as product for consumers. As Habermas emphasised, while public science had a commercial manifestation it also had the potential to foster intellectual empowerment by encouraging a broader rational critical debate.[8]

Post-modernist, post-structural and feminist scepticism towards traditional historical narratives has stimulated greater attention to the complexities, contingencies and divergences of phenomena such as identities and affiliations, which has had a major impact on – and indeed was partly driven by – recent work in the history of science. Older assumptions concerning participation in science and scientific objectivity have been questioned as the involvement of the working classes and women in scientific culture and the constructedness of science have been demonstrated. The older emphasis on the importance of conscious, rational and practical applications of scientific knowledge to technological and industrial innovation has been replaced by greater recognition of subconscious motivation, wider social and cultural influences, the breadth and subtlety of the scientific 'audience', the importance of objects, instrument and agency, the irrational nature of the creative process and the multiplicity of scientific sites. There has also been a very strong reaffirmation of scientific objectivism from 'realists' in the face of constructivism and social studies of science, although cultural historians of science have not really re-engaged with this.[9] Vladimir Jankovic, David Livingstone, Jan Golinski, Charles Withers and others have underscored the cultural construct-edness of spaces, places, institutions, identities and science itself as well as the geographies of scientific knowledge and interrelationships between natural philosophy and other aspects of intellectual endeavour and practices, such as chorography, geography and meteorology.[10]

Studies of Derby scientific culture have played an important role in changing cultural and historical studies of science. During the 1950s, work by Robinson and Schofield on the Derby Philosophical Society as much as on the Lunar Society helped to direct attention towards the relationship between science and industry, partly because of the geographical proximity of Derby to the pioneering textile industries of the Derwent Valley. In Klingender's influential study *Art and the Industrial Revolution* (1947) and to a lesser extent Nicholson's *Joseph Wright of Derby* (1968), Darwin, Wright and by implication their Lunar and Derby philosophical associates were interpreted as glorifiers of science and industry even though large parts of their work were neither concerned with – nor inspired by – these subjects.[11] Similarly, Robinson's study of the Derby Philosophical Society in 1953, which preceded Schofield's and his own publications on the Lunar Society, helped to re-establish the view that science played an important role in British industrialisation. Robinson's work still provides a useful introduction to the Philosophical Society, despite its tendency to exaggerate the industrial concerns of the Derby *savants*. However, these methodological presuppositions mean that less attention was given to scientific culture between 1800 and 1850.[12] Recent work on the chronology of industrialisation, the importance of proto-industry and nineteenth-century industrial growth and development, stretches the time-span in both directions and challenges such neglect. As Beckett and Heath suggest for the east midlands, 'the real breakthrough' into modern industry employing 'significant technological innovation, a fully-fledged factory system, and greater capitalisation, was a product of the years from about 1840 onwards rather than the classic industrial revolution period'.[13]

Subsequent work on Derby scientific culture has improved our understanding of the social and cultural uses of provincial science, notably King-Hele's studies of Darwin, Paul Sturges' examination of Derby Philosophical Society membership, Maxwell Craven's study of John Whitehurst and recent analyses of Joseph Wright. This work has greatly expanded upon Robinson's article, providing a much fuller analysis of Philosophical Society membership and the uses and contexts of natural philosophy, facilitating comparison with other British scientific communities. Using prosopographical analysis of the Philosophical Society, Sturges emphasised that interest in the industrial utility of science seems to have provided little incentive for membership given that most of the members had few industrial connections. In fact King-Hele, Sturges and Ian Inkster agree that the social and intellectual aspirations of medical men attracted by Darwin's leadership seems to have been the primary *raison d'être* for the Derby Philosophical Society, as their domination of the membership reveals.[14] However, as Sturges contends, inferences concerning individual motivation are precarious when based upon bald biographical summaries rather than surviving statements of intention. Although these studies of the Derby Philosophical Society suggest that social aspiration provided an important motivation for joining, they also indicate that the desire for polite rational entertainment, theological edification and industrial utility were other inducements. In this sense, Robinson's approach retains some validity and, as we shall see, receives additional support from analysis of the industrial concerns of the Derby Literary and Philosophical Society.

Although British Georgian scientific culture has received much recent analysis, work on nineteenth-century urban scientific culture has tended to concentrate upon Victorian civic science. According to Robinson, 'the growing attraction of the metropolis', the 'centralising tendency' in learned publications, and 'improved methods of transport' weakened the Derby Philosophical Society. However, as we shall see, the relationship between national, metropolitan and provincial scientific culture was more complex. As Jankovic, Golinski and others have argued, some aspects of natural philosophy such as meteorology and chemistry began to lose part of their local inspiration and character under the impetus of metropolitan centralisation. National scientific societies such as the Royal Institution (1799) and the British Association for the Advancement of Science (1831) provided influential models of elite, patriotic and civic science. These were adopted in provincial towns stimulated by relative economic revival after the post-war recession, political and religious campaigning and a more public-centred middle-class culture motivated by rational recreation, economic utility and convivial sociability.

While various phases can be discerned in the historiography of British scientific culture, as this book demonstrates, it can hardly be claimed that a single overarching dominant picture of British provincial scientific culture has emerged. The idea prevalent during the 1950s and 1960s of a causal relationship between the 'Scientific' and 'Industrial' revolutions, which Robinson's well-known paper on the Derby Philosophical Society helped to support, was rejected by later historians of science such as Porter, although Margaret Jacob and Larry Stewart have subsequently re-emphasised the interconnections between practical mathematics, engineering and industry. Porter's influential arguments concerning science and politeness and the relationship between metropolis and provinces have been questioned by urban historians, cultural and historical geographers and others. Furthermore, cultural and historical geographers, educational and feminist historians and others have challenged the excessive emphasis upon formal institutions and the perception of male domination. Likewise, as we have seen, the relationship between dissent, social marginality and scientific communities has also been called into question by work on dissent and the history of science. In this context, Derby makes an interesting case study as, unlike in Manchester and Leicester, Presbyterians, Congregationalists and Unitarians held prominent office in the corporation between 1700 and 1835. The Derby philosophers cannot straightforwardly be designated as social marginals, while, as we shall see, women and the working class played an important role in Derby scientific culture as audience and participants, as demonstrated by manuscript institutional records, subscription lists and other evidence, although this has tended to be overlooked in previous studies.

The first two chapters of the book examine the relationship between economic growth, the Georgian urban renaissance and the concept of British Enlightenment, particularly the role of science. Turning to Derby, they consider how and why natural philosophy held such a prominent position within enlightenment culture, including its importance in terms of natural theology, industrial utility and as polite activity manifest in the popularity of scientific literature, public lectures, schools and academies and philosophical associations. The importance of economic growth during the eighteenth and early nineteenth centuries and the manifestations of public culture and the urban renaissance are emphasised. The two chapters then explore the relationship between engineering, practical mathematics and scientific culture manifest in the activities of engineers such as George Sorocold, philosophical lecturers and teachers of the mathematics such as John Arden and William Griffis, and the relationship between dissent and natural philosophy and the established church, particularly the work of the Findern Academy at Derby and Findern.

Literary and philosophical societies were one of the most important manifestations of British enlightenment culture and the third and fourth chapters examine the circle around Joseph Wright, John Whitehurst and Earl Ferrers, the foundation of the Derby Philosophical Society, and other scientific coteries between the 1760s and 1790s. Like their counterparts in Birmingham and Manchester, the Derby *savants* saw science not merely as an isolated activity but as part of progressive European enlightenment culture. As the fifth and sixth chapters demonstrate, this was most clearly manifest in their political campaigns, which included those for urban improvement, against the Test and Corporation Acts and against the American and French wars. The campaigns to enclose common lands, the creation of the Derby Society for Political Information and the dispatch of an extraordinary delegation to the French assembly aroused considerable opposition, exacerbated local social divisions, split the Presbyterian congregation and resulted in the expulsion of one Derby Philosophical Society member.

During the nineteenth century, the Derby philosophers supported the foundation of a number of institutions intended to improve society and promote scientific education and rational recreation. With its novel industrial organisation and technology, as the eighth chapter demonstrates, the Derbyshire General Infirmary (1810) provided a major model for public institutions in Britain and abroad, while creating an institution that would satisfy the demands of local manufacturers and elite for an efficient and economical institution that would isolate, control and treat the sick from a rapidly expanding industrialised workforce, and tend to their moral and spiritual needs. Encouraged by the success of the Infirmary, the Derby philosophers made a determined effort to forge a public platform for science and promote scientific education for both sexes as a means of fostering social improvement, although renewed public political campaigning aroused hostility to their activities. Through the Derby Literary and Philosophical Society, they organised annual public lectures, collected scientific instruments and encouraged original scientific research, forming a subscription laboratory. Commercial, political and religious connections that spread to Europe and the USA were exploited and underpinned by love of science, hostility to international conflict and belief in intellectual progress. The progressive national and internationalist outlook of the members and the contacts with other philosophical communities encouraged them to transcend localism and provinciality.

The continuity in the ideas of the Derby philosophers is most evident in the progressive enlightenment of their social and educational philosophies, which embraced an evolutionary worldview inspired by Darwin. Darwin's scheme was the first in world history to combine biological evolution, associational psychology, evolutionary geology and cosmological developmentalism. The ninth chapter contends that through the activities of his father William George Spencer and the Derby philosophers, Herbert Spencer's evolutionism was primarily stimulated by his immersion in this culture. The global significance of Herbert Spencer's promotion of evolutionary ideas has been, of course, as the familiarity of his phrase 'survival of the fittest' indicates, generally recognised for over a century, yet accounts of the origins of his system still tend to follow too closely the author's own description, written when he was an old man many decades later. Spencer's interpretation of his own intellectual development gives an inadequate impression of the debt that he owed to provincial scientific culture and its institutions, which, as we shall see, was highly significant.

In their different ways, the final two chapters consider the degree to which the Derby philosophers were able to involve women and the working class in public scientific culture, encouraged by notions of rational recreation and the utility of scientific education. As we shall see, although women remained excluded from most Derby scientific societies by convention, there is considerable evidence that some participated in scientific culture though lectures and patronage. Through the Derby Mechanics' Institute, exhibitions by the museum and the mechanics' institute of 1839 and 1843, and Derby Arboretum, some women and working classes were able to gain and enjoy a scientific education, although this access remained circumscribed, contingent and negotiated. The Arboretum became one of the most successful of all mid-Victorian provincial public attractions and a regional rather than a merely local institution, serving as a civic and patriotic centre and inspiring the foundation of parks, arboretums and botanical gardens in Britain, Europe and the USA. However, as the conclusions demonstrate, important changes occurred in the nature of Derby public scientific culture around the mid-nineteenth century. Problems surrounded the combined scientific and popular status of the mechanics' institute and the Arboretum, represented at the latter by the remarkable, systematic collection of over a thousand varieties of trees and shrubs, and uses and appropriation of park spaces for public leisure activities, particularly anniversary festivals. For a variety of reasons, science came to occupy a less prominent position in Derby public culture. However, this should not be exaggerated, and museums, libraries, colleges and other institutions continued to popularise and teach scientific subjects in the late-Victorian and Edwardian periods.

Notes

1 G. J. Holyoake, 'Obituary for Herbert Spencer', *Literary Guide and Rationalist Review*, 91 (1 January 1904), 1.

2 R. E. Schofield, 'Industrial orientation of science in the Lunar Society of Birmingham', *Isis*, 48 (1957), 408–15; E. Robinson, 'The Lunar Society and the improvement of scientific instruments', *Annals of Science*, 12–13 (1956–57), 296–304, 1–8; R. E. Schofield, *The Lunar Society of Birmingham* (Oxford: Clarendon Press, 1963); A. E. Musson and E. Robinson, *Science and Technology in the Industrial Revolution* (Manchester: Manchester University Press, 1969), 190–9.

3 A. Thackray, 'Natural knowledge in cultural context: the Manchester model', *American Historical Review*, 79 (1974), 672–709; I. Inkster, 'Studies in the Social History of Science in England during the Industrial Revolution, c.1750–1850' (PhD thesis, University of Sheffield, 1977); I. Inkster and J. Morrell, *Metropolis and Province: Science in British Culture, 1780–1850* (London: Hutchinson, 1983); C. Russell, *Science and Social Change, 1700–1900* (London: Macmillan, 1983); I. Inkster, *Scientific Culture and Urbanisation in Industrialising Britain* (Aldershot: Ashgate, 1997); P. Elliott, 'Provincial urban society, scientific culture and sociopolitical marginality in Britain in the eighteenth and nineteenth centuries', *Social History*, 28 (2003), 394–442; P. Elliott, 'Towards a geography of English scientific

culture: provincial identity and literary and philosophical culture in the county town, *c*.1750–1850', *Urban History*, 32 (2005), 391–412.

4   J. T. Desaguliers, *A Course of Experimental Philosophy*, second edition, 2 vols. (London, 1745).

5   R. H. Kargon, *Science in Victorian Manchester: Enterprise and Expertise* (Manchester: Manchester University Press, 1978); A. D. Orange, *Philosophers and Provincials: The Yorkshire Philosophical Society from 1822 to 1844* (York: Yorkshire Philosophical Society, 1973); J. Morrell and A. Thackray, *Gentlemen of Science: The Early Years of the British Association for the Advancement of Science* (Oxford: Clarendon Press, 1981).

6   P. Borsay, *The English Urban Renaissance* (Oxford: Clarendon Press, 1989); R. J. Morris, 'Voluntary societies and British urban elites, 1780–1850: an analysis', *The Historical Journal*, 26 (1983), 95–118; P. Clark (ed.), *The Cambridge Urban History*, II: *1549–1840* (Cambridge: Cambridge University Press, 2000); P. Borsay, 'The London connection: cultural diffusion and the eighteenth-century provincial town', *London Journal*, 19 (1994), 21–35; C. B. Estabrook, *Urbane and Rustic England: Cultural Ties and Social Spheres in the Provinces, 1660–1780* (Manchester: Manchester University Press, 1998); P. Clark, *British Clubs and Societies, 1580–1800* (Oxford: Oxford University Press, 2001); J. Habermas, *The Structural Transformation of the Public Sphere*, trans. Thomas Burger (Cambridge: Polity Press, 1989); R. Porter, 'Science, provincial culture and public opinion in enlightenment England', *British Journal for Eighteenth Century Studies*, 3 (1980), 20–46.

7   T. Broman, 'The Habermasian public sphere and science in the enlightenment', *History of Science*, 36 (1998), 123–49; H. Chiswick, 'Public opinion and political culture in France during the second half of the eighteenth century', *English Historical Review*, 107 (2002), 48–77.

8   A. N. Walters, 'Science and politeness in eighteenth-century England', *History of Science*, 35 (1997), 121–54.

9   B. Latour and S. Woolgar, *Laboratory Life: The Construction of Scientific Facts* (New Jersey: Princeton University Press, 1981); P. R. Gross, N. Levitt and M. Lewis (eds.), *The Flight from Science and Reason* (New York: Academy of Sciences, 1996); P. R. Gross, *Higher Superstition: The Academic Left and its Quarrels with Science* (Baltimore: Johns Hopkins University Press, 1998); A. D. Sokal and J. Bricmont, *Fashionable Nonsense: Post-modern Intellectuals' Abuse of Science* (New York: Picador, 1999).

10 N. Jardine, J. A. Secord and E. C. Spary (eds.), *Cultures of Natural History* (Cambridge: Cambridge University Press, 1996); J. Golinksi, *Making Natural Knowledge: Constructivism and the History of Science* (Cambridge: Cambridge University Press, 1998); M. Ogborn, *Spaces of Modernity: London's Geographies, 1680–1780* (London: Guilford Press, 1998); C. Yanni, *Nature's Museums: Victorian Science and the Architecture of Display* (New York: Princeton Architectural Press, 1999); W. Clark, J. Golinski and S. Schaffer (eds.), *The Sciences in Enlightened Europe* (Chicago: Chicago University Press, 1997); V. Jankovic, *Reading the Skies: A Cultural History of English Weather, 1650–1820* (Chicago: Chicago University Press, 2000); V. Jankovic, 'The place of nature and the nature of place: the chorographic challenge to the history of British provincial science', *History of Science*,

38 (2000), 80–113; S. Alberti, 'Placing nature: natural history collections and their owners in nineteenth-century provincial England', *British Journal for the History of Science*, 35 (2002), 291–311; C. Withers, *Geography, Science and National Identity: Scotland Since 1520* (Cambridge: Cambridge University Press, 2002); S. Naylor, 'The field, the museum, and the lecture hall: the places of natural history in Victorian Cornwall', *Transactions of the Institute of British Geographers*, 27 (2002), 494–513; D. N. Livingstone, *Putting Science in its Place* (Chicago: Chicago University Press, 2003); S. Naylor, 'Nationalising provincial weather: meteorology in nineteenth-century Cornwall', *British Journal for the History of Science*, 39 (2006), 1–27; C. Withers, *Placing the Enlightenment: Thinking Geographically about the Age of Reason* (Chicago: Chicago University Press, 2007); see also the essays on the historical geographies of science in the special issue of *British Journal for the History of Science*, 38 (2005), part 1.

11 F. Klingender, *Art and the Industrial Revolution*, revised edition (London: Paladin, 1972); B. Nicolson, *Joseph Wright of Derby, Painter of Light*, 2 vols. (London: Paul Mellon Foundation, 1968).

12 S. Smiles, *Lives of the Engineers: Boulton and Watt* (London: John Murray, 1904), 336–57; E. Halévy, *History of the English People in the Nineteenth Century*, 6 vols. (London: Ernest Benn, 1924), II, 538–72; T. S. Ashton, *The Industrial Revolution* (Oxford: Oxford University Press, 1948). Some important source material was unavailable to Robinson, notably the journal book of the Derby Literary and Philosophical Society.

13 J. V. Beckett and J. E. Heath, 'When was the industrial revolution in the east midlands?', *Midland History*, 13 (1988), 77–93.

14 R. P. Sturges, 'The membership of the Derby Philosophical Society, 1783–1802', *Midland History*, 4 (1978), 212–29; Inkster, 'Studies in the Social History of Science', 469–513; D. King-Hele (ed.), *The Collected Letters of Erasmus Darwin* (Cambridge: Cambridge University Press, 2007); M. Craven, *John Whitehurst of Derby: Clockmaker and Scientist 1713–88* (Ashbourne: Mayfield, 1996); D. King-Hele, *Erasmus Darwin: A Life of Unequalled Achievement* (London: Giles de la Mare, 1999), 196–9; J. Egerton, *Wright of Derby* (London: Tate Gallery, 1990); S. Daniels, *Joseph Wright* (London: Tate Gallery, 1997).

1

Politics, religion, urban culture and natural philosophy, *c.*1700–1770[1]

Introduction

Many English towns were transformed during the late seventeenth and eighteenth centuries, most obviously through the construction of numerous brick and stone private and public buildings, which denoted and encouraged a parallel acceleration in the variety and intensity of urban sociability. Provincial towns came to be perceived as both cultural and economic centres, in what Peter Borsay defined as the Georgian 'urban renaissance'. This chapter examines how natural philosophy came to form an important part of urban culture in provincial towns such as Derby in the period. It begins by examining the political, religious and economic circumstances that allowed public culture to develop and flourish in the relative political stability of the eighteenth century encouraged by Whig dynasts such as the Cavendish family. It argues that scientific culture was just one part of Georgian urban renaissance culture, and explores the relationship between the transformation of the built environment and the development of Derby scientific culture. The chapter then considers the forms that this culture took, comparing these to other kinds of public culture that flourished simultaneously. It contends that through natural theology, as the activities of philosophers such as Benjamin Parker and dissenting schools such as the Findern Academy demonstrate, religion continued to stimulate scientific culture.

Political stability, economic prosperity and urban culture

The growth of public culture in Derby and many other provincial towns in the later Stuart and early Hanoverian periods undoubtedly owed much to what Plumb famously described as the 'age of stability'. After the serious political, economic and social dislocation of the Civil War and Protectorate periods, the restoration of the monarchy, and especially the so-called Glorious Revolution, eventually produced greater political stability that allowed trade to flourish and British imperial power to develop. Religious and political tensions were never, of course, entirely abated during the reigns of George I and George II, as the activities of the Jacobites and hostility towards dissenters demonstrate, but an expectation of peace and calm grew, providing encouragement for the growth of retail, trade and manufacture, sometimes designated the Georgian consumer revolution. While these trends are most evident, of course, in the post-Restoration metropolis, they are manifest in the English provinces where aristocratic families augmented their power and influence.[2]

In Derbyshire, the Cavendish family greatly benefited from supporting the accession of William of Orange. Their subsequent enthusiasm for the Hanoverian succession further augmented their power and influence, symbolised by the rise to ducal status, the magnificence of Chatsworth and the extensive Derbyshire estates. Chatsworth was radically reconstructed in the late seventeenth century as a palatial mansion with French-inspired formal gardens. Although there were independent Tory challenges from some Derbyshire aristocrats and gentry, the county remained politically dominated by the Cavendish family during the eighteenth and early nineteenth centuries, despite the different personalities of individual dukes and duchesses. This meant that while serious party political disagreements tended to be suppressed, there were fewer contested elections than in some other counties. Equally, the powers gained by the Derby burgesses during the seventeenth century were also held in check, to some extent, by Cavendish hegemony. The Duke of Devonshire retained the office of steward of the corporation, and the choice of the two members of parliament for the borough was generally split, by agreement, between the two.[3] The Derby Corporation consisted of a mayor, a high steward, nine aldermen, a recorder, fourteen brothers, fourteen capital burgesses and a town clerk, with the ordinary burgesses, who numbered about 900 in 1715, having the right to vote. The corporation had the power to create freemen as required, which provided them with the ability to outvote the residents in their own or the Cavendish interest.[4]

The degree of political and religious conviviality should not, however, be exaggerated. Important political and religious divides remained, notably between dissenters and Anglicans, and there was opposition towards the Whig hegemony. Occasional hostility is represented by the attack upon Whigs and dissenters made by Tory clergyman Dr Henry Sacheverell in sermons at All Saints, Derby and at St Paul's Cathedral, which singled out supporters of the Glorious Revolution as betrayers of church and constitution and dissenting academies as 'seminaries wherein atheism, deism, tritheism, socinianism with all the hellish principles of fanaticism, regicide and anarchy are openly professed and taught to corrupt and debauch the youth of the nation'. After Sacheverell was impeached and received a lenient sentence, this was greeted by rejoicing in many towns, including Derby, where 'the bells of All Saints clashed out a triumphant peel, which was shortly taken up by every steeple in the town, whilst huge bonfires were piled up and set on fire in the centre of the Market Place and on Nun's Green'.[5]

There is evidence for Jacobite sympathies among many Derby burgesses during the first half of the eighteenth century, although some of this is anecdotal. In 1715 some Derby clergymen, and apparently a large proportion of the population, openly espoused the cause of the Stuart claimant and there were riots on the accession of George I. According to William Hutton, Samuel Sturges, Vicar of All Saints' Church in Derby, 'prayed publicly for King James and then quickly changed this to King George', which may have simply been a slip of the tongue, as subsequently avowed. The 'congregation became tumultuous; the military gentlemen drew their swords, and ordered him out of the pulpit, into which he never returned'. Harris of St Peter's had to be called to order by the magistrates. Similarly, Hutton claimed that Henry Cantril of St Alkmund's 'drank the pretender's health upon his knees', a fact seemingly confirmed by his diary, although 'the wiser' William Lockett of St Michael's 'chose rather to amuse himself' mowing his lawn 'rather than meddling in politics'.[6] Support for the Jacobites is also evident from a sermon preached by the Whig clergyman Dr Michael Hutchinson at All Saints on 30 January 1717, when troops, whom he praised for their 'fidelity, courage and loyalty', were still quartered at Derby to prevent public demonstrations against the new monarch. Hutchinson had been selected by the corporation because of his loyalty to the 'illustrious House of Hanover', and his sermon was given on the anniversary of the martyrdom of Charles I. The purpose of the sermon was to show that 'tho' our excellent constitution in church and state is strong enough to resist all ... our enemies', yet it could be undermined by 'the indirect artifices of dissimulation and hypocrisie'. Derby continued to be disturbed by 'rage and fury', particularly on 30 January, when the Jacobite clergy had caused men to 'come reeking hot from prayers and sermons' containing covert statements of support for the Stuart claimant. Hutchinson blamed 'our common enemy the papists', who had 'administered fewel to the flame' and set 'all true Britons and Protestants ... biting and devouring one another'. He claimed that, supported by foreign rulers, George would secure the Protestant interest against the Roman Catholics.[7]

Nonconformists also, of course, tended to welcome the Hanoverian succession for the prospect of religious toleration that it offered, and Ferdinando Shaw, minister at the Friargate Presbyterian Chapel from 1699 to 1745, gave a sermon in support of the new king. His contemporary Ebenezer Latham, Presbyterian minister at Findern and teacher at the local dissenting academy, was equally forthright in celebrating the defeat of Jacobitism in 1746.[8] Though the Tory challenge remained between 1715 and 1745, especially from independent gentry and poorer burgesses, support for Jacobitism declined. The Tories usually lost in elections, though results were frequently close, as in 1741, when the creation of large numbers of honorary burgesses was necessary to secure the success of the Duke of Devonshire's son-in-law, Viscount Duncannon, and John Stanhope against the Tory, German Pole of Radbourne. Lists compiled by Pole's election agent, William Turner of Derby, and surviving poll books indicate that most of the Tories were gentry, artisans, craftsmen and small shopkeepers, while wealthier shopkeepers, merchants and professionals such as lawyers supported the Whigs.[9]

When in December 1745, Charles Edward Stuart reached Derby during the second Jacobite attempt to seize the crown and stayed there for two nights, his 5,000-strong army did not meet with much encouragement. Hanoverian loyalists were in some disarray, Cumberland's army was in Staffordshire, most of the civic leaders fled, the gentry sheltered on their estates, and the Duke of Devonshire marched off to Nottinghamshire with the Derbyshire Blues, a militia hastily concocted for the defence of the county. Despite some evidence of a good reception for Charles and the good behaviour of his army, only a handful of men from Derby actually joined the rebels, a pattern repeated in most English towns apart from Manchester. This can partly be explained by the fact that the Hanoverian dynasty had been in power for thirty years and become the 'legitimate' dynasty, partly through the fear of a civil war which would dislocate trade, and partly because adherence to the Jacobite cause had become more emotional than real and practical. As the sentimental Tory Samuel Johnson once admitted, 'if England were fairly polled, the present King would be sent away to-night', but 'people would not … risk anything to restore the exiled family. They would not give twenty shillings apiece to bring it about.' The lighting of bonfires, the ringing of bells and the crowds of people to witness an unusual occasion, which apparently occurred at Derby, were far removed from joining an army. They cost little. Romantic sympathy for the underdog and knowledge of the cruel and brutal conduct of 'Butcher' Cumberland cannot be allowed to obscure the major historical point that when Derbyshire Tories could perhaps have changed the course of British history by supporting the rebels, they failed to do so. By 1745, Jacobitism in Derbyshire was weaker than it had been in 1715, and there was no incentive in a period of obvious urban improvement to threaten social stability and dislocate trade by taking a leap into the unknown and to risk dragging Britain into a bloody civil war. In the face of growing economic prosperity, consumption and the encouragement towards greater social unity provided by urban public culture, apart from occasional marks of discontent, there was little open hostility towards the Cavendish Whig hegemony. Although always in a minority, Tories were well represented in the corporation, and the Cavendish family with their Whig and dissenting supporters never strove for complete political domination.[10]

The transformation of the urban environment

Smaller than its close neighbour Nottingham which had 10,000 people by 1740, the population of Derby nevertheless grew from around 4,000 in 1712 to some 6,000 by 1745 and 9,563 by 1788. As Penelope Corfield and others have shown, population size was not necessarily the main factor in determining the cultural and economic significance of Georgian towns, but a growing population usually demonstrated a thriving economy.[11] The growth of the Derby economy was founded upon agriculture, brewing, and the success of the lead mining industry in the Peak, which was at its height between 1600 and 1750. The Keuper Marl lowlands to the south of the county and the Trent Valley provided rich, well-watered agricultural land, while the carboniferous limestone and shale and gritstone uplands of the north, east and centre of the county with their coal measures and lead veins supported extractive industries and hill farming.[12] According to Daniel Defoe, 'good malt and good ale' were made, much of which was apparently consumed locally, though the Derby folk generously spared 'some to their neighbour too'. Similarly in 1721 the town was described as 'near, large, well-built and populous' with trade 'not very considerable, tho' it be a staple for wool'.[13] The historian William Woolley agreed that malting was the 'principal trade' and that Derby was 'very famous for very good ale, which the brewers send to London … to good advantage'. He also recognised the importance of the baking trade and the fact that there was a 'considerable manufacture of stockiners' carried on in Derby and adjacent areas. The Market Place was 'very handsome – a square with good buildings about it', there were markets on Wednesdays and Fridays which drew in

2 East prospect of Derby by G. Moneypenny.

crowds from the surrounding countryside, and six major fairs per year which attracted even more. Derby was surrounded with a 'very large district of excellent good arable, pasture and meadow lands belonging to it', while adjoining common lands had other economic functions including whitening cloth.[14]

Borsay has recognised the degree to which the creation of new kinds of town environment helped to create distinctive urban identity, and between 1660 and 1800 the Derby townscape was transformed from vernacular timber to brick, tile and stone, even if this sometimes took the form of fashionable re-fronting rather than reconstruction. Encouraged by the growth of trade and the relative political calm, the opportunities for organised elite sociability increased as the physical urban environment was transformed. In 1700, Celia Fiennes remarked that Derby was built mostly of brick apart from the five stone churches, yet over the ensuing decades, as the population increased, stone Palladian fronts and classical elegance replaced vernacular wood and brick. Already Fiennes noted that Derby was 'a fine, beautiful, and pleasant town' that was 'populous, well built' and had a 'large market place, a fine town-house, and very handsome streets'. There were 'more families of gentlemen in it' than was 'usual in towns so remote' and there was therefore 'a great deal of good and some gay company'. Due to the proximity of the 'inhospitable' and 'wild' Peak, which dominated much of the county, the gentry 'choose to reside at Derby, rather than upon their estates, as they do in other places'. Woolley and Defoe agreed that there were 'more families of gentlemen' living in Derby than was 'usual in towns so remote', so that there were 'many persons of good quality and a great number of coaches kept in it'. Although the Peak was indeed relatively 'wild' and remote in late seventeenth- and eighteenth-century terms, it supported a burgeoning lead mining industry at its height in the period and, combined with the agricultural land in the south of the county, Derbyshire estates were able to support a growing Derby-centred culture exemplified by the residence of urban gentry.[15]

Woolley provides an instructive example of the opportunities available from the county economy under Cavendish patronage. Woolley's family, who were greatly indebted to Cavendish favour, became wealthy through the gold and silk trades in London and Derbyshire and became landed gentry through the purchase of land in the south-west of the county. William retired from the Spanish trade in London and after becoming rich acquired interests in the Derbyshire lead mines, purchasing property in the county, Leicestershire and London. Although he suffered during the War of Spanish Succession, losing over £2,000, this seems to have provided him with an incentive to retire, and he was living in Derby by 1708 and bought an estate at Darley Abbey in 1709, where he lived until his death in 1719. His son William junior inherited the Darley estate and served in the corporation, becoming Mayor of Derby in 1722. Although he seems to have remained in the established church, his wife

3 Friargate Unitarian Chapel from S. Glover, *The History, and Gazetteer, and Directory of the County of Derby*, 2 vols. (1829, 1833), II.

attended the Friargate Presbyterian chapel, though she bore 'a sincere affection to all real Christians of what persuasion soever'.[16]

Fashionable townhouses were constructed in Derby, culminating in the creation of the elegant thoroughfare of Friargate, built on common land from the Nun's Green made available by improvement acts.[17] New public buildings appeared, including the County Hall (1660), the Full Street Assembly Room (1714), Bold Lane Theatre (1773), and the Guildhall (1731), while churches such as All Saints (1723–25) and St Werburgh's (1699) were rebuilt.[18] Individual private munificence gave way to collective effort as the middling sort assumed greater responsibility from the corporation for water supply, bridges, street paving and lighting, assembly rooms and hospitals. A regular system of post and stage-coaches was established, which utilised the first turnpike roads linking Derby to the rest of England. As elsewhere, many taverns were either re-fronted or rebuilt in the classical style, including coaching houses such as the King's Head, the New Inn and the Bell.[19] The new All Saints' Church was designed by James Gibbs, whose *Book of Architecture* (1728) had an important general influence on Derby buildings, and certain features such as Gibbs's surrounds, rustication and tripartite windows reappeared on town buildings well into the nineteenth century.[20] It would be wrong to assume that money was easily found for such major construction projects. Michael Hutchinson, the Whig Vicar of All Saints, had to resort to pulling down the main body of the medieval church and various unpopular measures to secure the construction of the new church. However, although it is an exaggeration to call this a Derby style of architecture, a pattern emerges of local architects transforming the built environment while favouring somewhat backward-looking Palladian models taken from architectural publications, adapted to local tastes.[21]

Polite culture and consumerism in Georgian Derby

The demand for fashionable goods in polite society at Derby is evident from the promotion and sale of luxury consumer items advertised in the newspapers, frequently imported from London, but increasingly available and manufactured locally. William Parker, a Derby silversmith for instance, offered gold, jewellery, silver tankards, mugs, glasses, scissors, telescopes, coats of arms, crests and toys, from his premises opposite the King's Head.[22] As the Georgian consumer economy was established and came to challenge traditional forms of rank and status in provincial towns such as Derby, there was greater ideological acceptance of the desirability of material wealth, which, it was claimed, could even be the basis for moral advance.[23] The complaints of moralists and churchmen were side-stepped with the claim that, as Hume argued, luxury was not in itself harmful and could be positively good by nurturing 'refinement in the gratification of the senses' and encouraging trade and commerce. Success in industry and the mechanical arts was indissolubly associated with advances in the liberal arts grounded in middling-sort urban sociability. It was the property derived from trade and commerce which made the 'middling rank' 'the best and firmest basis of public liberty' against peasant slavery or monarchical tyranny.[24]

One of the most important manifestations of polite culture in Derby was the production, marketing and consumption of printed texts, which occurred on a scale never previously experienced. There was a significant increase in the market demand and availability of printed literature, from political broadsides to Bibles and from volumes of poetry to newspapers. Booksellers were at the heart of urban culture, advertising and selling books and newspapers, running book circulation schemes, and frequently branching out to sell other products such as medicines and prints.[25] The first Derby newspaper was probably Samuel Hodgkinson's *Derby Postman*, which ran from 1719 to about April 1731, with a name change to *The British Spy: or Derby Postman* and a new start in 1726. Hodgkinson, whose premises were at St Werburgh's Churchyard, offered to print 'all manner of books' and included descriptions of Derby, local affairs and trading matters in his newspaper. At the same time the local market grew to the extent that, usually by subscription, booksellers were able to produce their own editions of works, especially sermons and religious pamphlets, illustrating the importance of religious disagreement in stimulating public culture, and also expanded into other fields.[26] Between the

1760s and early 1800s, for instance, the bookseller Francis Roome specialised in music and musical instruments, while another business owned by Samuel Fox, Rowland Almond and William Pritchard sold prints and mezzotints.[27] Booksellers also established circulating libraries, which played a major role in making books affordable by subscription and, as we shall see, supported and published works on natural history and natural philosophy.[28]

The role of the Georgian newspapers in facilitating public political debate is evident from the success of the *Derby Mercury*, founded by Samuel Drewry (d.1769) in 1732, the short-lived *Derby Courier* (1738) and the *Derby Herald*. The *Mercury* survived for almost two hundred years, remaining in the family for half this time and until the Whig *Derby and Chesterfield Reporter* was founded in 1823, had no successful rival.[29] Items derived from London newspapers initially dominated the pages of these newspapers; however, later in the century the column of local news expanded and became a regular feature.[30] Newspaper advertisements offer an excellent guide to the increased size and sophistication of consumerism. By the 1770s, fifty-three different magazines were advertised in the *Derby Mercury* and some 300 book titles including forty-five titles on religious subjects, twenty-five on law, twenty-five medical works, and twenty-one grammar and school books. Other popular subjects (apart from novels) included husbandry, gardening and field sports, cookery, travel and topography, jest and song books, histories, music, poetry, politics and biography.[31]

Provincial towns, as local capitals, brought together many different types of people from miles around, making varied social groupings quite common at markets, in taverns and on festival days, and urban leisure activities were more socially inclusive than they became by the end of the eighteenth century. In 1732, 'Gillenoe, the Frenchman' performed gymnastic feats on a high wire which stretched from the top of All Saints' Church to the bottom of St Michael's Church, watched by large crowds in the streets below. In December 1733, Herbert's Company of Comedians performed the tragedy of 'Tamerlane the Great' at Tyrrel's dancing room. The infamous Derby Shrovetide football, which often resulted in injury and sometimes death, and in which the only rules were that there were no rules, was generally supported by the urban elite until the 1790s.[32] Travelling entertainers vied with more traditional public festivals and celebrations, such as the fireworks and illuminations which took place on the monarch's birthday, or after a military victory, when beer was often distributed to the populace.[33] A wild beast show in the yard of the White Hart in April 1750, which included a rhinoceros and a live crocodile, was visited by the local gentry, the Judge of the Assize, several members of the Bar and the Unitarian minister James Clegg, who had come into town for an ordination with the Derby ministers Ebenezer Latham and Josiah Rogerson.[34]

Encouraged by advertising in the newspapers, new forms of public culture appeared which, because of their subscription basis, were less socially inclusive. These included coffee-houses where newspapers and magazines could be read, or the new assembly rooms opened in 1765 and designed by Earl Ferrers and Joseph Pickford.[35] There was even an attempt to establish a spa with pleasure walks by Dr William Chauncey at Abbey Barns in 1733, although this quickly foundered in the face of much more popular Derbyshire spas at Matlock, Buxton and elsewhere.[36] Another example was the theatre established in 1773, which replaced the more informal regular plays performed in taverns, outdoor booths or public buildings in races week and on other special occasions by visiting strolling players. The new theatre in Bold Lane presented performances, usually for two weeks at a time, and offered the opportunities for subscribers to reserve the best seats or boxes, it was triumphantly claimed, after the manner of metropolitan theatres, thus elevating the theatre to the highest form of polite entertainment.[37] Although the churches remained important in the process, the establishment of a distinctive polite secular public culture is also evident from the number of vocal and instrumental concerts in Derby, encouraged by Charles Denby (d.1771), his son Charles (1735–93) and booksellers such as Francis Roome. Some of these succeeded in attracting metropolitan performers, but they also served to further differentiate the leisure activities of the middling sort and gentry from the labouring classes.[38]

Natural philosophy and polite consumption

The 1730s and 1740s seem to be significant in the development of public literary and scientific culture in provincial areas such as Birmingham and the west midlands and the larger east midland towns.[39] The increasing importance of natural philosophy in polite culture is evident in the demand for scientific goods, as well as through more traditional inducements to study the subject, such as natural theology, which we shall examine shortly. The purchase of scientific books and paintings, subscriptions to lecture courses and the acquisition of drawing-room scientific apparatus such as orreries and telescopes were consumer manifestations of this culture. Derby instrument makers such as John Stenson offered barometers, hygrometers, thermometers, 'electrical pistols' and other scientific instruments of their own manufacture, for practical use and recommended as 'furniture for genteel apartments'.[40] This was also, to some extent, as we shall see, reflected in the demand for minerals, rocks and fossils as ornaments, encouraged by the proximity of Derby to the Peak District, and satisfied by local geologists and mineral collectors.

The publication of chorographical and county studies of natural history and antiquities was another manifestation of this interest in natural philosophy from gentry, aristocracy and the middling sort. This is evident in Derbyshire from the 'Wonders of the Peak' tradition and attempts to study and extol the benefits of spring waters, especially those at Buxton and Matlock. Charles Leigh's *Natural History of Lancashire, Cheshire and the Peak in Derbyshire* (1700) included detailed discussions of the natural history, waters and antiquities of these counties, while John Floyer considered Derbyshire springs in his *Enquiry into the Right Use of Baths* (1697).[41] William Woolley subscribed to Leigh's *Natural History* and used both works in his *History of Derbyshire*, which, although not published in his lifetime, included discussions of the natural history and landscape of the county. This parallels the interest taken in history and natural history at Nottingham by the German physician Charles Deering, who produced a botanical catalogue during the 1730s and published *Nottingham Vetus et Nova*, a town history with a large natural historical content.[42]

Benjamin Parker: natural philosophy and physicotheology

Natural philosophy also remained closely intertwined with natural theology and, as the career of Benjamin Parker (*c*.1700–47?) and the activities of the local dissenting academy demonstrate, religious disputes provided considerable impetus for different interpretations and appropriations of Newtonianism. Religious and political polemicists were able to utilise the development of bookselling and print markets to disseminate their ideas and enlarge audiences and Parker's works demonstrate that, utilising the patronage of local gentry and aristocracy, ambitious philosophical works were being composed at Derby by the 1730s. Parker was a Derby stocking maker who lived next to St Mary's Bridge, Derby and published a work on calculating the longitude in 1731, dedicated to the Earl of Chesterfield and Sir Nathaniel Curzon. Parker claimed to have taken to writing on physicotheology and natural philosophy to 'advance the spiritual welfare of others' by combating the 'vice and profaneness' of the age and because of his 'mean circumstances ... thro' misfortunes in trade'. He hoped that his works, 'chiefly the product of my own experience and speculations' and the Bible, would 'induce the curious to search into the beauty of nature, and to adore, love, fear, serve, and obey its author'. Parker published four more books at Derby on natural theology and philosophy and a treatise on personal education, these meeting with a 'more than expected entertainment and approbation'. In 1739 he moved to London, selling quack remedies, delivering lectures and publishing further works on natural philosophy, but apparently died in poverty.[43]

Parker's method of longitude calculation involved determining the lunar altitude at the time of its southing and comparing this with the known position of latitude at sea, hardly an original suggestion, having been made by Halley and others for decades. This required a precise knowledge of the moon's motion, something that Halley and Newton had tried to determine and which Parker was still urging in 1745, remarking bitterly that it was 'seldom nowadays, that the first discoverers of things, useful to the publick, meet with the encouragement they deserve'. The problem of the moon's orbit posed a special difficulty for Newtonian astronomy, which had literally given Newton a headache, because of the difficulty of reconciling the Lunar rotation on its axis with its orbit with the earth around the sun. Parker approached Halley to back his plan because of his position as Astronomer Royal and his interest in the Lunar method. William Whiston (1667–1752) had proposed a scheme with Humphry Ditton that involved firing rockets from ships stationed at regular intervals along trade routes. Though the scheme proved to be far too impractical, Whiston succeeded in creating enough interest to encourage the government to offer a prize of £20,000 for a successful method to be judged by a Board of Longitude. But the prize encouraged many 'longitude lunatics' to suggest a variety of bizarre proposals.[44]

Inspired by William Whiston's physicotheology and William Derham's *Astro-Theology* (1715) Parker strove during the 1730s and early 1740s to apply 'Newtonian' methods and ideas to natural theology in support of established Anglicanism, defending trinitarianism against the arianism openly espoused by Whiston and some other Newtonians. Much of Parker's work consisted of variations upon the argument for design intended to furnish 'proofs' for the existence of God from creation. Reconciling the Bible with his Anglicanism and physicotheology provided a challenge and a source of explanation for natural phenomena such as fireballs, the aurora borealis, the position of the planets, the location of minerals and the apparent evidence of purpose in the natural world. Parker sought and seems to have obtained patronage from university professors and both Whig and Tory gentry and aristocracy, including the Earl of Chesterfield, the Tory Sir Nathaniel Curzon and the Duke of Devonshire. He dedicated the second volume of his *Philosophical Meditations* to the 'reverend doctors and masters of colleges of the famous University of Oxford', while the third edition was 'revised and corrected by a gentleman of the University of Oxford'.[45]

The importance of religious inspiration in Parker's work is evident in his strong belief in the plurality of worlds and the prevalence of other beings throughout the universe. The idea was inspired by the work of Whiston and Derham, who both followed Richard Bentley in believing that Newtonian philosophy, empirical evidence and natural theology supported this notion. It was argued that the vast realms of creation would apparently be rendered incomprehensible in the divine plan unless they supported other beings, although Newton had been cautious and refrained from speculation in print concerning the origins of the universe and plurality of inhabited worlds. An idle and wasteful deity could not be credited. In 1734 Parker referred to 'the late discoveries from the principles of Sir Isaac Newton, concerning the fixed stars being so many suns, with planets revolving round them' and his belief in the inhabitants of the moon, ideas he was to develop during the 1740s.[46]

Parker considered the question extensively in his *Survey of the Creation*, citing correspondence with a learned friend which referred to the sublimity, uniformity and grandeur of the universe in which 'so many suns, each with their peculiar planets and, some of them, with satellites, existed'. There were 'worlds beyond worlds; systems beyond systems regularly interspersed, thro' the vast and unfathomable space'; this spoke eloquently for the 'design, contrivance, wisdom, power and goodness in the being, who is author of this fabric'. The empirical and mathematical basis of a plurality of inhabited worlds was tenuous, pushing the problem both intellectually and physically to the extreme limits of knowledge. However, following Derham, Whiston, Bentley and Newton himself, Parker contended that the laws of gravitation and attraction applied to 'all other material globes in the universe'. Hence from the separation of the 'grosser from the finer parts of their mixed and chaotic substances', the 'different materials of the chaos began to assume their proper essences' and 'places of residence' with 'parts of the air', for instance, naturally ascending to form atmospheres. The universality of God's 'established laws' in nature, the assumption that the stars were other suns, and empirical evidence apparently drawn from planetary observations allegedly supported the plurality of worlds hypothesis. With natural theology requiring purpose and meaning in creation, barren lifeless worlds were an impossible conception for many Christian philosophers. Hence by the 1750s, when Kant and Thomas Wright of Durham conjoined the nebular hypothesis with the above, acceptance of extraterrestrial life had become quite common among enlightenment philosophers.[47]

In his *Survey of the Creation*, Parker considered the case for life on all the planets of the solar system, transferring his anthropocentric arguments from the earth. The sun at the centre communicated light, heat and 'consequently life' to the planets, a position similar to the fixed stars, which had the 'same uses of affording light and heat ... to distant opake planets, or globes, to us invisible'. These distant planets had uses which were 'proportion'd by infinite wisdom, as their bulks and powers have render'd them capable of being serviceable'. Parker accepted that the moon was composed of the same elements of the earth and was covered with hills, mountains, rivers and seas 'discovered by all our modern observers'. The 'light, eclipses, monthly revolutions, progressions' all had their 'several admirable uses and services to us', while the earth 'mutually returns her services to the inhabitants'. This was an opportunity, exploited by Parker, to promote his method of determining the longitude using the position of the moon as an example of the services rendered by that body to mariners of the world. Reasoning from his universe of divine purpose with much imagination, astrology and a little empirical observation, Parker examined each of the supposed characteristics of the citizens of the solar system. The Martians were said to be given to 'wars, confusions, disorders and bloodshed', which was why they appeared to have no moon 'by whose light they might lengthen their disorders'.[48]

Facing the problem that Jupiter would have, according to Derham, some twenty-seven times less light and heat from the sun thereby imperilling Jovian life, Parker argued that the Jovian atmosphere might 'augment his apparent magnitude' like 'lenses concentrating light'. The amount of light obvious from the earth suggested that it was certainly not twenty-seven times colder despite the additional distance from the sun. Parker thought that the inhabitants of Jupiter, from their own astronomical observations (using telescopes), would think themselves with their own moons and Saturn to be 'the only globes belonging to the whole system of the sun'. Assuming that they were not more technologically advanced (which he accepted was a possibility), Parker suggested that because of their physical position, being further removed from the sun than the earth, the 'distinction between realities and common appearances' would be 'more difficult' for the Jovians to ascertain than it was for those placed near the middle of the solar system. Parker speculated that the Jovians might have recognised that the motion of Jupiter on its own axis caused the apparent motions of many astronomical bodies, and hence overcame the problem of reconciling appearances with nature's laws.[49]

The Findern Academy

While Parker strove to defend Anglican orthodoxy, Derbyshire dissenters saw natural philosophy through natural theology as providing support for their distinctive theologies and, after 1714, the Whig Hanoverian settlement. After the Act of Uniformity (1662) and the Five Mile Act (1665), educational provision began to split between the established church and the dissenters, reinforced by restrictions placed upon the award of degrees. Dissenting academies played an important part in the provision of scientific education in the later seventeenth and eighteenth centuries, although this was in a context in which grammar schools, church schools and the two universities were also beginning to introduce some teaching of natural philosophy.[50] Robert Merton, inspired in part by Max Weber, famously argued that dissenters and ascetic Protestants were disproportionately well represented in the new scientific community of seventeenth- and early eighteenth-century England and that the puritan ethic was both a direct expression of dominant values, and an independent source of new motivation.[51]

It has further been claimed that dissent is disproportionately represented among the leading eighteenth-century scientific and technological innovators, and dissenting academies evolved as differentiated educational institutions with their own distinguishing practices. According to Orange, certain features of rational dissent such as scripturalism, intellectualism, individualism and catholicity, tended to encourage a broad education which included scientific subjects.[52] Scientific and economic progress was accommodated in nonconformist theologies as the realisation of a divine plan, in contrast to the more static hierarchical Platonism of older organised religions, which tended to find social and intellectual innovation more difficult to accommodate. As Fitzpatrick has made clear, for the rational dissenters in particular, most notably Joseph Priestley – natural philosopher and theologian – scientific truths were accorded a similar status to the fundamentals of theology.[53] The well-known religious heterodoxy of Newton himself offered some support to the association.[54]

The most important Derbyshire nonconformist educational institution was the 'well known' dissenting academy for training ministers that had premises at Derby from 1745 to 1754 and had previously been at the south Derbyshire village of Findern, where it seems to have returned after Latham's death.[55] The Academy may have originated in 1693 with a Derby grammar school, from where it moved about 1710 to Hartshorn and then to Findern. Thomas Hill (d.1720) was the tutor before 1714 in Derby, and at Hartshorn and Findern. Ebenezer Latham (1688–1754), who succeeded Hill at the academy in 1720, was a colleague of Josiah Rogerson (1680–1763) at the Friargate chapel from 1745, who had edited John Platts's *Rational Account of the Principles of Christianity* (1737).[56] Latham's obituary referred to his 'uncommon learning and fine taste' and his 'eminent abilities both as a Divine, a Tutor, and a Physician'. His 'humanity and sweetness of Temper' had endeared him to friends and procured him a 'great reputation' and 'gain'd him the general Esteem of Persons of all Denominations'.[57] Latham was joined at the academy from 1721 by John Gregory, the author of a *Manual of Modern Geography* in 1739 that ran to four editions by 1760, who continued the school after Latham's death. Gregory's pupils included William Strutt, who attended the school at Findern before eventually becoming at pupil at Wilkinson's academy in Nottingham.[58] Gregory was also probably the author of a detailed astronomical letter to the *Gentleman's Magazine* in 1742 which attempted to calculate the pattern of the moon's orbit around the sun.[59]

The main subjects in the four-year course of study at Findern were medicine, divinity, classics, logic, mathematics, shorthand, natural philosophy, anatomy, chronology and Hebrew antiquities. A manuscript book entitled 'Exercitatis Physiologia' by Latham has survived which contains notes and exercises in Latin on natural philosophy, geometry and mathematics. There are discussions of chemistry and the elements, the earth and first principles, the nature of the body, the physics of fluids and solids, minerals, metals and rocks, animal sensation and astronomy. The astronomical notes include a comparison between the Ptolemaic and Copernican systems with diagrams drawn by compass.[60] Authorities used as textbooks at the Academy included Locke *On the Understanding*, Whiston's *Euclid*, Le Clerc's *Physics* and 's Gravesande's *Mathematical Elements of Natural Philosophy*, presumably in the translation by Desaguliers. Jean Le Clerc (1657–1736) was a French Protestant theologian who knew Locke and settled in Amsterdam, and W. J. 's Gravesande (1688–1772) was a Dutch philosopher and mathematician who became a friend of Newton and a Fellow of the Royal Society. His works on natural philosophy also played an important role in the dissemination of Newtonian physics in England.

More light is shed on the academy's curriculum by Gregory's *Manual*, composed using his teaching notes and utilised for his own students, which illustrates how studies of natural history were regarded as having patriotic as well as religious and economic value for the dissenters. Gregory's reasons for encouraging the study of geography combined pleasure with utility and theological purpose, recognising the value of polite education and conversational skills. He wanted his scholars to

> take off their minds from those follies and vanities to which youth is generally addicted: To fit them for conversation; to lead them to contemplate and admire the power and wisdom of the supreme being ... To give them some knowledge of the several nations of mankind that inhabit it: to shew them the peculiar happiness of Great Britain their native country ...[61]

They would then appreciate the value of Britain's geographical position 'where a surrounding sea secures us from the sudden inroads of a foreign enemy; and an excellent constitution preserves all our rights and liberties'. Gregory was also aware of the commercial market demand and repeatedly emphasised the usefulness, brevity, comprehensiveness and relative cheapness of his work, noting that other geographical treatises were bulky, unsuitable for the young, and costly. The *Manual*, and therefore Gregory's curriculum, included a glossary of geographical terms, an appendix on hydrography, 'Sketches of History and Curiosities', and civil and political information. The main body of the work contained descriptions of the countries of the world organised according to the four continents and featuring information on the 'situation, extent, product, government, religion, customs, &c. of every country'. Citing Locke's *Thoughts Concerning Education*, Gregory recommended that the sciences should be taught early in tandem with Latin, and he thought geography was the ideal science to begin, given that, as Locke said it was 'only an exercise of the eyes and memory, a child with pleasure will learn and retain [it]'. He recommended pinning maps to the walls and describing the geography to young people for at least half an hour each day.[62]

Conclusion

In provincial towns such as Derby, the relative peace and stability of the first half of the eighteenth century and the urban renaissance that it stimulated, encouraged the development of a polite public and consumer culture. This was reflected in the transformation of the urban environment and the new wealth, confidence and aspirations of the middling sort. Natural philosophy was an important component of this new public culture, manifest in the growing demand for the material goods of polite science, including scientific instruments and books. Although open conflict was curtailed, continuing religious and political differences also helped to stimulate public scientific culture and encourage the promotion of natural theology and natural philosophy in polemical publications and institutions such as Findern academy.

Notes

1 Sections of this chapter are based upon P. Elliott, 'The birth of public science in the English provinces: natural philosophy in Derby, c.1690–1760', *Annals of Science*, 57 (2000), 61–100, and I am grateful to the editors for permission to use this material here.

2 J. H. Plumb, *The Growth of Political Stability in England, 1689–1725* (London: Macmillan, 1973).

3 G. Turbutt, *A History of Derbyshire*, 4 vols. (Cardiff: Merton Priory Press, 1999), III, 1113–20; J. Pearson, *Stags and Serpents: The Story of the House of Cavendish and the Dukes of Devonshire* (London: Macmillan, 1983).

4 W. Woolley, *History of Derbyshire*, ed. C. Glover and P. Riden (Chesterfield: Derbyshire Record Society, 1981), 41; A. Davison, *Derby: Its Rise and Progress* (London: Bemrose, 1906), 254.

5 J. C. Cox and W. St. J. Hope, *The Chronicles of the Collegiate Church ... of All Saints, Derby* (London, 1881), 221; G. Holmes, *The Trial of Doctor Sacheverell* (London: Eyre Methuen, 1973).

6 W. Hutton, *The History of Derby: from the Remote Ages of Antiquity* (London, 1791), 243; Davison, *Derby*, 66.

7 M. Hutchinson, *Counterfeit Loyalty Displayed* (Derby, 1717), 3–4, 10–22.

8 F. Shaw, *Condolence and Congratulation: A Sermon on the Death of Queen Anne and the Happy Accession of King George to the Throne* (Derby, 1714); E. Latham, *Great Britian's Thanks to God ... a Sermon on Occasion of the Publick Thanksgiving, October 9, 1746* (Derby, 1746).

9 L. E. Simpson, *Derby and the Forty-Five* (London: Allan, 1933), 16–19; R. Simpson, *A Collection of Fragments Illustrative of the History and Antiquities of Derby*, 2 vols. (Derby, 1826), I, 205.

10 *Derby Mercury* (13 December 1745); 'The Chronicle of the Derbyshire Regiment ...', satire by "Nathan Ben Shaddai a priest of the Jews'", in Simpson, *Derby and the Forty-Five*, 165–7, appendix E, 267–71; Hutton, *History of Derby*, 243; Simpson, *History and Antiquities*, I, 196; Davison, *Derby*, 66–90; M. Craven, *Derby: An*

Illustrated History (Derby: Breedon Books, 1988), 74–81; P. K. Monod, *Jacobitism and the English People, 1688–1788* (Cambridge: Cambridge University Press, 1989); L. Colley, *Britons: Forging the Nation, 1707–1837* (London: Pimlico, 1994), 71–85.

11 Simpson, *Derby and the Forty-Five*, 4–5; P. Corfield, 'Small towns, large implications: social and cultural roles of small towns in eighteenth-century England and Wales', *British Journal for Eighteenth-Century Studies*, 10 (1987), 1225–38.

12 Turbutt, *History of Derbyshire*.

13 D. Defoe, *A Tour thro' the Whole Island of Great Britain*, sixth edition, 4 vols. (London, 1762), II, 155–7; *Derby Postman* (15 June, 1721).

14 Woolley, *History of Derbyshire*, 24–31.

15 Defoe, *Tour*, II, 155–7; C. Morris (ed.), *The Illustrated Journeys of Celia Fiennes, 1685–1712* (London: Macdonald, 1982), 149.

16 Woolley, *History of Derbyshire*, introduction, xvi–xix.

17 *An Act for Selling Part of Green called Nun's Green in the Borough of Derby* (8 Geor. III, 1768).

18 *An Act for Paving, Cleansing, Lighting, and otherwise Improving the Streets, Lanes and other Public Passages and Places, within the Borough of Derby* (32 Geo. III, 1792).

19 N. Pevsner, *The Buildings of England: Derbyshire* (London: Penguin, 1953), 110–22; *Derby Mercury* (4, 18 November 1768); M. Craven, *The Illustrated History of Derby's Pubs* (Derby: Breedon, 2002).

20 J. Gibbs, *A Book of Architecture: Containing Designs of Building and Ornaments* (London, 1728).

21 Cox and Hope, *Collegiate Church of All Saints, Derby*.

22 *Derby Mercury* (26 May 1791).

23 J. H. Plumb, *The Commercialisation of Leisure in Eighteenth-Century England* (Reading: University of Reading, 1973).

24 D. Hume, 'Of refinement in the arts', *Essays and Treatises on Several Subjects*, 2 vols. ([1752]; Edinburgh, 1825), II, 265, 268, 274–5.

25 I. Rivers (ed.), *Books and their Readers in Eighteenth-Century England* (Leicester: Leicester University Press, 1982); J. Feather, *The Provincial Book Trade in Eighteenth-Century England* (Cambridge: Cambridge University Press, 1985).

26 H. Cantrell, *The Royal Martyr; A True Christian* (Derby, 1716); J. Redwood, *Reason, Ridicule and Religion: The Age of Enlightenment in England, 1660–1750* (London: Thames & Hudson, 1996).

27 A. Wallis, 'A sketch of the early history of the printing press in Derbyshire', *Derbyshire Archaeological Journal*, 3 (1881), 137–56; R. P. Sturges, 'Cultural Life in Derby in the late Eighteenth Century, *c.*1770–1800' (MA thesis, University of Loughborough, 1968), 5.

28 *Derby Mercury* (8 January 1795, 30 April 1795, 12 May 1796); R. P. Sturges, 'Context for library history: libraries in eighteenth-century Derby', *Library History*, 4 (1976), 44–52.

29 N. Taylor, 'Derbyshire printers and printing before 1800', *Derbyshire Archaeological Journal*, 70 (1950), 38–69; Sturges, 'Cultural Life in Derby', ch. 2; Sturges, 'Context for library history'.

30 J. Brewer, *Party Ideology and Popular Politics at the Accession of George III*

(Cambridge: Cambridge University Press, 1976); H. Barker, 'Catering for provincial tastes: newspapers, readership and profit in late eighteenth-century England', *Historical Research* (1996), 44–61.

31 Sturges, 'Cultural Life in Derby', 5.

32 Hutton, *History of Derby*, 39.

33 Davison, *Derby*, 75–7; *Derby Mercury* (14 June 1787, 11 June 1795).

34 Davison, *Derby*, 108; V. Doe (ed.), *The Diary of James Clegg of Chapel-en-le-Frith*, 3 vols. (Chesterfield: Derbyshire Record Society, 1978–81), III, 741.

35 *Derby Mercury* (4, 18 November 1768); E. Saunders, *Joseph Pickford* (Stroud: Alan Sutton, 1993), ch. 5.

36 Simpson, *History and Antiquities*, II, 531–2; Craven, *Illustrated History of Derby's Pubs*, 140–2.

37 *Derby Mercury* (12 September 1777); Davison, *Derby*, 108–9, 147; Sturges, 'Cultural Life in Derby', ch. 3.

38 *Derby Mercury* (12 August 1748, 8 September 1785, 16 August 1787, 28 August 1788); Davison, *Derby*, 144–6; S. Taylor, 'Musical life in Derby in the 18th and 19th centuries', *Derbyshire Archaeological Journal*, 67 (1947), 1–54; R. P. Sturges, 'Harmony and good company: the emergence of musical performance in eighteenth-century Derby', *Music Review*, 39 (1978), 178–95; S. Wollenberg and S. McVeigh (eds.), *Concert Life in Eighteenth-Century Britain* (Aldershot: Ashgate, 2004); J. Brewer, *The Pleasures of the Imagination: English Culture in the Eighteenth Century* (London: HarperCollins, 1997).

39 J. Money, *Experience and Identity: Birmingham and the West Midlands, 1760–1800* (Manchester: Manchester University Press, 1977), 151, n. 51.

40 *Derby Mercury* (22 July 1784, 21 June 1787, 3 May 1792, etc.); M. Craven, *Derbeians of Distinction* (Derby: Breedon Books, 1999), 191–2; A. Walters, 'Science and politeness in eighteenth-century England', *History of Science*, 35 (1997), 121–54.

41 T. Brighton, *Discovery of the Peak District* (Chichester: Phillimore, 2004).

42 C. Deering, *A Botanical Catalogue of Plants about Nottingham* (Nottingham, 1738); C. Deering, *Nottingham Vetus et Nova* (Nottingham, 1751); A. C. Wood, 'Dr. Charles Deering', *Transactions of the Thoroton Society*, 14 (1941), 24–39.

43 B. Parker, *Parker's Projection of the Longitude at Sea* (Nottingham, 1731); B. Parker, *A Journey Thro' the World: in a View of the Several Stages of Human Life*, second edition (Birmingham, 1738), preface, viii–xi.

44 B. Parker, *A Survey of the Six Days Work of the Creation* (London, 1745), 118; E. Halley, 'A proposal of a method for finding the longitude at sea within a degree, or twenty leagues', *Philosophical Transactions*, 37 (1731–32), 185–95; W. Whiston and H. Ditton, *A New Method for Discovering the Longitude both at Sea and Land* (London, 1714); L. Stewart, *The Rise of Public Science: Rhetoric, Technology and Natural Philosophy in Newtonian Britain, 1660–1750* (Cambridge: Cambridge University Press, 1992); W. J. H. Andrewes (ed.), *The Quest for Longitude* (Cambridge MA: Harvard University Press, 1996); A. Cooke, *Edmond Halley: Charting the Heavens and the Seas* (Oxford: Clarendon Press, 1998), 395–8.

45 W. Whiston, *A New Theory of the Earth, from its Original to the Consummation of All Things* (London, 1696); W. Derham, *Astro-Theology: or a Demonstration of the Being and Attributes of God from a Survey of the Heavens* (London, 1715); B. Parker,

Philosophical Meditations with Divine Inferences (Birmingham, 1738); *A Second Volume of Philosophical Meditations* (Birmingham, 1738); *Philosophical Dissertations with Proper Reflections Concerning the Non-Eternity of Matter* (London, 1738); *A Survey of the Wisdom of God in the Works of Creation* (London, 1745); Stewart, *Rise of Public Science*; R. Porter, *The Making of Geology: Earth Science in Britain, 1660–1815* (Cambridge: Cambridge University Press, 1977), chs 2, 3 and 4.

46 S. J. Dick, *Plurality of Worlds: The Origins of the Extraterrestrial Life Debate* (Cambridge: Cambridge University Press, 1982); M. J. Crowe, *The Extraterrestrial Life Debate, 1750–1900* (Cambridge: Cambridge University Press, 1986); Parker, *Philosophical Meditations*, 26.

47 Parker, *Survey of the Six Days Work of the Creation*, xvi–xix; T. Wright, *An Original Theory or New Hypothesis of the Universe* (London, 1750).

48 Parker, *Survey of the Six Days Work of the Creation*, 81–4, 98, 116, 117–20, 138.

49 Parker, *Survey of the Six Days Work of the Creation*, 140–1, 157.

50 I. Parker, *Dissenting Academies in England* (Cambridge: Cambridge University Press, 1914); M. McLachlan, *English Education under the Test Acts* (Manchester: Manchester University Press, 1931); N. Hans, *New Trends in Education in the Eighteenth Century* (London: Routledge & Kegan Paul, 1966); A. E. Musson and E. Robinson, *Science and Technology in the Industrial Revolution* (Manchester: Manchester University Press, 1969).

51 R. K. Merton, *Science, Technology and Society in Seventeenth Century England* (New York: Harper & Row, 1970); I. B. Cohen (ed.), *Puritanism and the Rise of Modern Science: The Merton Thesis* (New Brunswick: Rutgers University Press, 1990).

52 D. Orange, 'Rational dissent and provincial science: William Turner and the Newcastle Literary and Philosophical Society', in I. Inkster and J. Morrell (eds.), *Metropolis and Province: Science in British Culture, 1780–1850* (London: Hutchinson, 1983), 224.

53 M. Fitzpatrick, 'Heretical religion and radical political ideas in late eighteenth-century England', in E. Hellmuth (ed.), *The Transformation of Political Culture* (Oxford: Oxford University Press, 1990), 339–72.

54 L. Stewart, 'Seeing through the scholium: religion and reading Newton in the eighteenth century', *History of Science*, 34 (1996), 123–65.

55 McLachlan, *English Education*, 131–4.

56 John Platts (1693–1735), minister at Ilkeston, Derbyshire 1708 to 1735.

57 *Derby Mercury* (18 January 1754).

58 Doe (ed.), *The Diary of James Clegg*, appendix 2, 945–60; E. Strutt (attrib.), 'Memoir of William Strutt, esq., FRS', *Derby Mercury* (12 January 1831).

59 J. Gregory, *A Manual of Modern Geography*, third edition (London, 1748); letter by 'J. G.' of Derby, 17 July 1742, *Gentleman's Magazine*, 12 (1742), 369–71.

60 E. Latham, 'Exercitatis Physiologia', Latin manuscript book, DLSL, ms. 3368.

61 Gregory, *Manual of Modern Geography*, preface.

62 Gregory, *Manual of Modern Geography*, preface.

2

Natural philosophy and utility[1]

Introduction

The Derby silk mill is generally regarded as the first British modern manufactory, but what was the relationship between the textile mills and scientific culture? While natural theology provided an important incentive for the study and consumption of natural philosophy in Derby between 1690 and 1760, belief in the utility of science in mechanics and industry was equally important in the establishment of public science. The Royal Society and provincial philosophical societies encouraged the pursuit of natural philosophy and natural history for the benefits they were thought to confer in practical mechanics, navigation, cartography, surveying, and military and naval campaigning. Similarly, scientific ideas and practices were thought to benefit the development of British commercial, naval and imperial interests, along with exploitation of country estates and colonies through enclosure, horticulture, arboriculture and mining. Trade and industry created a demand for improved travel communications, and were themselves stimulated by better waterways and roads, while the demand for mechanical and surveying knowledge and skills was also encouraged by important changes in the outlook and ambitions of the landed classes, who turned their attention to the creation of formal landscape gardens in addition to estate improvement and agricultural exploitation.[2]

This chapter examines how the development of manufacture, industry and mechanics in Derby encouraged the creation of a public scientific culture which placed a premium upon natural philosophy for its supposed economic utility. It argues that the development and celebration of the silk mills, the water-supply system and other local engineering works, particularly by George Sorocold, helped to encourage the study of natural philosophy, while the rhetoric of scientific utility was employed to promote philosophical lectures in Derby from the 1720s. The relationship between practical mechanics and natural philosophy is also evident in the career of John Whitehurst, who built upon a successful clockmaking business to pursue mechanical and philosophical interests.

Water supply, navigation and the silk industry

Although some of the details surrounding the career of George Sorocold are as yet unclear, including the date of his death, he was certainly the most important mechanic in Derby between the 1690s and 1720s, and his work had a major impact upon industry and culture in the town.[3] Sorocold came from a Lancashire family of lesser gentry, members of which had been at Cambridge, and married Mary Francis at All Saints, Derby on 7 December 1684.[4] Where Sorocold gained his education and practical experience is unclear, but if he read Latin, he would probably have read some of the works on hygrography published in Europe since the Renaissance. He was employed on a water-supply system in Macclesfield during the mid-1680s, and between 1687 and 1688 to hang the bells of All Saints. In March 1692, he took control from the Derby Corporation of Gunpowder Mill, Byfleet Island and two sluices in the Derwent, which were to be used for the construction of a town water-supply system. This was to be returned back to the corporation and remained the main water supply for the town into the nineteenth century. A series of pipes carried water to twenty-three streets from the water wheel over Markeaton Brook in three places, over Sadler Gate Bridge and over the Gaol Bridge to St Peter's Street. The wheel raised the water on to the top of St Michael's Church about 35 or 40 feet above, with the height creating enough pressure to circulate it around the pipes estimated at four miles in length. The pressure was such that Erasmus Darwin and his Derby philosophical friends later utilised the principal pipe of the waterworks for their frigorific experiments on the expansion of air in the 1780s. The pipes varied in thickness of bore from one inch to four and a half inches thick and were cut from elm. The water engine could be raised or lowered according to the height of the Derwent, ensuring continuous supply, and simultaneously powered a malt mill and a cutting machine which bored new elm pipes for the system – as Woolley remarked, 'all managed by only one man'.[5] Water was piped only to subscribers, but profits went towards the support and maintenance of eighty lamps for street lighting in the borough.

Although full documentation is lacking, it is clear that Sorocold designed water-supply systems for a number of other towns, including the London Marchant's Waterworks (1696) and the London Bridge Works (1701–3). The former was in association with John Hadley and described by Sir Godfrey Copley as 'the best piece of work I have seen'. The latter was described as the product of 'that great English engineer, Mr. Sorocold', and raised water from the north end of the bridge to a great height, creating the necessary pressure to serve London subscribers. The flux and reflux of the Thames worked the engine, an account of which was published by the engineer Henry Beighton in 1732. Beighton remarked upon the fact that the London engine was 'admirably' contrived for strength and usefulness, and that the barrels, trunks and apparatus were subject to as little friction as possible, so that they were 'far superior' to the engines at the waterworks for the gardens of the palaces of Louis XIV at Marly constructed during the 1680s.[6]

Sorocold also worked on waterworks for the gardens of country houses such as Melbourne, Calke Abbey and Sprotborough, commissions that were stimulated by the relative peace and prosperity of the late Stuart and Hanoverian periods, especially estate wealth and changes in landscape gardening fashion. The formal gardens were also inspired by the waterworks at Marly. At Melbourne Hall for Thomas Coke, Sorocold collaborated with George London and Robert Bakewell, the Derby ironsmith, to redesign the gardens, providing a new water-supply system, an ornamental pool and a lake.[7] Sorocold may also have been involved in draining Derbyshire lead mines, and he was consulted by John, Earl of Mar concerning the drainage of his Clackmannonshire collieries in 1709. The use of Sorocold's engines for mine drainage had been suggested as early as 1694 and he probably came into contact with the early steam engines utilised for this purpose. Around 1709, Sorocold was involved with the Mines Royal and Mineral and Battery Works along with Thomas Holland and John Harris, the author of the *Lexicon Technicum* (1704–10).[8] The fire engines of Thomas Newcomen were being applied to mine drainage from about 1710 and one was built by George Sparrow and associates for the lead mine at Yatestoop, Winster, Derbyshire in 1716–17, where Henry Beighton also appears to have become involved, and others followed. The fire engine was also being employed in water-supply systems, which were Sorocold's speciality.[9]

The importance of the carriage of goods and the incentive this provided to improve road and river communications from Derby are evident from a petition of 1702, which noted that Derbyshire abounded with 'great store of heavy commodities, as lead, iron, marble, plaister, millstones, etc.' As these were usually transported through Derby to the Trent, this made the town, in Woolley's view, 'a thoroughfare, or storehouse'. Repeated attempts were made during the second half of the seventeenth century to reduce carriage costs by making the Derwent navigable south of the town and constructing wharves, which would, according to Woolley, be 'much for the benefit of the trade of this town'. However, despite the support of most of the inhabitants of Derby and the corporation, these did not succeed until 1720, because of opposition from 'the towns of Nottingham, Bautry and Chesterfield' and 'gentlemen whose estates lie upon that river'.[10]

The impact of improved river communications and turnpike roads is evident in the stimulus that these provided to trade and industry, including iron working. Bakewell the ironsmith, who had been apprenticed in London, was encouraged by the demands for gates, screens and other iron structures for private houses, gardens and churches to open a workshop and forge at Oake's Yard, St Peter's Street, Derby. In 1734 William Evans established iron rolling, slitting and flattening mills on the Morledge in the Derwent, which prepared iron for smelting and copper sheathing for the navy. Bakewell's commissions included a wrought-iron abour and the terrace balustrade for the gardens at Melbourne Hall laid out by Coke and London, where Sorocold had also worked.[11]

In addition to the scheme to make the Derwent navigable from Derby to the River Trent, Sorocold's other notable Derby work involved the design and construction of a silk mill financed by Thomas Cotchett, and work on the more ambitious silk mill created afterwards by the Lombe brothers. Cotchett came from an old Derby family and was born at Mickleover in 1640. In 1661 he was admitted a barrister at Gray's Inn, but little is known of his silk venture except that it was a small factory containing eight Dutch mills which continued to function under Lombe as the 'Old Shop'.[12] Water rights were leased on the Derwent in 1704 and Woolley refers to the mill in 1712. The larger mill was erected by John Lombe and his half-brother Thomas Lombe (1685–1739) in 1721 adjacent to the old mill, which became part of the same business venture.

At first imported to Europe from China, silk came to be produced in the medieval period in the Italian city states, which came to dominate the trade. Silk had to be brought into Britain already wound, spun and twisted, at first by hand-powered machines, then from water-powered versions. The Cotchett and more successfully the Lombe venture allowed the silk to be wound, thrown and strengthened by 'doubling' to produce a workable thread. The 'Italian works' housed twelve circular throwing machines and seventy-eight winding machines in the three upper stories. The machinery was powered by Sorocold's single undershot water wheel 23 feet in diameter with the hand-powered process of doubling being performed in a separate building. The key to the eventual success of the business was technological, and machinery components and ideas were initially imported from Europe. Thus Cotchett had imported Dutch mills and John Lombe famously journeyed to Italy to learn the details of the Piedmontese throwing machine, an act of industrial espionage which was punishable by death if the perpetrator was caught, and the Italians offered a reward for his capture. Continental technology retained its superiority, which explains why Lombe was prepared to risk his life in Italy to gain practical knowledge of a water-powered throwing machine in action, although much of the technology was well established, details being easily available in publications. The exact role of Sorocold is unclear, though it must have been important, and he appears to have realised the machinery from Lombe's description, drawings and his own knowledge and experience. A letter of 2 March 1717 refers to him as 'the ingenious, unfortunate, mathematician' and in his *Tour thro' the Whole Island of Great Britain*, Defoe described an accident where Sorocold fell into the mill sluice while demonstrating his engineering work and was pulled under the water wheel, without, however, receiving any 'hurt at all'.[13]

The silk mill became celebrated as a model of industrial improvement gained through the application of practical mechanics. Furthermore, Lombe's mill was a prototype for all subsequent textile mills, notably the pioneering silk mills of Macclesfield and the cotton manufactories of the Derwent Valley and Manchester. The Derby silk mills also had an influence beyond the textile industry, as demonstrated by the degree of celebrity they attained as a wonder of technology in, for example, Defoe's *Tour*. Sir Thomas Lombe's application to parliament for the extension of his patent in 1732 attracted considerable attention, though Lombe was keen to exaggerate the difficulties he had in receiving good-quality Italian silk to ensure his continued monopoly.[14] The *Gentleman's Magazine* recorded that the engine contained

> 26,586 wheels and 97,746 movements, which work 73,726 yards of silk thread every time the water wheel goes round ... and 318,504,960 yards in one day and night. One water wheel gives motion to all the rest of the wheels and movements, of which any one may be stopped separately; one fire engine conveys warm air to every individual part of the machine, and one regulator governs the whole work.[15]

This 'fire engine' was a furnace rather than a Newcomen steam engine. In parliament the silk mill was described as 'a most curious and intricate structure' which had improved the whole County of Derby and allowed the poor to be employed – a common justification for improvements rendered by practical Georgian industry. According to Alderman Perry, who in contrast highlighted the greater industrial efficiency, it was

> a very large engine, which is first moved by water, by means of which first motion a great many wheels and spindles are set a moving, and thereby great quantities of silk are twisted in a much finer manner, and by much fewer hands, than can possibly be done by any engine that was ever yet invented.[16]

Though the House refused to renew the patent, a committee granted Lombe £14,000 'as an encouragement for his useful invention', which the act called 'this wonderful piece of machinery' and which it attributed to the direction of Sorocold. Knowledge of the engine spread, partly because Lombe had to provide models of it under the terms of the act, which only after much prevarication were later deposited in the Tower of London following his death. John Lombe had had to assemble and specially train workers by hiring 'various rooms in Derby, and particularly the town hall, where he erected temporary engines, turned by hand'.[17] There must have been a shortage of suitably skilled local workers, because Lombe brought some Italians over who had the requisite experience with silk engines, and Sorocold's knowledge and experience must have been invaluable. Exceptions were mechanics and clockmakers who had prior knowledge of gear systems, and presumably men who had already worked with Sorocold on his water-supply engines. Bakewell, the ironsmith, produced a pair of elaborate gates and probably some of the iron components, although most of the machinery was constructed from wood. Other writers followed Defoe in describing Lombe's mill. In 1748 the Rev. James Gatt, a Scottish minister at Gretna, composed a Latin poem in celebration of the 'wonderful' creation, which emphasised that the Derby silk mill was 'constructed with such skill that in my judgement it is certain that nothing finer has been found in the whole world'. Who, asked Gatt, standing beside it was 'not astounded by so many movements, and as many wheels, and the brilliance of mind by which such a machine is produced?' For Gatt, if the art of man was so great then the mind of God must be so much the greater; thus the silk mill led him to celebrate the 'divine perfection of his majesty'.[18]

Natural philosophy was frequently justified because of its apparent utility in the arts and sciences in addition to its fashionableness in polite culture. Sorocold's water-supply system and the silk mill were precursors of other engineering ventures. The water-supply system was mentioned by John Houghton when still under construction in 1693, and Celia Fiennes described it, remarking upon the fact that the engine could be raised or lowered according to the height of the Derwent.[19] Though, ironically, the silk mill throwing-machinery design had been stolen from Italy, its engine was perceived as a brilliant application of reason to industry. Contemporaries regarded the complexity, intricacy, neatness, size, precision and degree of mechanisation as being the most 'wonderful' aspects of the engine, and celebrated the heating system from the fire engine and the fact that any part of the 'wheels and movements ... may be stopped separately'. Lombe's estate was worth at least £120,000 on his death in 1739 – a great encouragement to others to study mechanics, mathematics and natural philosophy, which had apparently been so important, regardless of the actual role that these may have had in the creation of the mill. Mathematics, hydrostatics, surveying, geometry and drawing were staple subjects of the lecturers at Derby in the same period, with models of machinery such as pumps being another common feature.[20] Lombe's Derby silk mill and, to a lesser extent, Sorocold's water-supply system became celebrated as technological wonders, the applications of rational thought and mathematics to industrial production, whether or not they actually demonstrated the application of philosophy to technology. It is likely that local philosophers such as Whitehurst were inspired by this and scientific lecturers seem to have alluded to the Sorocold machinery, particularly when discussing hydrostatics, which was included in the syllabuses of Arden and Griffis. Whitehurst's friend James Ferguson included models of water mills and pumps in his lectures, and when he came to Derby from the 1760s made sketches of the silk mill machinery.[21]

The application of Newtonian natural philosophy to practical mechanical problems encouraged mathematical study. The importance of mathematics in scientific culture is evident in the correspondence contained in newspapers and periodicals such as the *Gentleman's Diary*, the *Ladies' Diary* and the *Gentleman's Magazine*. Featuring mathematical correspondence and problems from around Britain, these publications helped to create and satisfy the desire for mathematics and natural philosophy at local and national levels. Founded by the mathematician and mechanic Henry Beighton (1636–1743), the *Ladies' Diary* has been described as 'one of the main' Georgian 'channels for the diffusion of mathematical science'. Despite the title, the magazine had a wide male and female readership, from schoolboys to clergy, schoolmasters and mistresses, and contained astronomical information and mathematical questions, many in rhyme, with prizes for successful solutions to problems.[22]

Thomas Simpson, the astronomer and mathematician who became editor from 1754 and professor of mathematics at the Woolwich Academy, taught mathematics in Derby between 1733 and 1735. An avid reader of the *Diary*, Simpson sent in articles and problems under his own name and a variety of pseudonyms. A weaver and astrologer from Market Bosworth, Leicestershire, like Benjamin Parker, Simpson deliberately curried favour with the Cavendish family by writing and publishing an election poem in their praise (1733). He composed his *New Treatise of the Fluxions* at Derby and carried it to London in 1735, leaving his family behind in Derby. The simultaneous development of the calculus by Newton and Leibnitz provided the opportunity to manage mathematical information that was theoretically infinitely small by the differential equation, and allowed mathematicians and mechanics to utilise the fluxions to analyse the geometry of curves and surfaces, while astronomers could ascertain the pattern of ellipse orbits using Kepler's laws. The *Treatise* featured demonstrations of practical examples of the application of calculus to a series of astronomical and mathematical problems. According to Charles Hutton, Simpson published the work because he 'did not know of any English book founded on the true principles of fluxions that contained any thing material, especially the practical part'. Therefore, practical applications of the calculus were given with examples of solutions to various problems 'rendered plain to ordinary capacities', including, the 'use of fluxions in some of the sublimest branches of physics and astronomy … done in a method quite different from any thing' already published. Simpson also considered pendulums, centripetal force, the paths of shadows, the motion of projectiles in different mediums and the attractive powers of bodies in various forms.[23]

Other mathematical and philosophical pieces were apparently sent to the *Ladies' Diary* from Derby during the 1730s, including some signed by Richard Lovatt and Christopher Hale, although these may be pseudonyms of Simpson. According to Lovatt:

When mighty Newton the foundation laid,
Of his mysterious art none coul'd invade
Nor take from him the honour which was due,
Great Britain's sons will long his works pursue.
By curious theorems he the moon coul'd trace,
And her true motion give in every place ...

In another poetical question of 1738 Lovatt described how with a friend he had seen the 'celestial orbs', 'wond'rous works by nature wrought', 'heav'nly bodies which to us appear, like blazing tapers, wand'ring in the air'. Another mathematical question from Hale required readers to find the height of All Saints' Church, Derby.[24]

Courses of lectures in natural philosophy

By increasingly situating their courses of lectures in towns, often in taverns, itinerant philosophers were forming and satisfying a distinctly urban demand, and from the 1730s and 1740s lecturers such as Richard Griffis (or Griffiths) and John Arden gave courses on natural philosophy at Derby, demonstrating the audience already existing for the subject.[25] Derby's north midland location placed it on the routes between London and northern towns, making it a destination on the lecture touring circuit. The improvement of the road network allowed the conveyance of ever more elaborate apparatus which added to the drama and spectacle. There was no hard and fast division between natural philosophy lecturers and other public entertainments; indeed, itinerant lecturers had to have something of the showman in them – scientific instruments serving as their theatrical props. Quacks and showmen offered entertainment and sold potions and nostrums, but so did more 'serious' scientific lecturers, and while philosophical institutions such as the Royal Society strove to forge a legitimate science, the market for curiosities remained insatiable.

In 1728 John Grundy senior (1696–1748), 'land-surveyor and teacher of the mathematics', advertised in the *Derby Postman* his services as cartographer and surveyor of gentlemen's estates, using a 'method entirely new, which never fails to justly represent, both in shape, position, and similar proportion every furlong and single land'. He promised that any gentleman ignorant of surveying methods might 'by inspection, tell whether his survey be justly done' by instruments including the theodolite, Beighton's plain table and Gunter's chain'. During the winter Grundy offered to teach subjects including merchant's accounts, algebra, geometry, 'both plane and solid, the doctrine of triangles, both plane and spherical, and either orthographic or stereographic projection of the sphere, gauging, dialling, surveying, navigation, astronomy, optics, and drawing'.[26] Grundy senior was one of the most important English river engineers and his philosophical interests are underscored by his membership of the Spalding Gentlemen's Society. He came from Leicester-shire and lived at Congerstone and later Spalding, working in Leicestershire, Cheshire, Nottinghamshire and Lincolnshire on various canal, river naviga-tion and drainage schemes. In the early 1730s, he was working on the River Welland and Bladeslade's Canal near Chester, which was intended to improve the navigation to that town. Later he worked on making the Trent navigable near Newark, while his son John junior was also an engineer who collaborated with John Smeaton.[27]

Griffis, who was 'well known' at Derby, gave private tuition at gentle-men's houses and philosophical courses there and in other midland towns such as Lichfield, Birmingham, Burton and Wolverhampton. By 1757 he had lectured in Bath, Bristol, Frome, Warminster and Salisbury to 'great numbers of people' on natural philosophy, chemistry, pneumatics, optics, geography and astronomy and the 'application of these principles to the improvements of trade' using a 'large and noble set of instruments'.[28] In 1739 Griffis had for 'some time' been delivering a course of experimental philosophy in Derby and had begun a second subscription. In 1743 another course in Derby was adver-tised and lectures were offered at private houses if twenty or more of either sex would subscribe. His equipment included a 'vast variety of instruments' such as a pump that worked without a water wheel, a bucket engine, a 'curious air gun, a planetarium, a cometarium, a Ptolemaic sphere, and an improv'd orrery, telescopes and microscopes of all sorts, particularly a new invented one by Mr. Lindsey'. This 'celebrated philosophical apparatus' was sold to the scientific lecturer Adam Walker in 1766.[29]

Arden, who came from Bromsgrove, was resident in Derby between 1739 and 1752, where he taught mathematics from a schoolroom in Irongate. Adver-tisements provide some details. In 1739 it was stated:

> At Mrs Milward's large room upon the Sadlergate Bridge in Derby (late Mr. Adam's school-room) will be completely taught, reading, writing, in all its usual hands, arithmetic in all its parts, viz. vulgar, decimal, logarithmetrical, instru-mental and algebraical, geometry, speculative, and practical, ... together with the use of the globes, and other mathematical instruments.[30]

Additionally, Arden offered to teach subjects including projection of the sphere, geography, practical and theoretical, the 'calculation of the celes-tial motions, particularly of the eclipses of the sun, moon, and satellites of Jupiter'.[31] He also contributed mathematical questions and other pieces to the *Mercury* and the *Gentleman's Diary*, including some that dealt with various problems.[32] It is possible that Richard Oldacre(s) may have been one of Arden's pupils at this time. Later the proprietor of a successful academy at Woodbor-ough near Nottingham, Oldacre went to Derby in 1741, taking day and evening classes, and 'made himself master of almost every part of mathematics and natural philosophy'. He later taught natural philosophy and related subjects at his school, including geometry, surveying, navigation and astronomy using scientific instruments.[33]

Arden's courses of natural philosophy were in direct competition to those of the popular Griffis, and both tried to attract more subscribers by offering new and curious subjects and novel machines and equipment. In 1744, in collaboration with Francis Midon, Arden announced a course of eight lectures, seven on the 'nature and properties of air' and the eighth to be given by Midon on 'all the most surprising phaenomena of attractions and electricity'. The lectures would be continued three times a week and it was assured that they would be intelligible in a short time even to those 'without any previous application to books'.[34] Arden continued to give scientific lectures in Derby until 1751. In 1749 he offered another course in natural and experimental philosophy as soon as forty or more had subscribed. The lectures exhibited 'a great variety of experiments', many of which, it was claimed, were 'entirely new' and would elucidate 'natural philosophy in general, or the properties, affections, and phenomena of natural bodies hitherto discovered, and the laws by which they act'.[35] Subjects covered included hydrostatics, pneumatics, the 'surprising properties of air', and astronomy 'or the phaenomena arising from the motion of the heavenly bodies, geography, and the use of the globes', the attraction of natural philosophy as polite learning being emphasised here, particularly for women, rather than its utility. The course cost half a guinea and catalogues of the experiments were offered free by Arden from his schoolroom, along with private tuition for 'young ladies and gentlemen'.[36] Arden offered another similar course in January 1752 costing half a guinea, emphasising that 'the greatest care will be taken to make experiments as plain and intelligible as possible' even to beginners.[37]

The content of Griffis's and Arden's lectures was similar to those of other itinerant teachers around the country.[38] Caleb Rotherham, for example, of the Kendal Academy, offered similar subjects in Manchester, while Benjamin Martin (1704–82) included the subjects of physics, mechanics, hydrostatics, hydraulics, pneumatics, phonics, optics, astronomy and geography in his lectures. James Ferguson (1710–76) began his lecturing career in London during the 1740s with courses on the 'use of the globes', geography and astronomy, which included the 'motions of the moon and Venus' shown on one of his early four-wheeled orreries. Ferguson charged one shilling per lecture, about the same as Arden, who required half a guinea for a course of ten lectures, and as the value of a guinea was set in 1717 at 21 shillings (though it did fluctuate), this gives a price of just over one shilling for each of Arden's Derby lectures.[39]

The main difference between Arden's and Ferguson's lectures was that the former included a larger range of subjects such as astronomy, pneumatics, mathematics (geometry, algebra and logarithms), geography and hydrostatics, whereas the latter originally focused mainly on astronomy. At Bristol in 1749, Ferguson advertised five astronomy lectures using the orrery and 'several other machines and diagrams'.[40] Demonstration equipment was a very important feature and included steam engines, electric machines, globes and, of course, various kinds of orrery. Arden's advertisements do not mention an orrery, though his lectures included 'the calculation of the caelestial motions', 'eclipses of the sun, moon, and satellites of Jupiter', and 'astronomy, or the phaenomena arising from the motion of the heavenly bodies'. This suggests that he probably utilised one from 1739, as Ferguson was doing, although the first definite reference to the use of an orrery at Derby is in the Griffis lectures of 1743.[41]

Arden left Derby abruptly before the end of January 1752, just after advertising another lecture course. He seems to have sold the business as an ongoing concern including schoolroom and instruments to George Leighlar, who advertised that quadrants, geometry, arithmetic, navigation, and the use of the globes were to be taught in the 'Iron Gate, at Mr. Arden's late school room'. This was probably because Arden had been offered a position at the Heath Academy near Wakefield by Joseph Randall, the founder, 'whose attainments in the mathematics and the science of calculation was considerable'. Arden taught at the academy between 1752 and 1754, when it failed.[42] The mathematician George Gargrave (1710–85) also taught there before setting up a mathematical school at Wakefield, and one of the most famous pupils was Dr William Hey, who met Arden in 1778 when he was giving a course of lectures at Bath. Hey later helped to found a philosophical society at Leeds in 1783 and acknowledged that it was from Arden that he had 'acquired a taste for natural philosophy'. Arden moved to Beverley, then in Yorkshire, where he lived between 1754 and 1776 or 1777, when he moved to Bath.[43] He lectured at many towns, including Manchester (1756, 1762, 1774 and 1776), Nottingham, Liverpool (1740, 1782), Shrewsbury and Newcastle (before 1764), Derby (1762), Rotherham (1761), Bristol and Birmingham (1765). The course of twenty lectures at Liverpool in 1782 was typical and included a series of demonstration experiments intended to show the properties of matter and laws of motion, mechanics, astronomy, hydrostatics, pneumatics, geography and 'the use of the globes'.[44] At Bath, he resided at St James's Street, where he had a lecture room and gave philosophy courses assisted by his son James, helping to found the Agricultural Society and the Bath Philosophical Society. He returned to Beverley between 1782 and 1784, where he died in died in May 1791.[45]

Between 1762 and 1764, Arden constructed an electric orrery, which underscores the requirement for original and striking apparatus for philosophical lectures. This was described in correspondence with Charlewood Lawton and Thomas Birch, secretary of the Royal Society. Arden claimed that the electric orrery had been a great success in his lectures at Shrewsbury and Newcastle and hoped that Birch would print his description in the *Philosophical Transactions*, though he never did.[46] The device took the form of a mobile supported on a vertical axis and composed of gilt balls mounted on four well-balanced and freely rotating wire arms. When connected to a Leyden jar and electrical machine, the arms were driven round by electrical discharges from pointed conductors at the ends of the arms, and thereby represented the rotation of the sun and the orbital motions of Mercury, Venus and the earth.[47] In a similar way, the rotation of the earth and the moon was shown. Arden emphasised that this type of orrery would be cheap to produce, suggesting that if it were made public he doubted not that it would 'soon be so much improv'd as to answer the end of explaining the sistem of the world full as well if not better than the large expensive orreries of 100 or 110 guineas price.'[48]

Ferguson also created electrical machines for his lecture courses, introducing electricity to his syllabus by early 1768. Arden introduced electricity into his lectures much earlier and probably mastered the subject in the later 1740s or 1750s to obviate the need to split course earnings with Midon. The Leyden jar, an early form of static-electrical capacitor comprised of a metal-lined glass jar half-filled with water, allowed frictional charge to be stored and released, making it ideal for experimentation and public demonstration. Ferguson described various electrical machines in his *Introduction to Electricity* (1770), many of which were little more than toys and included clocks, an orrery, a grinding mill and a pump. They were all powered by what Ferguson referred to as the 'electrical wind' and utilised the lightest possible materials – usually card, cork and wire – to facilitate animation. Ferguson's electric orrery was different from Arden's device. Instead of showing the orbits of Mercury, Venus, the earth and the moon, Ferguson's device showed only the diurnal rotation of the earth and the phases of the moon. Arden had used a well-balanced wire mobile with pointed conductors at the ends and a Leyden jar for the structure. Ferguson, however, used gears, wheels and two pinions to turn the moon and earth on two axes. In all his electrical devices, motion was provided from a large cardboard wheel that looked similar to a paddle wheel, which was turned by the action of electricity emanating from a pointed conductor.[49]

Arden returned to Derby in 1762 to give another course of experimental philosophy with a syllabus little different from previously, consisting of twenty lectures on astronomy, mechanics, geography, hydrostatics and pneumatics. The only significant change was the incorporation of 'optics: or the science of vision'. Newtonian optical experiments, like electricity, were ideal for striking public demonstration and could easily be repeated at home.[50] It is unlikely that Arden had read the original works of Newton, but he would probably have read one of the works such as Desaguliers's *Course of Experimental Philosophy* (1734) or Robert Smith's *Complete System of Optics* (1738). He also included

4 John Whitehurst.					5 James Ferguson.

'pneumatics ... the surprising properties of the air ... pressure and elasticity ... rarefaction, condensation, etc.', which would have been demonstrated using an air pump. The equipment was 'very extensive, neat, and elegantly finished, according to the latest improvements', and the experiments were to be 'as plain and intelligible as possible even to those who have not apply'd anytime or study this way'.[51] The globes and orrery were used, the latter likely to have been a mechanical geared model, although it could have been the electric variety that he was developing.

John Whitehurst

The career of John Whitehurst at Derby and London also underscores the interconnections between practical mechanics and natural philosophy. Whitehurst moved from Congleton, Cheshire to Derby in 1736 and became a town burgess in 1737 on the presentation of a clock for the town hall. He married Elizabeth, daughter of the Rev. George Gretton, in 1745 and developed a clockmaking business which employed a number of apprentices. While the business thrived from its provincial base, from the 1750s Whitehurst assiduously strove to develop a network of commercial and philosophical contacts, friends and patrons. Retaining his Derby business interests, he moved to London in 1775 to become Stamper of the Money Weights for the increasingly unpopular North administration. Perhaps the most important of these networks were those in London and Birmingham centred upon what later became known as the Lunar Society, and Whitehurst was elected FRS in 1779. By the 1770s, he had become most known for his geological work, exemplified by the publication of his *Inquiry into the Original State and Formation of the Earth* (1778). With the important exception of surviving clocks, much less is known of his earlier career; however, this was crucial in the establishment of scientific culture in Derby between the 1730s and 1770s. According to Charles Hutton, his school education was 'very defective'. Whitehurst was brought up by his father, who showed him 'mechanical and scientific pursuits' and 'encouraged him in every thing that tended to enlarge the sphere of his knowledge'. He accompanied him on 'subterraneous researches' that had been inspired by 'the many stupendous phenomena in Derbyshire'. Prior to 1778 he had been 'consulted in almost all undertakings in Derbyshire, and the neighbouring counties, where the aid of superior skill in mechanics, pneumatics and hydraulics was required.'[52]

A quarrel caused by the publication of an account of a pyrometer constructed by Whitehurst reveals the importance of practical mechanics in stimulating his philosophical activities. The device measured the extension or contraction of a length of metal when heated and the expansion of 'other rods, even tobacco pipes and glass tubes'. In 1749 when the pyrometer was in the possession of Arden, an illustrated description appeared in the *Gentleman's Magazine*. The device incorporated a large plate of whited brass in which was engraved a circular calibration and a lengthy frame with a screw at one end to fasten the rod of metal to be measured. When the rod expanded it pushed a bar of metal that by means of a watch spring, a small pulley and a larger pulley moved the needle to record the amount of expansion that had taken place. The pyrometer was thought to be able to measure the expansion of rods with a new degree of precision.[53]

The publication of the description resulted in a quarrel with E. Sparke, who accused Whitehurst of dishonestly copying the idea, claiming that John Ellicott had already designed, constructed and published an account of a similar device in the *Philosophical Transactions*. Sparke also claimed that Whitehurst's design was inferior to Ellicott's machine, even though, as Whitehurst pointed out, these two accusations tended to contradict each other. The major difference between Ellicott's pyrometer and Whitehurst's device was that the former used a weights-and-pulley system rather than a spring. Ellicott's measured the amount of heat that was being applied to the metal while also measuring the expansion. Thus, as Sparke pointed out,

> The principal use to which these instruments are applicable being to measure the quantity of expansion produced in different metals, with any certain degree of heat; the excellency of such an instrument consists not only in rendering the least quantity of expansion perceptible, but likewise in determining with exactness the degree of heat made use of in each experiment.[54]

Whitehurst was explicitly criticised for having 'copied the most essential parts of his instrument.'

From the reply Whitehurst made to the charges, printed after Sparke's letter, we learn that he only submitted the design for publication at the instigation of his Derby friend John Arden. Whitehurst explained that he wrote the reply only 'to justify myself from the imputation of having either pirated it, or copied any thing from it'.[55] He argued convincingly that no one who realised that Ellicott's machine had been printed in the *Transactions* and copied it would have been foolish enough to publish it in the *Gentleman's Magazine*. He contended that as solutions to the problem were limited to two methods, 'either wheels, and pinions, or pulley and bands' so 'when a man is convinced of the imperfection of the former, would he not very naturally put the latter in practice?'[56] The advantage of Ellicott's attempt to standardise by measuring the amount of heat being applied to the rod was acknowledged by Whitehurst, who did not question the priority of his invention. His main quarrel was with Sparke, who should have shown 'a little tenderness' for his reputation.

As Whitehurst and Arden were not aware of the Ellicott pyrometer, it would appear that they did not read the *Philosophical Transactions*. As Whitehurst stated indignantly, 'Mr. Sparke might have thought, that there are many mechanics who never saw the ... Transactions. This was really my case.' Arden also 'did not think there was anything like the pyrometer anywhere else', which was why he advised Whitehurst to publish. Although it is possible that Arden may have overlooked Ellicott's article, it is significant that despite being a teacher of mathematics and natural philosophy, Arden does not seem to have read, or to have read regularly, the *Philosophical Transactions*. Although Whitehurst became a philosopher of national repute by the 1770s and an FRS, during the 1740s he was still primarily a clockmaker and practical mechanic, and the *Philosophical Transactions* was an expensive publication, even in abridgements, beyond the reach of many individuals without wealthy patronage or access to philosophical libraries. The disparity between the cost of scientific books at some four to ten shillings, and the cost of literary, historical and theological titles, averaging only two shillings, remained considerable prior to 1750, providing a barrier to membership of the philosophical community. Subscription libraries were only just coming into existence and one of the major reasons for the foundation of philosophical societies was, of course, library provision.[57]

Limitations to the Whitehurst pyrometer may suggest a lack of theoretical scientific knowledge at this stage in his career. The device employed his knowledge of clock mechanisms, allowing him to develop a dial and calibration system sufficiently exact to record minute expansion of heated substances. This was probably stimulated by the need to devise accurate clock mechanisms resistant to heat, and Whitehurst incorporated temperature-compensating pendulums in his clocks. However, the failure to measure and control all major variables and the implicit assumption that expansion under heat would be uniform over the whole surface were limitations, as both the amount of heat being applied and the amount of expansion needed to be measured. Without a measurement of the quantity of heat it is hard to see what use the pyrometer could have been beyond the demonstrative and educational (which may explain Arden's interest). However, Ellicott's system of weights was inferior to Whitehurst's springs and pulleys and was hardly portable, whereas Whitehurst's device was easier to carry for use anywhere.

Whitehurst continued to work on his pyrometer. In 1758 he reported to Matthew Boulton that he had 'at last' constructed a pyrometer 'to please me'. It had 'all the perfection I could wish for, and will, I think, ascertain the expansion of metals with more exactness than the machine extant'. He continued to conduct trials with Boulton, and another letter from this period describes the new pyrometer and a hygrometer of Boulton's design, requesting that rods be prepared for experiments on the effects of heat expansion. Whitehurst also described some joint research to 'stop the vibrations of bells in chime music'.[58] The work on the measurement of heat expansion reveals that, like some of his Lunar colleagues, Whitehurst took a keen interest in attempts to define, quantify and measure phenomena with the greatest exactitude. It also demonstrates the relationships between clockmaking, practical mechanics, industry and natural science in the period. In clockmaking, of course, Whitehurst was concerned to measure time accurately and consistently and, in London from 1775, as Controller of Weights and Measures, he tried to calculate the smallest units using a pendulum and published an account of his work.[59]

Conclusion

In *The Gifts of Athena*, the economic historian Joel Mokyr has emphasised the importance of creative communities in the development of the British industrial economy, centred upon scientific societies, coffee houses, scientific lectures and other contexts. Distinguishing between propositional knowledge or beliefs concerning natural phenomena and instructional or prescriptive knowledge, he has also argued that in Britain the gap between those who engaged in the former and those applying it to production was the narrowest of any country by 1700, becoming narrower during the eighteenth century. Beyond the traditional question of causation between science and industry, he is more interested in the 'strong complementarity' or 'continuous feedback' between engineers, artisans and natural philosophers which made the British economy distinctive and hastened the industrialisation process.[60] As we have seen, the growth of this culture is fully evident at Derby between 1690 and 1760, where, as the careers of Sorocold, Whitehurst and Arden demonstrate, there were close connections between practical mechanics and natural philosophy. As the advertisements and demonstrations provided by philosophical lecturers demonstrate, belief in the industrial utility of natural philosophy through the operation of practical mathematics and mechanics provided one of the most important incentives for scientific study, helping to create a public urban scientific culture.

Notes

1 Part of this chapter is based upon P. Elliott, 'The birth of public science in the English provinces: natural philosophy in Derby, *c.*1690–1760', *Annals of Science* 57 (2000), 61–100, and I am grateful to the editors for permission to use this material here.

2 M. Jacob, *The Cultural Meaning of the Scientific Revolution* (New York: MacGraw-Hill, 1988); I. Inkster, *Science and Technology in History* (London: Macmillan, 1991); L. Stewart, *The Rise of Public Science: Rhetoric, Technology and Natural Philosophy in Newtonian Britain, 1660–1750* (Cambridge: Cambridge University Press, 1992); M. C. Jacob and B. J. Teeter Dobbs, *Newton and the Culture of Newtonianism* (Atlantic Highlands NJ: Humanities Press, 1995); I. Inkster, 'Technological and industrial change: a comparative essay', in R. Porter (ed.), *The Cambridge History of Science*, vol. IV: *Eighteenth-Century Science* (Cambridge: Cambridge University Press, 2003), 845–82; J. Mokyr, *The Gifts of Athena: Historical Origins of the Knowledge Economy* (Princeton: Princeton University Press, 2005).

3 F. Williamson, 'George Sorocold of Derby: a pioneer of water supply', *Derbyshire Archaeological Journal*, 56 (1936), 50; F. Williamson, 'Extent of Sorocold's water supply system at Derby', *Derbyshire Archaeological Journal*, 74 (1954), 100–11; M. Chrimes, 'George Sorocold', in *Biographical Dictionary of Civil Engineers in Great Britain and Ireland: 1500–1830*, I, ed. A. W. Skempton, R. W. Rennison, R. C. Cox, T. Ruddock, P. Cross-Rudkin and M. M. Chrimes (London: Thomas Telford, 2002), 641–5. M. Craven has argued, on the basis that the lease for the waterworks changed hands at the time, that Sorocold may have died in 1737–38.

4 One George Sorocold was admitted a fellow-commoner at Emmanuel College, Cambridge, 15 May 1684: John A. Venn, *Alumni Cantabrigienses*, 6 vols. (Cambridge: Cambridge University Press, 1954), IV, part 1, 44.

5 E. Darwin, 'Frigorific experiments on the mechanical expansion of air', *Philosophical Transactions*, 78 (1788), 46; W. Woolley, *History of Derbyshire*, ed. C. Glover and P. Riden (Chesterfield: Derbyshire Record Society, 1981), 26; W. Hutton, *The History of Derby: from the Remote Ages of Antiquity* (London, 1791), 14; S. Glover, *History and Directory of the Borough of Derby* (Derby, 1843); F. Nixon, *The Industrial Archaeology of Derbyshire* (London: David & Charles, 1969), 95–8. Hydraulic pipe-boring machines appeared at least as early as the fifteenth century and an engraving of one appeared in G. Bockler, *Theatrum Machinarum Novum* of 1661, reproduced in T. S. Reynolds, *Stronger than a Hundred Men: A History of the Vertical Water Wheel* (Baltimore: Johns Hopkins University Press, 1983), 75. The water wheel was similar to one patented by John Hadley in March 1693, suggesting a close connection between himself and Sorocold.

6 Williamson, 'George Sorocold'; H. Beighton, 'A description of the water works on London Bridge', *Philosophical Transactions*, 37 (1732), 5; N. Smith, *Man and Water:*

A History of Hydro-Technology (London: P. Davies, 1976); Reynolds, *Stronger than a Hundred Men*, 61, 78, 80, 187. Ramelli had described a water wheel that could be raised or lowered according to the height of the river in 1588, Agricola had pictured hydraulic mine drainage systems with triple suction pumps in his *De re metallica* (1556), and Vittorio Zonca had detailed various hydraulic mills in his *Novo Teatro di machine et edificii* (1607).

7 M. Craven and M. Stanley, *The Derbyshire Country House*, revised edition (Ashbourne: Landmark, 2004), 150.

8 Williamson, 'Extent of Sorocold's water supply system', 64–5; Stewart, *Rise of Public Science*, 239.

9 Stewart, *Rise of Public Science*, 244–5; R. L. Hills, *Power from Steam: A History of the Stationary Steam Engine* (Cambridge: Cambridge University Press, 1989).

10 Woolley, *History of Derbyshire*, 24; Petition of the Mayor and Burgesses presented to the Commons to render the Derwent navigable (17 November 1702), Williamson, 'George Sorocold', 50.

11 Hutton, *History of Derby*, 211; Glover, *History and Directory* (1843), 80; Craven and Stanley, *Derbyshire Country House*, 150.

12 Williamson, 'George Sorocold'; F. Williamson, 'The Cotchett family of Derby', *Derbyshire Archaeological Journal*, 60 (1939), 172–6; W. H. Chaloner, 'Sir Thomas Lombe and the British silk industry', *History Today*, 3 (1953), 778–89; A. Calladine, 'Lombe's mill: an exercise in reconstruction', *Industrial Archaeology Review*, 16 (1993), 82–99; F. Nixon, *Notes on the Engineering History of Derbyshire* (Derby: Derbyshire Archaeological Society, 1956), 6–7.

13 Williamson, 'George Sorocold', 59–64; Calladine, 'Lombe's mill', 87; D. Defoe, *A Tour thro' the Whole Island of Great Britain*, first edition, 3 vols. (London, 1727), III, 38.

14 Calladine, 'Lombe's mill', 87; D. Bentley-Smith, *A Georgian Gent. and Co.: The Life and Times of Charles Roe* (Ashbourne: Landmark, 2005).

15 *Gentleman's Magazine*, 21 (1732), ii, 719, 940, 985–6.

16 A Newcomen engine was installed at Yatestoop lead mine, Winster, Derbyshire in 1716–17: Stewart, *Rise of Public Science*, 244; *The History and Proceedings of the House of Commons from the Restoration to the Present Time* (London, 1742), VII, 140–5.

17 Chaloner, 'Sir Thomas Lombe', 782.

18 J. G. Macqueen, 'An unpublished poem on the Derby silk mill', *Derbyshire Miscellany*, 10 (1983–85), 146–8.

19 J. Houghton, *Collection for Improvement of Husbandry and Trade*, 38 (21 April 1693); C. Fiennes, *Through England on a Side Saddle in the time of William and Mary* (London, 1888), 140, quoted in Williamson, 'George Sorocold', 68.

20 *Gentleman's Magazine*, 11 (1739), 47.

21 J. Ferguson, *Tables and Tracts* (London, 1767), 172–3.

22 A. E. Musson and E. Robinson, *Science and Technology in the Industrial Revolution* (Manchester: Manchester University Press, 1969), 47; N. Hans, *New Trends in Education in the Eighteenth Century* (London: Routledge & Kegan Paul, 1966), 155–7.

23 T. Simpson, *A New Treatise of Fluxions* (London, 1737); C. Hutton, 'Some account of the life and writings of Mr. Thomas Simpson', *Annual Register* (1764), 29–38;

biography from unidentified nineteenth-century Derby newspaper, Derby Local Studies Library (hereafter DLSL), news cuttings file, vol. 1 (acc. 8359); F. M. Clarke, *Thomas Simpson and his Times* (New York, 1929); R. V. and P. J. Wallis, *Biobibliography of British Mathematics and its Applications*, part II: *1701–60* (Newcastle: Newcastle University, 1986), 221–4; R. S. Westfall, *The Life of Isaac Newton* (Cambridge: Cambridge University Press, 1994), 40–6, 98–101.

24 *Ladies' Diary, or the Woman's Almanack* (1733), 20–1 (1738), 19.

25 John Flamsteed (1646–1719), the first Astronomer Royal, came from Denby near Derby and was educated at Derby Free School. Although he carried out some early observations from Derby and was stimulated by contact with a local group of astronomers that included Immanuel Halton of South Wingfield Hall, who lent him philosophical and mathematical textbooks, Flamsteed's work pre-dated the development of a wider scientific culture in Derby with astronomical interests. F. Bailey, *An Account of the Rev. John Flamsteed: The First Astronomer Royal* (London, 1835–37); L. Murdin, *Under Newton's Shadow: Astronomical Practices in the Seventeenth Century* (Bristol: Hilger, 1985).

26 *Derby Postman* (14 November 1728).

27 N. R. Wright, *John Grundy of Spalding, Engineer, 1719–1783* (Lincoln: Lincolnshire County Council, 1983), 2–5.

28 *Salisbury Journal* (12 April 1757), reference from Musson and Robinson, *Science and Technology*, 164.

29 *Derby Mercury* (22 September 1748); W. Griffis, *A Short Account of a Course of Mechanical and Experimental Philosophy and Astronomy* (London, 1748); Musson and Robinson, *Science and Technology*, 105.

30 *Derby Mercury* (14 June 1739).

31 *Derby Mercury* (28 June 1739); J. Arden, *A Short Account of a Course of Philosophy and Astronomy* (1744).

32 *Derby Mercury* (12 July 1739, 28 February 1740).

33 P. Elliott, '"Food for the mind and rapture to the sense": scientific culture in Nottingham, 1740–1800', *Transactions of the Thoroton Society*, 109 (2005), 1–13.

34 *Derby Mercury* (6 January 1744).

35 *Derby Mercury* (29 June 1749).

36 *Derby Mercury* (29 June 1749).

37 *Derby Mercury* (3 January 1752).

38 Hans, *New Trends*, 148; Musson and Robinson, *Science and Technology*, 101, 103.

39 Hans, *New Trends*, 145, 152; J. R. Millburn, *Wheelwright of the Heavens: The Life and Work of James Ferguson, FRS* (London: Vade Mecum Press, 1988), 301–6; Musson and Robinson, *Science and Technology*, 111, 120, 144.

40 Millburn, *Wheelwright of the Heavens*, 66–7.

41 *Derby Mercury* (22 September 1743).

42 Musson and Robinson, *Science and Technology*, 156.

43 Musson and Robinson, *Science and Technology*, 156.

44 F. W. Gibbs, 'Itinerant lecturers in natural philosophy', *Ambix*, 6 (1960), 113–14; H. Torrens, 'Bath philosophical societies', *Notes and Queries*, 231 (1986); H. Torrens, 'The four Bath philosophical societies, 1779–1959', in R. Rolls, J. Guy and J. R. Guy (eds.), *A Pox on the Provinces: Proceedings of the 12th Congress of the British Society*

for the History of Medicine (Bath: Bath University Press, 1990), 180–8; I. Inkster, 'Scientific culture and scientific education in Liverpool prior to 1812: a case study-ing the social history of education', in M. D. Stephens and G. W. Roderick (ed.), *Essays on Scientific and Technical Education in Early Industrial Britain* (Notting-ham: University of Nottingham, 1981).

45 Torrens, 'The four Bath philosophical societies'; Gibbs, 'Itinerant lecturers', 113–41.

46 H. C. King and J. R. Millburn, *Geared to the Stars* (Bristol: Hilger, 1978), 192–34, copy of the letter to Lawton (10 March 1764) kindly supplied to the author by Professor H. S. Torrens of Keele University.

47 King and Millburn, *Geared to the Stars*, 193.

48 British Museum ms., add. 4440 fos. 270v–274r, King and Millburn, *Geared to the Stars*, 192–3.

49 J. Ferguson, *Introduction to Electricity* (London, 1770); Millburn, *Wheelwright of the Heavens*, 225.

50 *Derby Mercury* (9, 14 April 1762); S. Schaffer, 'Glass works: Newton's prisms and the uses of experiment', in D. Gooding, T. Pinch and S. Schaffer (eds.), *The Uses of Experiment: Studies in the Natural Sciences* (Cambridge: Cambridge University Press, 1989), 67–104.

51 *Derby Mercury* (15 April 1762).

52 *Derby Mercury* (24 June 1748, 22 September 1749); J. Whitehurst, *Tracts; Philosoph-ical and Mechanical* (London, 1792); R. E. Schofield, *The Lunar Society of Birming-ham* (Oxford: Clarendon Press, 1963); M. Craven, *John Whitehurst: Clockmaker and Scientist* (Ashbourne: Mayfield, 1996); C. Hutton, 'Authentic memoirs of the life and writings of the late John Whitehurst FRS', *Universal Magazine* (November 1788), 224–9; *European Magazine* (1788), ii, 316–20; *Gentleman's Magazine* (1788), i, 182–3; Hutton, *History of Derby*, appendix.

53 *Gentleman's Magazine*, 19 (1749), 361–2; 20 (1750), 118–22.

54 J. Ellicott, 'The description and manner of using an instrument for measuring the degree of the expansion of metals by heat', *Philosophical Transactions*, 39 (1735–36), 297–9; *Gentleman's Magazine*, 20 (1750), 119.

55 *Gentleman's Magazine*, 20 (March 1750), 121.

56 *Gentleman's Magazine*, 20 (March 1750), 120.

57 G. S. Rousseau, 'Science books and their readers in the eighteenth century', in I. Rivers (ed.), *Books and their Readers in Eighteenth Century England* (Leices-ter: Leicester University Press, 1982), 197–255; R. P. Sturges, 'Context for library history: libraries in eighteenth-century Derby', *Library History*, 4 (1976), 44–52.

58 Schofield, *Lunar Society*, 22–3; E. Robinson, 'The Lunar Society and the improve-ment of scientific instruments', I and II, *Annals of Science*, 12, 13 (1956–57), 296–304, 1–8.

59 J. Whitehurst, *An Attempt toward Obtaining Invariable Measures of Length, Capac-ity, and Weight* (London, 1787); Schofield, *Lunar Society*, 256–61.

60 Mokyr, *Gifts of Athena*, 4–8, 65–6.

3

Joseph Wright, John Whitehurst and Derby philosophical culture, *c*.1760–1783

Introduction

Joseph Wright of Derby has long been celebrated for his candlelight paintings from the 1760s and early 1770s, particularly 'An Experiment on a Bird in the Air Pump' and 'A Philosopher giving that Lecture on the Orrery in which a Lamp is put in Place of the Sun'. Despite their inherent ambiguities, these two works of art have been regarded as archetypal celebrations of British and European enlightenment culture, scientific education and especially Newtonian natural philosophy. Although the undue emphasis upon these few, untypical paintings and upon their scientific aspects has recently been corrected and more attention given to Wright's later creations, as we have seen from the number of philosophical lectures taking place at Derby, the paintings can still be perceived to have been inspired by local scientific culture. There are also illuminating parallels between the development of audiences for science and for works of art in the context of urban renaissance consumerism, and middling sort and gentry identities and aspirations. It was Derbyshire middle-class families and gentry such as the Mundys of Markeaton Hall who provided important sources for many of his commissions. Indeed, engraved reproductions of paintings such as 'The Orrery', which were quickly in circulation and helped to establish Wright's national and international reputation, can be perceived as scientific commodities in their own right, which helped to promote the pursuit of Newtonian natural philosophy.

Rather than re-examining these paintings, already analysed and interpreted so often, this chapter primarily focuses upon the establishment of philosophical communities centred upon Derby during the 1760s and 1770s and the role that they played in the development of public scientific culture. It argues that Wright's 'scientific' paintings were primarily inspired by a philosophical coterie whose most important members were Washington Shirley, fifth Earl Ferrers, Peter Perez Burdett and John Whitehurst. Other philosophical groups were also active in the town, attesting to the growing audience for natural philosophy and the links between different kinds of associations such as freemasonry, including the first institution in the town that apparently described itself as a

6 Joseph Wright, 'An Experiment on a Bird in the Air Pump', mezzotint by V. Green (1769) from W. Bemrose, *The Life and Works of Joseph Wright* (1885).

philosophical society. As Whitehurst's career demonstrates, although centred upon Derby, these groups sustained important links with other provincial and metropolitan philosophical circles, providing for the first time a small but continuous semi-public scientific culture.

Wright's scientific paintings

The 'Orrey' and 'Air-Pump' each show a philosopher dramatically demonstrating the power of enlightenment science demonstrated by the Newtonian system to a (mostly) entranced audience. Wright covered similar philosophical themes in 'A Philosopher by Lamplight' (1768), 'The Alchymist' (1771) and the 'Philosopher by Lamplight' which was soon changed to 'Democritus' or 'A Hermit Studying Anatomy'.[1] 'The Orrery' and the 'Air Pump' are strongly reminiscent of religious art: young and old alike emerge from the darkness of superstition and ignorance to bathe in the golden light of rational understanding. Both paintings utilise a concealed light source, thus allowing Wright to experiment with shadow and light on figures. The paintings have been variously seen as inspired by religious art, enlightenment philosophy and aesthetic sensibility. They encapsulate the spirit of inquiry engendered by the

7 Joseph Wright, self-portrait from W. Bemrose,
The Life and Works of Joseph Wright (1885).

dissemination of natural philosophy and are usually associated with the Lunar Society and itinerant lecturers. But as Daniels has argued, they should not be perceived purely in narrow empiricist or utilitarian terms, representing rather 'a Rosicrucian theatre of enlightenment in which the rituals of material and spiritual transformation are combined'.[2]

Public lectures are not being depicted, which usually had audiences of at least twenty or thirty subscribers, but a more intimate and domestic form of tuition to a small group of family and friends. Various identities for the lecturer have been suggested, including Whitehurst, Denby the musician, organist and composer at All Saints' Church, and various visiting lecturers.[3] However, the paintings should still be associated with the semi-public and private courses of lectures on natural philosophy that had, as we have seen, been given at Derby and other midland towns for decades. As courses of lectures became more frequent and popular, so the range of subjects and apparatus increased, although astronomy remained popular. John Warltire, for instance, lectured at Derby in 1771, 1781 and 1798 and Mr Pitt in 1773, 1778 and 1785, the latter offering subjects including 'electricity, mechanics, astronomy, magnetism, hydrostatics, optics, pneumatics' in 1778. Lecture advertisements promised that language would be kept as simple as possible, and both sexes were encouraged to attend as 'ladies, as well as gentlemen without any previous knowledge of these subjects may readily understand the principal phenomena of nature'.[4]

Astronomy was, of course, one of the most fashionable sciences in the eighteenth century, as the popularity of James Ferguson's astronomical works reveal.[5] The first chapter of Ferguson's *Astronomy Explained* (1756) presented a dramatic view of the whole universe and a plurality of life worlds, an immense three-dimensional heavens with other planetary systems inhabited by intelligent travelling beings.[6] Children were also encouraged to be interested in astronomy and learn the basics of the Newtonian system, textbooks such as 'Tom Telescope's' *Newtonian System of Philosophy* (1761) going through many editions.[7] Many astronomical letters and articles appeared in the Derby newspapers, many of which assumed that readers possessed or aspired to own expensive scientific instruments.[8] As we have seen, instrument makers such as John Stenson, an apprentice of Whitehurst, offered a range of scientific equipment for ornamental or practical purposes.[9]

Wright's 'Experiment on a Bird in the Air Pump' is obviously not a depiction of the latest natural philosophy but of elementary scientific subjects being taught to a small and intimate audience. The interest is really on the reaction of the audience rather than natural philosophy in itself, although the lecturer

8 Thomas Haden by Joseph Wright, from W. Bemrose,
The Life and Works of Joseph Wright (1885).

may be based upon Whitehurst. Wright again made dramatic use of the effects of light and shadow. The air-pump is being used to suck out air from a glass retort in which a bird is struggling for life, while a young man with a watch on the left is timing both the convulsions of the bird and its revival. A skull in the foreground sits in a glass bowl, serving as a symbol of mortality along with the struggles of the bird, and there are two Magdeburg hemispheres on the right. Ferguson's Derby lectures in 1764 included pneumatics, although he opposed the practice of suffocating birds and animals in experiments as 'a most bitter and cruel death'. Normally he substituted a lungs glass, as the experiment was otherwise too shocking to spectators who had 'the least degree of humanity'.[10]

Philosophical groups in Derby

Associations were at the heart of Georgian urban culture and helped to define the nature of enlightenment society. A bewildering variety of clubs and societies existed, from charitable to drinking and sex clubs, music and gambling societies, subscription libraries and literary clubs, antiquarian socie-ties, gentlemen's societies and Masonic lodges.[11] Hume argued that in contrast to the solitude of 'barbarous nations', the 'middling rank of men' refined in the arts and sciences flocked into towns, where they loved 'to receive and communicate knowledge; to shew their wit or breeding; their taste in conver-sation or living, in clothes or furniture'. He noted how clubs and societies 'are every where formed', both sexes meeting 'in an easy and sociable manner; and the tempers of men, as well as their behaviour' being refined 'apace'.[12] Hume explicitly associated social and economic improvement with 'middling rank' urban organisational culture – where 'industry, knowledge, and humanity' were 'linked together by an indissoluble chain'.

While many Georgian clubs and societies were quite private associa-tions often associated with meals, drinking and taverns, they helped to form urban public culture, offering important opportunities for social networking, convivial conversation, and enjoyment, as well as intellectual discourse. As Clark has shown, more formal Georgian associations were in some ways modelled on older forms of urban association such as trades guilds, religious denominations, local corporations and vestries, which, in some respects, they replaced.[13] At Derby the 'Town Chronicle', for example, which recorded the names of the borough's chief officers for each year during the sixteenth and seventeenth centuries, and some important local and national events, ceased in 1698.[14] There was also a decline in the importance of the town guilds and a reduction of the proportion of inhabitants who were burgesses. The town tradesmen's Mercer's Company, for example, re-established in 1674, stopped prosecuting rogue traders in the 1730s and stopped meeting in 1740, when minute books records cease.[15] Though a new guildhall was constructed in 1731 it was, significantly, shortly afterwards used for events such as borough assemblies and for housing the town records.[16]

Clubs with interests in natural philosophy rapidly appeared during the late seventeenth and eighteenth centuries, partly inspired by the Royal Society and the Society of Antiquaries (1707). These included the gentlemen's societies at Spalding, Peterborough and Stamford, whose scientific interests were closely allied to antiquarianism, art and mechanics.[17] 'An Experiment on a Bird in the Air Pump', 'The Orrery' and Wright's other philosophical candlelight paintings of the 1760s and early 1770s were inspired by a similar small intellectual coterie centred upon Derby under the patronage of Washington Shirley, fifth Earl Ferrers. This included the cartographer and mathematician Peter Perez Burdett (1734–93), John Whitehurst, John Pickering, and Benjamin Bates, a physician who lived in Derby until 1774. Bates purchased 'An Experiment on a Bird in the Air Pump' and two of Wright's other paintings from this period, 'The Gladiator' (which included a portrait of Burdett) and a portrait of Galen. Joseph Cradock said that he and Bates were 'both intimate with Wright the painter', the 'rising genius', whose work was not yet fully appreciated.[18] Pickering was treated by Erasmus Darwin when ill for three weeks at Bates's apartment near the Market Place, Derby. In a letter to Bates, Wright described how in correspondence from 'our friend' Burdett he had been told that Bates would give him £130 for a picture, probably the 'Experiment on a Bird in the Air Pump', on condition that Bates would never reveal how little he had paid for it.[19]

Ferrers, the most influential member of the group, advanced through naval ranks to become captain and admiral, seeing action in various engagements, and continued to design and race ships. After the fourth Earl was tried and executed by the House of Lords for murder, Shirley unexpectedly came back to Staunton to become fifth Earl in 1760.[20] He was also an amateur architect whose designs seem to have included his own house at Staunton Harold and some aspects of Derby's new assembly rooms, where he probably employed the local builder and architect, Joseph Pickford. He was also close to Whitehurst, acting as a patron and subscribing to his *Inquiry* in 1778. The degree to which Whitehurst was identified with the Earl is evident from a satirical work dated 1767 which described the latter as being the author of a 'theory of the earth' at a time when Whitehurst's geological work was in preparation.[21]

The group shared Masonic and philosophical interests, reflecting the role that the lodges played across enlightenment Europe as promoters of scientific ideas and culture.[22] As grand master of the grand lodge in London between 1762 and 1764, Shirley held the highest office in English freemasonry. He was most closely associated with the famous and influential original no. 4 'Horn Lodge', of which he was master in 1762 and which included many members of parliament, royalty and aristocracy among the membership and was closely associated with the armed forces.[23] At Derby, Whitehurst, Pickford and John Chatterton were also freemasons, and a Virgin's Inn Lodge (no. 104) existed from September 1732 until February 1777, coincidentally almost exactly the dates of Whitehurst's residence. This lodge, which met originally at the Royal Oak Inn, was replaced in March 1785 by a lodge called the Tyrian (no. 468).[24] Whitehurst stamped Masonic symbols on the workings of his clocks and joined a London lodge by 1772, while carvings on ornamental panels on Pickford's house have also been interpreted as Masonic symbols. The purchasing and probably commission of 'The Orrery' by Shirley was probably associated with his election as FRS on 14 December 1761 on the basis of his observations of the transit of Venus in Leicestershire and his gift of 'an orrery or transitarium' that he had invented 'to facilitate the conception of future transits'. The observations were made 'in the manner of a camera obscura; with a Telescope

A – All Saints Church.
B – St Werberghs
C – St Peters
D – St Allmonds
E – St Michaels
F – Silk Mills
G – China Works
H – County Hall
I – Town Hall
K – Assembly Room
L – Jail

9 Plan of the east side of Derby, detail from engraving by G. Moneypenny after P. P. Burdett (1791).

four feet in length, placed in a frame and fixed in a window shutter so as to exclude all light'.[25] Shirley described how, with his assistant the mathematician and instrument maker Davis, they took the observations by projecting the image of the sun on to paper behind the telescope. They had recorded the precise time of the passage of Venus across the face of the sun with a watch. He concluded that the latitude had increased since 1639 and that the transit was performed in a shorter time than had been determined by Halley. Furthermore, the horizontal sun's parallax could be determined by comparing these results with those of the observations at the East Indies and Hudson Bay.[26]

Burdett, probably Wright's closest friend during the 1760s, is seen taking notes in 'The Orrery' and featured in others of Wright's works. A talented surveyor, cartographer, artist and engraver, Burdett produced a prize-winning map of Derbyshire based upon a comprehensive survey carried out between 1762 and 1767, inspired by the prize offered by the Society of Arts. Like Whitehurst, Pickford and Shirley, he was probably also a freemason by the 1760s and lodged with Ferrers at Staunton Harold, claiming to be a kinsman.[27] Burdett probably commissioned Pickford, who also worked for Shirley, to design a fashionable Georgian gothic townhouse in All Saints' Churchyard, where he was living by June 1766, probably the earliest Derby neo-gothic edifice.[28] Wright accompanied him to Liverpool in 1768, where Burdett surveyed the county of Cheshire and became first President of a Liverpool Society of Artists. At Liverpool Burdett also produced the first aquatint to be published in England, though a scheme to transfer aquatinted designs for the decoration of porcelain and pottery to the firm of Wedgwood and Bentley collapsed in some acrimony after Burdett failed to perfect the process. After negotiations, which may have been facilitated by Shirley and international Masonic contacts, he left for Germany in 1774 in serious debt after the failure of a scheme to emigrate to America with the support of Franklin, to whom he had probably been introduced by Whitehurst. Burdett ended his life in the service of Markgraf Karl Friederich of Baden in southern Germany, for whom he produced a state survey, his talents as a mechanic and cartographer proving attractive.[29]

Burdett's map of Derbyshire (1767) was the first general map of the county to be based upon an original survey to appear since Christopher Saxton's map of 1577. It featured geographical information such as roads, towns, woodland, heath and moor, and landed seats, with positions calculated by triangulation, symbolically using as a central meridian the line of longitude running through Derby.[30] Roads were classified as turnpike and non-turnpikes or secondary, with distances calculated from All Saints' Church, Derby. Burdett had originally hoped to include the heights of mountains, lists of fairs and market days, and 'natural or antique curiosities', reflecting the scientific and antiquarian concerns of British enlightenment culture evident in the activities of his Derby-centred philosophical group.[31] Although Ferrers and Whitehurst assisted with the venture and there was support in the local newspaper, there were problems gaining enough subscribers, delaying publication.[32] Burdett used astronomical observations to determine latitude and meridian line and conducted other observations with the assistance of John Whyman of Aston. In 1764, for instance, they observed the solar eclipse at Derby using a spiral micrometer invented by Burdett, publishing an account in the *Gentleman's Magazine*.[33] Burdett's map was considered by John Mitchell to be much the best county map in existence and made Derbyshire one of the most accurately mapped English counties, although it failed to make Burdett much money.[34]

Shirley's philosophical group may be synonymous with a Derby philo-sophical club described by the dramatist Joseph Cradock, who was taught by the Rev. William Pickering, the Vicar of Mackworth near Derby. According to

10 Plan of the west side of Derby, detail from engraving by G. Moneypenny after P. P. Burdett (1791).

Cradock, after 1760 a 'weekly club' was held near All Saints' Church, 'consisting of philosophers then principally engaged on the subject of electricity'. Three of the 'friendly members' were 'Rev. Mr Winter originally a dissenter, Mr Pickering at least a Tory, and Mr Whitehurst supposed to be an infidel'. Furthermore, 'from this humble society ... the celebrated Mr Darwin and his ingenious pupil and associate Mr Watt derived no small advantages'.[35] Although the references are confused, this may be synonymous with another circle referred to by Thomas Mozley, which included Wright and Pickford. According to Mozley, this was 'a coterie contemporaneous and on friendly terms with the philosophical society founded by Erasmus Darwin, but with a difference of cast, for philosophers are, socially as exclusive as other people'.[36] Leaving aside the allusions by Cradock and Mozley to Watt and Darwin respectively, which emphasise the connections with the Lunar Society, these accounts confirm the existence of overlapping philosophical coteries in Derby prior to the foundation of the Derby Philosophical Society. They included Whitehurst, Pickford, the Rev. Joshua Winter, curate at All Saints for over fifty years and probably John Pickering, who taught mathematics to Cradock and was depicted by Wright with mathematical drawings.[37]

Wright's circle and the philosophical club described by Cradock may, in turn, be synonymous with a philosophical society that celebrated its anniversary at the New Inn, Derby in 1779. Although it is not clear which anniversary was being celebrated and the language is highly symbolic and figurative rather than literal, the existence of such a society before the formation of the Derby Philosophical Society in 1783 may be confirmed by a later account which refers to the philosophical society established in Derby 'about the year 1772'.[38] The tone of the poem, attributed to Samuel Pratt, is formulaic and highly reminiscent of Masonic language and the praise lavished on Newtonianism by Desaguliers and Akenside in his 'Hymn to Science'.[39] The anniversary address celebrates the superior wisdom that philosophical learning could bring for 'wisdom's sons' – 'great souls' inspired by 'the love of science' who prefer 'knowledge to the brilliant ore', while those beneath were 'grovelling slaves' devoted to 'sloth and lust' who wasted their lives. These 'Derby' philosophers exulted in the liberal arts such as astronomy and 'mounted high, surveying various regions in the sky', were 'enrapt with inextinguishable joy' as the deity's universal laws were revealed to them. Masonic ritual involved the process of moving through three stages or degrees, often shown as a ladder or steps towards the Deity, who, incidentally, was sometimes represented as Jupiter or Jove. Discoveries in natural philosophy were likened to Masonic stages, with increased knowledge of the 'sublime mysteries' as initiates ascended the degrees, revealing the laws of nature both at a universal level and in terms of the self.[40]

Whitehurst was, as we have seen, the most important figure in Derby philosophy between the 1750s and 1770s, as a member of the Lunar Society

and author of an *Inquiry into the Original State and Formation of the Earth*
in 1778. His appearance in later life was, according to Charles Hutton, very
striking, as he was 'above the middle stature, rather thin' and of a 'countenance
expressive at once of penetration and mildness. His fine grey locks, unpol-
luted by art, gave a venerable air to his whole appearance.'[41] William Hutton
described seeing him at Buxton in 1785 and wishing his acquaintance. He was
'near six feet high, straight, thin, and wore his own dark grey bushy hair: he
was plain in his dress and had much the appearance of a respectable farmer'.[42]
Encouraged by friends such as Franklin and Anthony Tissington FRS, the
mines agent with philosophical interests, Whitehurst gained a reputation as a
geologist. He sent Franklin an adumbration of his *Inquiry* in 1763, illustrating
the extent to which this was stimulated by the work of Burdett and other
members of Shirley's coterie, and entertained Franklin on visits to Derbyshire
between 1759 and 1774.[43] Whitehurst's geological reputation is symbolised in
Wright's portrait of 1774, which shows him seated with pen in hand holding a
page from his *Inquiry*, this being part of the 'Section of the Strata of High Tor'
with a fiery volcano in the background, and Wright famously wished for his
company on Mount Vesuvius.[44]

Whitehurst used philosophical and Masonic contacts to expand his business
and scholarly networks beyond Derbyshire. He may have joined the Virgin's Inn
Lodge in Derby during the 1730s and was a member of the Lodge of Emulation
no. 21 in London by 1770, doubtless encouraged by patrons such as Shirley.[45]
Although based in the provinces for decades, Whitehurst cannot be regarded
as simply a provincial natural philosopher. By the 1760s, he was a member
of metropolitan scientific circles as well as those in Birmingham and Derby,
including a London philosophical society in existence by 1766 and a society at
Old Slaughter's coffee house known as the 'Club of Thirteen'. Members included
Whitehurst, Daniel Solander, Day, Wedgwood and Bentley again, James
(Athenian) Stuart, Benjamin Franklin and Rudolph E. Raspe the geologist,
antiquarian and author of the *Adventures of Baron Munchausen*. Whitehurst
joined other London associations, including one that met at the Baptist Head
Coffee House, whose minute books between 1782 and 1787 survive. Members
included Nairne, Richard Kirwan, Boulton, Watt, Wedgwood, James Hutton,
William Nicholson, Benjamin Vaughan, Adam Walker and J. H. de Magellan.
Matters discussed included Watt's blowing engines at Wilkinson's iron works,
casting and iron-founding methods, the recovery of silver from copper and
other base metals, the nature of common air and, finally, electricity.[46]

Conclusion

Wright's philosophical candlelight paintings and the evidence concerning
various philosophical coteries attest to the appearance of a public scientific

culture in Derby by the 1760s. As the careers of Shirley, Whitehurst and Burdett and the anniversary address of the Derby philosophical society of 1779 demonstrate, freemasonry also provided an impetus for philosophical study and urban sociability. Just as Wright's candlelight paintings partly copied earlier religious art, so freemasonry employed language equally inspired by religion, such as the imagery of intellectual illumination triumphing over the darkness of ignorance, to help to forge an audience for natural philosophy. Through the provincial and metropolitan activities of Wright, Whitehurst, Ferrers and associates such as the members of the Lunar Society and Franklin, which helped to inspire the 'Experiment on a Bird in the Air Pump' and 'The Orrery', these philosophical circles, with their Tory and Anglican sympathies, operated both locally and nationally.

Notes

1 B. Nicolson, *Joseph Wright of Derby: Painter of Light*, 2 vols. (London: Paul Mellon Foundation, 1968), I, 52–3, 118–20; J. Egerton, *Wright of Derby* (London: Tate Gallery, 1990), 84–6, 91–2; S. Daniels, *Joseph Wright* (London: Tate Gallery, 1999), 26–30; P. Gay, *The Enlightenment: An Interpretation*, I (New York: Norton, 1967), 305–6.

2 S. Daniels, 'Loutherbourg's chemical theatre: Coalbrookdale by night', in J. Barrell (ed.), *Painting and the Politics of Culture: New Essays on British Art, 1700–1850* (Oxford: Oxford University Press, 1992), 195–230, 222.

3 J. Gandon, *The Life of James Gandon Esq.* (Dublin, 1846), 211–12; T. Mozley, *Reminiscences, Chiefly of Oriel College and the Oxford Movement*, 2 vols. (London, 1882), I, 65; W. Bemrose, *The Life and Works of Joseph Wright* (London, 1885), 55; Nicolson, *Joseph Wright*, I, 115; J. Cradock, *Literary and Miscellaneous Memoirs*, 3 vols. (1826), I, 10; H. Wright, 'Wright of Derby', obituary in *Gentleman's Magazine* (1797), 804; *Monthly Magazine* (1797), 289–94; F. Klingender, *Art and the Industrial Revolution*, ed. A. Elton (St Albans: Paladin, 1968); M. Craven, *John Whitehurst: Clockmaker and Scientist* (Ashbourne: Mayfield, 1996); D. Fraser, 'Fields of radiance: the scientific and industrial scenes of Joseph Wright', in D. Cosgrove and S. Daniels (eds.), *The Iconography of Landscape* (Cambridge: Cambridge University Press, 1988), 119–41; S. Daniels, *Fields of Vision* (Cambridge: Polity Press, 1993), 43–79; Egerton, *Wright of Derby*; Daniels, *Joseph Wright*.

4 *Derby Mercury* (14, 21 June 1771, 20 November 1778, 25 December 1778, 1 January 1779).

5 J. Ferguson, *Astronomy Explained on Sir Isaac Newton's Principles* (London, 1756); J. R. Millburn, *Wheelwright of the Heavens: The life and Work of James Ferguson, FRS* (London: Vade Mecum Press, 1988); H. C. King, *The History of the Telescope* (London: Charles Griffin, 1955).

6 Millburn, *Wheelwright of the Heavens*, 167–8, 184.

7 J. A. Secord, 'Newton in the nursery: Tom Telescope and the philosophy of tops and balls, 1761–1838', *History of Science*, 23 (1985), 127–51.

8 *Derby Mercury* (29 December 1775, 12 July, 1 October 1776, 8 June 1781).

9 *Derby Mercury* (22 July 1784, 21 June 1787, 3 May 1792); M. Craven, *Derbeians of Distinction* (Derby: Breedon, 1997), 191–2.

10 *Derby Mercury* (13 April 1764); J. Ferguson, *Lectures on Select Subjects* (London, 1760), 185; Nicolson, *Joseph Wright*, I, 113.

11 R. P. Sturges, 'Harmony and good company: the emergence of musical perform-ance in eighteenth-century Derby', *Music Review*, 39 (1978), 180.

12 D. Hume, 'Of refinement in the arts', *Essays and Treatises on Several Subjects*, 2 vols. (Edinburgh, 1825), II, 274, 268.

13 P. Clark, *British Clubs and Societies, 1580–1800: The Origins of an Associational World* (Oxford: Clarendon Press, 2001), 234–73.

14 R. Clark, 'The Derby "Town Chronicle" 1513–1698', *Derbyshire Archaeological Journal*, 118 (1998), 163–84.

15 A. W. Davison, *Derby: Its Rise and Progress* (London: Bemrose, 1906), 275.

16 M. Craven, *Derby: History and Guide* (Stroud: Sutton, 1994), 64.

17 G. Averley, 'English Scientific Societies of the Eighteenth and Early Nineteenth Centuries' (unpublished PhD thesis, University of Teeside Polytechnic, 1989); P. Elliott, 'Provincial urban society, scientific culture and socio-political marginality in Britain in the eighteenth and nineteenth centuries', *Social History*, 28 (2003), 394–442; P. Elliott, 'Towards a geography of English scientific culture: provincial identity and literary and philosophical culture in the county town, *c*.1750–1850', *Urban History*, 32 (2005), 391–412; R. E. Schofield, *The Lunar Society of Birming-ham* (Oxford: Clarendon Press, 1963); J. Uglow, *The Lunar Men: The Friends who Made the Future, 1730–1810* (London: Faber, 2003).

18 J. Cradock, *Literary and Miscellaneous Memoirs*, second edition, 3 vols. (1826), I, 20.

19 Bemrose, *Joseph Wright*, 12; Nicolson, *Joseph Wright*, I, 104–5; E. Darwin, letter to W. Withering, 25 February 1775, in D. King-Hele (ed.), *The Collected Letters of Erasmus Darwin* (Cambridge: Cambridge University Press, 2007), 130–2.

20 *Gentleman's Magazine* (1773), 464–5; W. L. Clowes, *The Royal Navy: A History*, 4 vols. (London, 1898), III, 566; Letter from P. P. Burdett to Earl Ferrers, 16 February 1766 (incomplete), Shirley family papers, Leicestershire Record Office, 26 D53/1943. Ferrers naval commissions have also been preserved among these papers.

21 J. Nichols, *The History and Antiquities of the County of Leicester*, 4 vols. (1804), III, 717; E. Saunders, *Joseph Pickford of Derby: A Georgian Architect* (Stroud: Sutton, 1993), 59; H. M. Colvin, *Biographical Dictionary of British Architects, 1660–1840* (New Haven: Yale Univesity Press, 1995).

22 M. C. Jacob, *Living the Enlightenment: Freemasonry and Politics in 18th Century Europe* (Oxford: Oxford University Press, 1991), 65; P. Elliott and S. Daniels, '"The school of true, useful and universal science"? Freemasonry, natural philosophy and scientific culture in eighteenth-century England', *British Journal for the History of Science*, 39 (2006), 207–30.

23 W. F. Gould, *History of Freemasonry*, 4 vols. (London: Caxton, 1953), IV, 242–5.

24 O. Manton, *Early Freemasonry in Derbyshire* (Manchester, 1913/1916), 6–10.

25 Journal Book of the Royal Society, 10 December, 1761, 221–4; Nichols, *History and Antiquities of Leicester*, III, part II, 717.

26 H. Woolf, *The Transit of Venus* (London, 1959); J. R. Millburn, *Benjamin Martin: Author, Instrument Maker and Country Showman* (Leiden: Noordhoff, 1976), 118–23; Millburn, *Wheelwright of the Heavens*, 128–37.

27 *P. P. Burdett's Map of Derbyshire, 1791 Edition*, ed. with introduction J. B. Harley, D. V. Fowkes and J. C. Harvey (Derbyshire Archaeological Society, 1975); Egerton, *Wright of Derby*, 87–91.

28 *Derby Mercury* (19 June 1766); Saunders, *Joseph Pickford*, 28–9.

29 *Derby Mercury*, obituary, 1797; P. P. Burdett (attrib.), 'Fragments of a philosophical institution in which the Earl Ferrers was enlisted as patron', Shirley family papers, Leicestershire Record Office, 26 D53/2654; Egerton, *Wright of Derby*, 87–91; J. Whitehurst, letter to Franklin, Derby, 25 February 1767, in W. B. Willcox (ed.), *The Papers of Benjamin Franklin*, 38 vols. (New Haven: Yale University Press, 1959–2006), XIV, 41.

30 *Derby Mercury* (27 August 1762); R. H. Bird, 'Notes on the Derbyshire map surveyed and produced by Peter Perez Burdett, 1762–1767', *Bulletin of the Peak Mines Historical Society*, 4 (1969), 75–81; *P. P. Burdett's Map of Derbyshire, 1791 Edition*, introduction by Harley, Fowkes and Harvey; M. O'Sullivan, 'Derbyshire', in C. R. Currie and C. P. Lewis (eds.), *A Guide to English County Histories* (Stroud: Sutton, 1997), 107–14.

31 *P. P. Burdett's Map of Derbyshire, 1791 Edition*, introduction, 3; and compare the original claims in the prospectus and *Derby Mercury* (27 August 1762) with the final advertisement, *Derby Mercury* (23 October 1767).

32 *Derby Mercury* (19 June, 14 November 1766, 26 June 1767).

33 *Derby Mercury* (27 August 1762, 9 April, 24 June 1766, 17, 24 April, 8 May, 26 June 1767); *Gentleman's Magazine*, 34 (1764), 193.

34 Nichols, *History and Antiquities of Leicester*, III, part II, 717; *P. P. Burdett's Map of Derbyshire, 1791 Edition*.

35 Cradock, *Literary and Miscellaneous Memoirs*, I, 10.

36 Mozley, *Reminiscences of Oriel College*, I, 65.

37 Nicolson, *Joseph Wright*, I, 117–18; Egerton, *Joseph Wright*, 90; C. Cox and W. St. J. Hope, *The Chronicles of the Collegiate Church ... of All Saints, Derby* (London, 1881), 26–9; *Derby Mercury* (8 April 1774).

38 J. Pigot, *National Commercial Directory [for the Midlands, the North and Wales]* (Manchester, 1835), 39.

39 *Derby Mercury* (30 July 1779) and *Harrison's Derby and Nottingham Journal* (29 July 1779); M. Akenside, 'Hymn to science', *Gentleman's Magazine* (1739).

40 W. K. MacNulty, *Freemasonry: A Journey through Ritual and Symbol* (Thames & Hudson, 1991), 18–32.

41 John Whitehurst, *The Works of John Whitehurst F.R.S. with Memoirs of his Life and Writings* (London, 1792), 18.

42 W. Hutton, *The History of Derby: from the Remote Ages of Antiquity* (London, 1791), 294.

43 Willcox (ed.), *Papers of Benjamin Franklin*, X, 226–31; XIV, 41–2; Craven, *John Whitehurst*, 41–2.

44 Nicolson, *Joseph Wright*, I, 294.

45 *Derby Mercury* (24 June 1748, 22 September 1749); obituary, *Gentleman's Magazine*,

58 (1788), 182; Craven, *John Whitehurst*, 42; C. Hutton, 'Authentic memoirs of the life and writings of the late John Whitehurst FRS', *Universal Magazine* (November 1788), 226, and 'memoir' in J. Whitehurst, *Tracts; Philosophical and Mechanical* (London, 1792), 9–10.

46 E. Robinson, 'R. E. Raspe, Franklin's Club of Thirteen, and the Lunar Society', *Annals of Science*, 11 (1955), 144; A. E. Musson and E. Robinson, *Science and Technology in the Industrial Revolution* (Manchester: Manchester University Press, 1969), 58, 74, 126–27, 141, 391; T. H. Levere and G. L. Turner (eds.), *Discussing Chemistry and Steam: The Minutes of a Coffee House Philosophical Society, 1780–1787* (Oxford: Oxford University Press, 2002).

4

The Derby Philosophical Society

Introduction

Scientific associations were important centres for the exchange of ideas and the development of public enlightenment scientific culture, linking smaller towns with the great European cultural centres. In England, literary and philosophical societies such as those at Manchester, Liverpool and Newcastle helped to develop and disseminate scientific ideas as part of a broad progressive culture that was able to attract many Tories and Anglicans as well as dissenters and reformers until the 1790s. The Derby Philosophical Society, primarily the inspiration of Erasmus Darwin, was a significant British scientific association and served as a regional organisation, bringing together professionals, especially medical men, gentry, industrialists, manufacturers and other individuals well into the nineteenth century. It formed a large scientific library and served as a forum for scientific discussion, especially concerning the latest developments in chemistry, electricity and chemistry, which were well represented in the collection.

However, some of the ambitious intentions ennunciated in Darwin's opening address to the members of the Derby Philosophical Society in 1784 failed to be realised, such as the idea of publishing transactions in the manner of the Manchester Literary and Philosophical Society, and the society generated little publicity. Nevertheless, inspired by enlightenment progressivism, the pursuit of natural philosophy and the leadership of Darwin, the Derby philosophers were at the forefront of local improvement and national political campaigns such as the measures to enclose common lands, the campaign against the Test and Corporation Acts and the Derby Society for Political Information.

The foundation of the Derby Philosophical Society

The main impetus for the foundation of the Derby Philosophical Society came from Darwin's move from Birmingham to Radbourne in 1781 and Derby in 1783. On 26 December 1782, he complained to Matthew Boulton about being 'cut off' in Derby, 'from the milk of science, which flows in such redundant streams from your learned lunations'.[1] As we have seen, various philosophical groups were active in Derby during the 1760s and 1770s, so these comments should be taken as referring to Darwin's personal feelings at moving further from the centre of Lunar meetings rather than as a general dismissal of Derby public culture. Intended as a gentlemanly forum for the discussion of philosophical ideas and the acquisition of a library, the first full meeting of the Derby Philosophical Society was held at the King's Head, one of Derby's leading coaching inns, on 7 August 1784, although this was preceded by a series of informal domestic gatherings from around February 1783.[2] Subsequently, it was arranged that members were to meet at the King's Head on the first Saturday of every month at 6 p.m., while two full meetings took place each year in April and October.

The entrance fee to the society was one guinea and subscription one guinea annually, with a fine of one shilling for absence at a monthly meeting and half a crown for absence from the biannual meetings.[3] The only officials, elected at the biannual meetings by majority vote, were a president and a secretary, the latter being Richard Roe from 1784 until his death in 1814. Though the Philosophical Society members kept no records concerning the nature of their domestic discussions, brief reports offer glimpses of what took place. On 13 March 1783, Susannah Wedgwood wrote to her father Josiah that 'the philosophical club goes on with great spirit, all the ingenious gentlemen in the town belong to it, they meet every Saturday night at each others houses'.[4] Just before, on 4 March 1783, Darwin wrote to Matthew Boulton announcing,

> We have establish'd an infant philosophical society at Derby, but do not presume to compare it to your well grown gigantic philosophers at Birmingham. Perhaps like the free-mason societies, we may sometime make your society a visit, our number at present amounts to seven, and we meet hebdomidally. I wish you would bring a party of your society and hold one moon at our house. N.B. our Society intend to eclipse the moon on the 18 of this month, pray don't you counteract our conjurations.[5]

In an address to the members at their first regular meeting, held at his house on Full Street on 18 July 1784 with the Derwent flowing rapidly past at the end of his back garden, Darwin presented a vision of the value of philosophical societies for the progress of knowledge, which must have been inspirational. The address, subsequently reprinted with editions of the rules and library catalogue, offered a powerful call to arms for secular science over torpor, ignorance and anarchy, emphasising the grand continuity inherent in their project, uniting the philosophers of antiquity with the greatest figures in contemporary science. Through the 'daring hand of experimental philosophy' they could transcend their feeble human frame and help to 'enrich the terraqueous globe', contributing to the useful arts, necessities and 'embellishments of life' from magnetism to cookery, which did 'honour to human nature'. The 'liberal and agreeable associations' that promoted it, like the freemasons, were a 'band of Wampum', or 'chain of concord', and they too could collect together the 'scattered facts' of philosophy and converge them 'into one luminous point … to exhibit the distinct and beautiful images of science'. He hoped that they would be able to gather together the publications of other societies and 'ingenious philosophers' in an 'increasing, and valuable library', perhaps enriching one day 'by our own publications … the common heap of knowledge' which would 'never cease to accumulate so long as the human footstep is seen upon the earth'.[6]

It all began optimistically enough. A balloon launch was chosen by the founders as a suitable means of marking the occasion of the birth of the Philosophical Society, as a symbol of technological progress wrought by reason and experiment. In June 1783, the Montgolifier brothers had launched the first balloon in France, acting upon a suggestion from the Scottish chemist Dr Joseph Black that if a bladder was filled with inflammable air (hydrogen) then it would form a mass lighter than atmospheric air and so rise up. The first human flight in a balloon took place at Paris on 27 August 1783, when J. F. Pilâtre de Rozier rose to 80 feet though he was secured by a rope. The first British flight was in November 1783, when 'Count' Francisco Zambeccari ascended from Cheapside in London. George III and his court were shown this remarkable invention and balloon fever gripped the country for the next year or so.

The Derby balloon was launched from Nun's Green on Saturday, 27 December, only six months after the Mongolifier invention and only one month after the Zambeccari flight in London. As Darwin joked to Boulton, the aim was to send a message to the members of the Lunar Society, symbolising the kinship between the two philosophical clubs, and it was 'calculated to have fallen in your garden at Soho'. The Derby Philosophical Society seems to have been the first British scientific society to make such an effort and a detailed report appeared in the *Derby Mercury*.[7] A label pinned to the balloon offered a reward for its recovery and the image of the balloon flying over the trees apparently caused some commotion among field labourers. However, as Darwin explained, 'the wicked wind carried it to Sir Edward Littletons' at Teddesley Park near Penkridge in Staffordshire.[8] According to the *Mercury*, the balloon was found by a labourer who 'saw it alight, and rebound again several times before he could catch it; and who thought that the time he found it was about noon; so that it seems to have passed with very great velocity'. Darwin described the making of the balloon in a letter to Wedgwood and a full description also appeared in the Derby newspaper complete with a sketch, although Drewry the proprietor seems to have quickly got tired of balloons, commenting that 'enough has been said ordinary and extraordinary, of air-balloons … a subject, however useless, which has excited much curiosity'.

11 Erasmus Darwin, engraving after the painting by J. Wright.

It was first to be a sphere of four feet in diameter, but Darwin proposed cutting it in two at the equator and the addition of 'a torrid zone of nine inches breadth' to ensure lift-off. The skin was made of silk 'all cover'd with oil boil'd with red lead in it'. A major innovation of early British industrialisation was the mass production of oil of vitriol or sulphuric acid, and the hydrogen was prepared by placing 'old iron turnings into a narrow-mouthed vessel and pouring upon them a certain quantity of oil of vitriol and common water'. This cased 'an effervescence' and the 'inflammable air' to 'rise in a plentiful stream', from whence it was conveyed by a pipe into the balloon, soon showing a 'disposition to ascend'.[9]

Balloon launches became a common European event from the 1780s, becoming, as well as the height of fashion, a symbol of the successful application of European enlightenment reason to nature, represented by their role in meteorological experiments. By August 1784, balloon flights of thousands of feet were being made in France and meteorological experiments were being conducted on board.[10] The English response differed from that of the French, who took ballooning more seriously and organised state-sponsored scientific experiments, whereas the English tended to view it as a trivial diversion, although many balloons were launched. Thus Joseph Banks, President of the Royal Society, who was kept informed of the French flights, had to be admonished by Benjamin Franklin for suffering 'pride to prevent our progress in science'.[11] This was partly because of the threat to public order that balloon flights posed and partly because the French had got there first.[12] Echoing the hostility of *The Times* and other newspapers, after one accident in 1784 the *Derby Mercury* urged that 'the philosophic gentlemen who launch almost every evening those dangerous machines, should give up an amusement which must afford but little pleasure' and 'may be the cause of doing great injury'.[13] The tendency of balloon flights to attract adventurers, pickpockets and rioters ensured that the craze declined, while many attempts failed and caused violence, such as the attempted ascent by Chevalier Moret in August 1784, when a crowd of disappointed subscribers tore the balloon to pieces on realising that it would never fly.[14] The fact that Banks and the Royal Society paid less attention to balloons makes the early response of provincial philosophical societies such as those at Birmingham and Derby all the more significant. The

Lunar Society member John Southern published a *Treatise upon Aerostatic Machines* (1785), one of the earliest English books on balloons. This contained a description of a process for manufacturing inflammable air, and a method of pasting the sheets together for making paper balloons given to him by Abraham Bennet, the curate electrician, who had been launching balloons at Wirksworth. Priestley used balloons to conduct experiments during the 1780s, and Darwin featured them and 'the bold Montgolfier' in his *Botanic Garden*, boasting that he was turning all French in chemistry just as he had in politics. The association between balloons, natural philosophy and political radicalism was later exploited by opponents of reform during the 1790s, who depicted Darwin, Priestley and other philosophers in cartoons suggesting there was little difference between experiments in airs and the hot air of speeches of sedition by foolish philosophers.[15]

Philosophical activities

Although it did not achieve all the objectives defined by Darwin in his opening address, the Derby Philosophical Society existed independently for some eighty years and managed to acquire a significant library of scientific and medical works, which came to dominate the proceedings. This collection later made a vital contribution towards the original Derby municipal library. The core group resident in Derby and regularly attending meetings remained small, although the Derby Philosophical Society functioned as a regional corre-

sponding association and library during Darwin's lifetime and during the first decades of the nineteenth century, when there were no equivalent organisations in the east midland region. Membership was dominated by middle-class professionals, especially medical men, gentry and manufacturers, reflecting Derby's county town status and manufacturing and industrial importance. There always seems to have been tensions between members who subscribed to Darwin's vision of an active scientific society, conducting philosophical experiments and publishing results, and those who saw the Philosophical Society primarily as a useful library through which expensive medical and scientific works could be obtained. This was to result, as we shall

12 Joseph Priestley, from S. Smiles, *Lives of the Engineers: Boulton and Watt* (1904).

see, in the formation of the Derby Literary and Philosophical Society after
Darwin's death. While taking an interest in natural philosophy, the majority
of members did not undertake original scientific work or publish science-
related literature.[16] Besides Darwin the two most active individuals in Derby
were William Strutt and Richard Roe, who served as secretary between 1783
and his death in 1814. Roe worked as a coach, sign and house painter, property
and land sale agent, and servant employment agent. He also served as writing
master at Repton School and kept an academy which taught various subjects,
including mathematics, grammar, French, drawing and geography, supported
by other members of the Philosophical Society, who sent their children.[17]

Although some Derby Philosophical Society members engaged in scientific
experiments and composed philosophical works, especially those closest to
Darwin, and experiments were demonstrated and discussed at meetings, the
society never became as active in promoting science as Darwin had hoped.
However, surviving membership and library records, the works of Darwin
and other sources provide some information concerning the philosophical
interests of members as a whole. It is clear that, encouraged by their medical
and industrial interests, other members besides Darwin took a keen interest
in chemistry, electricity, galvanism, meteorology and botany. Electricity, for
instance, the wondrous science of the European enlightenment, was regarded
as a dramatic and active power in meteorology, an agent in chemistry and
a form of medical treatment.[18] Works such as Tiberius Cavallo's *Treatise
on Electricity* (1786–95), George Adam's *Essay on Electricity* (1784–85) and
Richard Fowler's *Experiments and Observations on Animal Electricity* were
obtained and circulated between the members, while Roe supplied electrical
machines.[19] Darwin, Abraham Bennet, Whitehurst, Swanwick and Stenson
collected meteorological information, and Darwin conducted experiments
with William Strutt, Richard French and Samuel Fox, the results of which
were published in Darwin's paper on cloud formation and adiabatic expansion, which has been described as his greatest contribution to the physical
sciences.[20] The study of atmospheric electricity also gave the initial impetus for
Bennet's electrical research, which, with the encouragement of Darwin and the
Derbyshire *savants*, culminated in the invention of the gold-leaf electroscope
and later the doubler. (Although Bennet is not listed as a member of the Derby
Philosophical Society he probably attended some meetings.) Bennet's research
with Nicholson's doubler, described by Darwin as the greatest electrical
discovery since the Leyden jar, demonstrated the electromotive potential of
metals and, in inspiring Volta, made an important contribution to the invention of the pile.[21] Darwin utilised galvanism clinically and conducted galvanic
experiments with William Strutt, Thomas Swanwick and Henry Hadley.[22] As
we shall see, electricity also provided an analogy in Darwin's psychophysiology for the operation of the spirit of animation. This illustrates the tension

between the appeal of natural philosophy as objective progressive knowledge, and the political activities pursued by most of the leading members which it helped to inspire.

Medical men

Medical men made up about half of the resident and over half of the non-resident members of the Derby Philosophical Society in Darwin's lifetime. Dr John Hollis Pigot, Dr John Beridge and Dr Peter Crompton were physicians, the highest of the archaic tripartite medical structure in eighteenth-century society. Dr William Brookes Johnson was MB of Cambridge, and Henry Hadley, Thomas Haden and William Tancred Fowler were surgeons.[23] Surgeons were the middle rank of medicine and generally performed physical operations, while apothecaries formed the lowest rung as the suppliers of drugs. In practice, these divisions were not fixed, as physicians might perform operations, surgeons could diagnose disease, and apothecaries often, particularly for the poor, diagnosed disease and offered remedies. Among the non-resident members of the Philosophical Society in Darwin's lifetime, medicine was even more dominant, with eighteen or nineteen of the thirty-six being medical men. These included the physicians Dr Thomas Arnold of Leicester, Dr Robert Bree also of Leicester, Dr Robert Waring Darwin of Shrewsbury (son of Erasmus Darwin), Dr Snowden White of Nottingham, Dr John Storer also of Nottingham and Dr James Justyn Bent. Bent was a non-resident member based at Newcastle-under-Lyme and numbered members of the Wedgwood family, including Josiah – whose leg he had to amputate – among his patient for thirty years. With his younger brother William Bent he organised a private health scheme for Wedgwood's workforce at Etruria and in 1774 published a paper on a treatment which avoided amputation for an abscess on the shoulder.[24] Of the other medical men who were Philosophical Society members, four were physicians, Doctors Buck of Newark, Jones of Lichfield, Taylor of Ashbourne, and Wilson of Mansfield, while the rest were probably all surgeons, none being recorded as apothecaries alone, underscoring the gentlemanly status of the society.[25] Most of the physicians, including Darwin, Arnold, Bree, White and Storer, were also graduates of the Scottish universities.

John Storer was a leading Nottingham physician and prominent member of the local scientific community. He was later president of the Nottingham Subscription Library at Bromley House, which functioned like a philosophical society in its early years, arranging scientific lectures and discussion groups. He was also instrumental in the foundation of the Nottingham General Lunatic Asylum and one of the leaders of the campaign for a mechanics' institute in 1824.[26] Thus his career and scientific interests closely parallel some other Derby philosophers, as we shall see, such as Richard Forester. Dr Thomas Arnold was perhaps the best known of the Derby Philosophical Society physicians after Darwin himself during the 1780s and 1790s.[27] From 1789 he kept a private asylum at his house in Bond Street, Leicester and became acknowledged as a theorist on insanity, publishing a textbook that was the longest work on the subject in English at the time.[28] His Leicester associate Dr Robert Bree served as physician at the Leicester Infirmary established in 1771. It was usual for provincial madhouses to admit parish pauper patients, and during the 1780s Arnold admitted some charity cases and paupers. In January 1793 he was instrumental in helping to launch an appeal for subscriptions to add a lunatic asylum to the Infirmary. In September 1794, this resulted in the construction of a small building where ten cases were accommodated. Arnold had dissenting and radical sympathies similar to those of some of the other Derby philosophers; he was also a close friend of William Withering.

Membership of the Derby Philosophical Society offered access to an important scientific and medical library and the opportunity to associate with prominent physicians such as Darwin, who, despite remaining in the provinces had become widely acclaimed by the 1780s, to the extent that, as his grandson Charles noted, even George III had apparently expressed a desire for him to come to London to become his physician.[29] During the 1790s, Darwin also tried to cement his medical reputation by publishing *The Zoonomia*, dedicated to the members of the College of Physicians, philosophers, students, and 'all those who study the operations of the mind as a science, or who practise medicine as a profession'. The work was intended to 'reduce the facts belonging to animal life into classes, orders, genera, and species, and by comparing them with each other, to unravel the theory of diseases'.[30]

Commentators from the Rev. Thomas Gisborne to Charles Lyell emphasised the importance of medical men in provincial urban scientific culture, and this was manifest in their support for literary and scientific institutions. This enhanced their status in local communities as well as providing access to libraries and opportunities to attend lectures, all of which helped to differentiate them from quacks and healers who competed with them in the late-Georgian medical marketplace.[31] In his popular general study of the duties of the male middle classes and higher ranks, Gisborne, a founder member of the Derby Philosophical Society in 1783 and honorary member of the Manchester Literary and Philosophical Society, examined the duties of physicians in a lengthy chapter that drew upon his friendship with Darwin and the Manchester physicians Gregory and Percival. According to Gisborne, the British medical profession enjoyed 'that degree of estimation and credit, which a science conferring on mankind the greatest of all comforts except those of religion, justly deserves'. Physicians were 'almost invariably men of liberal education and cultivated minds' and the profession opened the way to 'reputation and wealth', raising him 'to a level, in the intercourse of common life, with the highest classes of society'.[32] That such medical men supported local literary and philosophical culture was hardly surprising given that the physician ought to 'bestow a due share of his time on other collateral pursuits and acquisitions, as chemistry, botany, and natural philosophy ... which ... have a close connection with the healing art'. Gisborne also recommended that the physician gain particular knowledge of 'pharmaceutical chemistry', and

Drawn by O. Oakley from
a Painting by J. Wright. R.A.

Engraved by H. Meyer.
No 3. Red Lion Square.

JEDEDIAH STRUTT ESQUIRE,

(late of Derby.)

The Eminent Inventor of that Important Machine

Called the DERBY RIB STOCKING FRAME.

To William Strutt Esq.r S.t Helens House Derby.
This plate is most respectfully dedicated by his obliged
Humble Servant Stephen Glover the Publisher.

13 Jedediah Strutt, engraving after the painting by J. Wright.

Latin, Greek and especially French. This was not merely because physicians as professionals were 'placed in the upper ranks of society', but also so that he might 'peruse with facility the valuable tracts on medical subjects published in the tongue'.[33] Thus, for Gisborne, physicians were most emphatically members of the social elite, men 'of reputation and wealth', who were 'placed in the upper ranks of society', and membership of philosophical societies helped to maintain this status.

With half the Philosophical Society membership being medical practitioners, it was hardly surprising that

14 Josiah Wedgwood, from S. Smiles, *Lives of the Engineers: Boulton and Watt* (1904).

medical subjects dominated the library. By the end of 1793, the library contained 440 volumes, of which at least 150 were on medical subjects, excluding philosophical journals containing articles on medical subjects. In the period between 1 January 1794 and 16 March 1795, the Society bought a total of eighty-eight volumes of which at least thirty-nine were on medical subjects; thus the proportion of medical works acquired increased.[34] These included Arnold's *Observations on Insanity*, various medical works by Beddoes, Clarke's *Essays on Diseases and Fevers*, Duncan's *Medical Commentaries*, five works by the surgeon John Hunter, Monroe's medical works including the *Medical Chemistry* (3 vols.) by 1793, and treatises on mineral waters.[35]

Industrial and manufacturing families

Although the domination of medical members suggests that the perception of the industrial utility of natural philosophy was not decisive, a group of industrial and manufacturing families stand out in Derby scientific culture. In this, they parallel the philosophical activities of the Greg, Henry, McConnel, Philips and the Robinson families in Manchester, who helped to found the Manchester Literary and Philosophical Society. The Strutts were probably the most influential political and scientific leaders in Derbyshire between the 1780s and 1840s. Their wealth was founded upon the cotton spinning ventures of Jedediah Strutt, who had partnered Richard Arkwright with Samuel Need, and although this concern had split during the 1780s, Strutt retained enough of a portion to serve as the basis for a major cotton spinning business at Belper

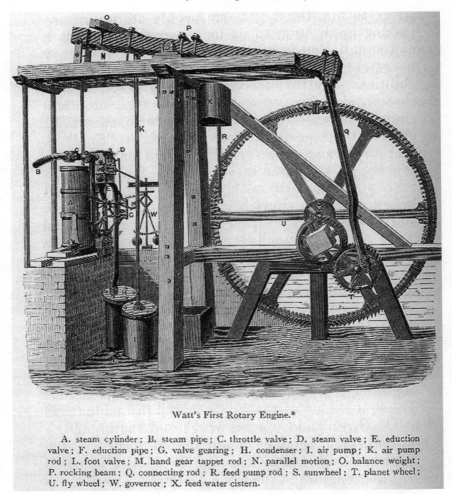

Watt's First Rotary Engine.*

A. steam cylinder; B. steam pipe; C. throttle valve; D. steam valve; E. eduction valve; F. eduction pipe; G. valve gearing; H. condenser; I. air pump; K. air pump rod; L. foot valve; M. hand gear tappet rod; N. parallel motion; O. balance weight; P. rocking beam; Q. connecting rod; R. feed pump rod; S. sunwheel; T. planet wheel; U. fly wheel; W. governor; X. feed water cistern.

15 Watt Rotary Steam Engine from S. Smiles, *Lives of the Engineers: Boulton and Watt* (1904).

and Derby, which was continued by his sons William, Joseph and George Benson. William and Joseph resided in Derby and took a keen interest in philosophical culture, the former becoming Darwin's closest scientific companion and serving as Philosophical Society president after Darwin's death in 1802. Like the Gregs and Henrys at Manchester, the Strutts were Unitarians and political reformers who rose to social and political eminence. The process was completed by Joseph Strutt, who served as first mayor of the reformed corporation, and Edward Strutt, who converted to Anglicanism and became Baron Belper.[36]

Other local industrial and manufacturing families also rose to prominence in Derby cultural and philosophical life, notably the Wedgwoods, Evanses and Duesburys. Originally iron-founders, the Evanses' concerns included cotton mills at Darley Abbey, a small village just to the north of Derby, and banking interests from the 1770s. They intermarried with the Strutt family, sharing their taste for land, property and natural philosophy if not, as Anglicans, their Unitarian affiliation. William Evans I, the brother-in-law of Joseph and William Strutt through his marriage to their sister Elizabeth (1758–1836), helped to found the Chesterfield Literary and Scientific Debating Society in 1786 and probably joined the Derby Philosophical Society. Walter, his half-brother, and William II, his son, also became members of the Philosophical Society by 1813 and subscribed to local scientific works such as White Watson's *Delineation of the Strata of Derbyshire* (1811).[37]

The Duesburys, who came to dominate china manufacturing in Derby, were also prominent supporters of scientific culture, and William Duesbury II joined the Philosophical Society and assisted Richard Roe in procuring books from London for the library.[38] Even more successful as businessmen, technological innovators and natural philosophers were the Wedgwoods who, while not residing in Derby, supported the Philosophical Society. Josiah Wedgwood (1730–95) was a member of the Lunar Society and close friend of Erasmus Darwin, though probably not a member of the Philosophical Society.[39] The J. Wedgwood referred to in the Philosophical Society records was probably his son John Wedgwood, the banker, of Etruria. Ralph Wedgwood, the cousin of Josiah, was another corresponding Philosophical Society member. He was also

an inventor and potter, whose most successful invention was supposed to have been carbon paper. The interrelationship between natural philosophy, technological innovation and industrial development is illustrated by the activities of the Wedgwoods, especially Josiah, and was celebrated by contemporaries such as Darwin, who contributed an obituary to the *Derby Mercury*, which praised his 'great public spirit', general intelligence, chemical knowledge and invention of a new industrial thermometer. Wedgwood had 'raised the manufacture of earthen wares, from the obscure state which he found them, to the degree of utility, elegance, and splendour, by which they are now

16 The Rev. Thomas Gisborne.

17 William Strutt's residence: St Helen's House in 2008.

distinguished in every part of the world.'[40] Overlooking the more oppressive aspects of industrial development, Darwin emphasised the degree to which science and industry as embodied by the activities of the Strutts and Wedgwoods were interconnected acts of 'great public spirit'.

Conclusion

At one level, for the first twenty years the Derby Philosophical Society was a fairly small scientific association which failed to achieve most of the objectives stated by Darwin in his opening address and did not establish a public platform for science in urban culture. Dominated by one natural philosopher of world standing, the Society published nothing except editions of its library catalogue, and only a few individuals met regularly to play a part in producing original scientific work or disseminating scientific ideas. It is also striking that some local individuals with clear philosophical interests such as the Rev. James Pilkington, Bennet and Thomas Swanwick are not listed as members, although membership records are lacking between the 1790s and 1813.

While there is some truth in this, the considerable funds necessary to publish meant that most Georgian philosophical societies did not publish transactions. The Manchester philosophers, in a much larger town, were an exception, but even they had major difficulties keeping their transactions going. The fact that most Derby Philosophical Society meetings were conducted by a small group of friends in domestic circumstances, which explains why the business of meetings usually went unrecorded, reflected the interdependency of public and private life prevalent in Georgian clubs, and helps to account for the durability of the society. The number, variety and geographical spread of the membership also suggest that the Philosophical Society was relatively successful compared to other contemporary such associations and can be regarded as a regional organisation. Furthermore, despite the lack of formal publications, the society was able to attract public attention for natural philosophy, as the publicity surrounding the balloon launch and the publication of Darwin's works demonstrates. The example of the Derby Philosophical Society also helped to encourage the formation of other scientific societies in Derby during the nineteenth century.

Although Darwin was dominant, as references in his work testify, he seems to have received considerable stimulus from other members, including William Strutt, Roe, Fox and Forester, as the galvanic experiments with glass demonstrate. Some of the non-resident members such as Arnold, Boothby, Bage, Wedgwood and Robert Darwin were also highly successful in various fields of intellectual endeavour. Finally, it should not be forgotten that, despite the political activities of the Derby philosophers, the Philosophical Society survived independently for almost sixty years after Darwin's death until it was absorbed by an amalgamation, when the much-augmented library helped to form the collection for a public museum, library and gallery in Derby. With its formal institutional structure and rules and regulations designed by Darwin and his friends, the Derby Philosophical Society long outlived many other Georgian philosophical groups such as the Lunar Society, Richard Phillips's Adelphi Society at Leicester, the first Bath philosophical society, and the London coffee-house philosophical clubs. As the list of nineteenth-century members demonstrates, it continued to attract a variety of Liberal and Tory individuals from across the north midland counties and beyond, including gentry and aristocracy, clergy, industrialists, medical men and other professionals. Inspired by Darwin, it was not actually dependent upon him and long outlived him. Situated in an important county town and major industrial and manufacturing centre, for the members of the Derby Philosophical Society, natural philosophy was simultaneously polite knowledge, form of improvement and utilitarian endeavour.

Notes

1 D. King-Hele, *Erasmus Darwin: A Life of Unequalled Achievement* (London: Giles de la Mare, 1999); R. Schofield, *The Lunar Society of Birmingham* (Oxford: Clarendon Press, 1963); J. Uglow, *The Lunar Men: The Friends who Made the Future, 1730–1810* (London: Faber, 2003).

2 E. Darwin, letter to M. Boulton, 26 December 1782, in D. King-Hele (ed.), *The Collected Letters of Erasmus Darwin* (Cambridge: Cambridge University Press, 2007), 214; S. Glover, *The History, and Gazetteer, and Directory of the County of Derby*, 2 vols. (Derby, 1829, 1833), II, 430; M. Craven, *The Illustrated History of Derby's Pubs* (Derby: Breedon, 2002), 90–1.

3 In 1999 King-Hele estimated that the subscription was equivalent to about £80 per year and the fine to about £4 per meeting in today's money, King-Hele, *Erasmus Darwin*, 198.

4 Quoted in A. E. Musson and E. Robinson, *Science and Technology in the Industrial Revolution* (Manchester: Manchester University Press, 1969), 191.

5 E. Darwin, letter to Matthew Boulton, 4 March 1783, King-Hele (ed.), *Letters of Erasmus Darwin*, 215.

6 E. Darwin, 'Address to the Philosophical Society delivered at their first regular meeting, held July 18, 1784', DLSL.

7 *Derby Mercury* (1 January 1784).

8 Letter to M. Boulton, 17 January 1784, King-Hele (ed.), *Letters of Erasmus Darwin*, 224.

9 Quoted in King-Hele, *Erasmus Darwin*, 196–7; *Derby Mercury* (8 January 1784).

10 L. C. T. Rolt, *The Aeronauts: A History of Ballooning, 1783–1903* (London: Longmans, 1966); R. Gillespie, 'Ballooning in France and Britain 1783–1786', *Isis*, 75 (1984), 249–68, pp. 249–63.

11 Gillespie, 'Ballooning in France and Britain', 261–8.

12 Rolt, *The Aeronauts*; Gillespie, 'Ballooning in France and Britain'.

13 *Derby Mercury* (23 December 1784); *The Times* (5 May, 3 September 1784) argued that ballooning was foreign, lower class and led to public disorder, therefore it should be banned.

14 Gillespie, 'Ballooning in France and Britain', 263.

15 J. Southern, *A Treatise upon Aerostatic Machines containing Rules for Calculating their Powers of Ascension …* (Birmingham, 1785); *Derby Mercury* (1, 8 January 1784); Letter to Matthew Boulton, 17 January 1784; King-Hele (ed.), *Letters of Erasmus Darwin*, 224.

16 Musson and Robinson, *Science and Technology*, 190–9; R. P. Sturges, 'The membership of the Derby Philosophical Society, 1783–1802', *Midland History*, 4 (1978), 212–29.

17 *Derby Mercury* (31 May 1787, 5 January 1792, 16 July 1795).

18 J. Priestley, *The History and Present State of Electricity, with Original Experiments*, first edition (London, 1767); W. Hackmann, *Electricity from Glass: The History of the Frictional Electrical Machine* (Alpen an den Rijn: Sijthoff & Noordhoff, 1978); J. Heilbron, *Electricity in the 17th and 18th Centuries: A Study of Early Modern Physics* (New York: Dover, 1999); M. Pera, *The Ambiguous Frog: The Galvani/Volta*

Controversy on Animal Electricity (Princeton: Princeton University Press, 1992); P. L. Mottelay, *A Bibliographical History of Electricity and Magnetism* (London: Griffin, 1922); S. Schaffer, 'Natural philosophy and public spectacle in the eighteenth century', *History of Science*, 21 (1983), 1–43; V. Jankovic, *Reading the Skies: A Cultural History of English Weather, 1650–1820* (Manchester: Manchester University Press, 2000); P. Elliott, 'Abraham Bennet FRS (1749–1799): a provincial electrician in eighteenth-century England', *Notes and Records of the Royal Society of London* (1999), 59–78; J. Golinski, *British Weather and the Climate of Enlightenment* (Chicago: Chicago University Press, 2007); P. Elliott, '"More subtle than the electric aura": Georgian medical electricity, the spirit of animation and the development of Erasmus Darwin's psychophysiology', *Medical History*, 52 (2008), 195–221.

19 *Rules and Catalogue of the Library of the Derby Philosophical Society* (Derby, 1793), with supplements (1795, 1798); Catalogue and charging ledger of the Derby Philosophical Society (1786–9), DLSL; *Derby Mercury* (31 May 1787).

20 E. Darwin, 'Frigorific experiments on the mechanical expansion of air', *Philosophical Transactions*, 78 (1788), 44–6; King-Hele, *Erasmus Darwin*, 226–8.

21 E. Darwin, *Zoonomia; or, the Laws of Organic Life*, first edition, 2 vols. (London, 1794–96), I, 120; Elliott, 'Abraham Bennet FRS'.

22 E. Darwin, letter to William Strutt (6 August 1801), letters to Georgiana, Duchess of Devonshire (6 August 1801, 29 November and November 1800), in King-Hele (ed.), *Letters of Erasmus Darwin*, 556–60.

23 Sturges, 'Membership', 228–9 and appendix.

24 E. Posner, 'Eighteenth-century health and social service in the pottery industry of north Staffordshire', *Medical History*, 18 (1974), 138–45; E. Posner, 'Josiah Wedgwood's doctors', *Pharmaceutical Historian*, 1 (1973), 6–8. J. Bent, 'An account of a woman enjoying the use of her right arm after the head of the *os humeri* was cut away', *Philosophical Transactions*, 13 (1774), 539–41.

25 Sturges, 'Membership', 228–9.

26 *Nottingham Journal* (13 October 1837, 27 March, 10, 17, 22 April 1824); I. Inkster, 'Scientific culture and education in Nottingham, 1800–1843', *Transactions of the Thoroton Society* (1978), 82, 45–50; P. Elliott, '"Food for the mind and rapture to the sense": scientific culture in Nottingham, 1740–1800', *Transactions of the Thoroton Society*, 109 (2005).

27 P. K. Carpenter, 'Thomas Arnold: a provincial psychiatrist in Georgian England', *Medical History*, 33 (1989), 199–216; E. R. Frizelle and J. D. Martin, *The Leicester Royal Infirmary, 1771–1971* (Leicester, 1971); P. K. Carpenter, 'The private lunatic asylums of Leicestershire', *Transactions of the Leicestershire Archaeological and Historical Society*, 61 (1987), 34–42.

28 T. Arnold, *Observations on the Nature, Kinds, Causes and Prevention of Insanity*, 2 vols. (Leicester, 1782, 1786); Carpenter, 'Thomas Arnold', 204; R. Hunter and I. MacAlpine, *Three Hundred Years of Psychiatry 1535–1860* (Oxford: Oxford University Press, 1963); R. Hoeldtke, 'The history of associationism and British medical psychology', *Medical History*, 11 (1967), 46–65.

29 C. Darwin, *Life of Erasmus Darwin*, ed. D. King-Hele (Cambridge: Cambridge University Press, 2003), 69; King-Hele, *Erasmus Darwin*, 172, 312.

30 Darwin, *Zoonomia* (1794), I, preface.

31 C. Lyell, 'Scientific institutions', *Quarterly Review*, 34 (1826), 172; T. Gisborne, 'On the benefits and duties resulting from the institution of societies for the advancement of literature and philosophy', *Memoirs of the Literary and Philosophical Society of Manchester*, 5 (1798), 70–88; 'Obituary of Rev. Thomas Gisborne', *Gentleman's Magazine* (1846), i, 643–5; Sir J. Stephen, *Essays in Ecclesiastical Biography*, fifth edition (London, 1867), 530–6; B. Nicolson, *Joseph Wright of Derby: Painter of Light*, 2 vols. (New York: Paul Mellon Foundation, 1968), I, 132–7.

32 T. Gisborne, *An Enquiry into the Duties of Men in the Higher Rank and Middle Classes of Society in Great Britain*, first edition (London, 1794), 383–5.

33 Gisborne, *Enquiry into the Duties of Men*, 389. Gisborne also encouraged James Fox (1760–1835), who worked as his servant, to develop his manufacturing skills and he went on to found an internationally renowned engineering company at Little Chester which developed some of the earliest industrial planing machines in the world.

34 *Laws and Regulations Agreed Upon by the Philosophical Society at Derby, Catalogue of the Library* (December 1793); supplement (March 1795).

35 *Derby Philosophical Society Rules and Catalogue* (Derby, 1793); letter to James Pilkington (5 February 1788), King-Hele (ed.), *Letters of Erasmus Darwin*, 303–4.

36 R. S. Fitton and A. P. Wadsworth, *The Strutts and the Arkwrights, 1758–1830: A Study of the Early Factory System* (Manchester: Manchester University Press, 1958).

37 *Derby Mercury* (16 March 1786, 18 May 1786); Derby Philosophical Society catalogue and charging ledger (1785–89), DLSL, BA 106, 9230; Derby Philosophical Society cash ledger (1813–45), DLSL, BA 106, 9229, 9230; J. Lindsay, 'An early industrial community: the Evans' cotton mill at Darley Abbey, Derbyshire, 1783–1810', *Business History Review*, 34 (1960), 277–301; W. Watson, *A Delineation of the Strata of Derbyshire* (Sheffield, 1811), list of subscribers.

38 Thanks to Judith Anderson for this information. Craven, *Derbeians*, 77–9.

39 E. Meteyard, *The Life of Josiah Wedgwood* (London, 1866); B. and H. Wedgwood, *The Wedgwood Circle, 1730–1897* (Studio Vista, 1980); R. Reilly, *Josiah Wedgwood* (London: Macmillan, 1992).

40 *Derby Mercury* (8 January 1795): I am indebted to D. King-Hele for the identification of this obituary as Darwin's work.

41 P. Elliott, 'Provincial urban society, scientific culture and socio-political marginality in Britain in the eighteenth and nineteenth centuries', *Social History*, 28 (2003), 361–87; P. Elliott, 'Towards a geography of English scientific culture: provincial identity and literary and philosophical culture in the English county town, *c.*1750–1850', *Urban History*, 32 (2005), 391–412.

5

Dissent, politics and natural philosophy: the campaigns of the 1780s and 1790s

Introduction

Inspired by enlightenment progressivism, their pursuit of natural philosophy and the leadership of Darwin, the Derby philosophers were at the forefront of local improvement and national political campaigns, especially through the activities of the Derby Society for Political Information. The relationship between philosophical culture and politics in Derby was coloured by the fact that many of the middle classes, particularly the dissenters, had become wealthy and influential locally but still faced civil and political restrictions nationally, with continuing aristocratic power, a very limited franchise, the Test and Corporation Acts and social prejudice. Women, of course, faced greater restrictions, and we will examine their involvement in scientific culture in the penultimate chapter. As well as apparently conferring direct utilitarian benefits, as Darwin emphasised in his address to the Philosophical Society, experimental science provided important inspiration for social and political improvement and an opportunity to enhance and confirm social status through public cultural activity that was – ostensibly at least – socially, religiously and politically inclusive.

Before examining the enclosure controversy, this chapter will explore the political activities of the Derby philosophers in the context of borough and county politics and religious and political affiliations, especially the inspiration provided by enlightenment natural philosophy. It will focus upon the Derby Political Society, including the delegation to France, the relationship between the Derby Philosophical Society and local political campaigning, and the nature of the political philosophy espoused by the Derby reformers as enunciated in a variety of polemical and literary publications. In order to appreciate the significance of the political campaigns, it will begin by considering the state of politics in Derby and Derbyshire during the Georgian period.

Politics in Derbyshire

The agrarian west of Derbyshire had proportionately greater representation than the more urbanised east and south. There were few contested elections prior to 1832, with the Duke of Devonshire as the most powerful aristocratic landlord and his agents exerting the dominant influence. The Duke controlled one county member, the other tending to be an independent Tory, while either a member of the Cavendish family or a ducal nominee served as one of the two borough members until 1848, the other being by tradition and agreement a corporation nominee until 1832.[1] Open bribery was rarely necessary to control the vote, as power and influence usually sufficed. Prior to the Reform Bill of 1832, the vote was the prerogative of forty-shilling freeholders in the county and Derby borough burgesses, most of whom were small tradesmen, shopkeepers and labourers who had been born to a burgess father, though it remained possible for resident aliens to be elected as freemen. The relative importance of the burgesses declined rapidly during the eighteenth century, so that they became less representative of the total population, probably numbering between 650 and 900. Assuming the higher figure, in 1680 that was about 30% of a population of approximately 3,000, while in 1791 this had fallen to just 10% of the total population, and in 1801 just 8.3% of the total population.[2]

Prior to 1836, under the terms of charters granted by Charles I and Charles II, the 'closed' Derby corporation consisted of a mayor, various town officials, nine aldermen, fourteen brethren, fourteen common council and an indefinite number of freemen, who voted in elections. Legal authority was vested in the common council made up of aldermen, brethren and capital burgesses, which could make by-laws and impose fines. The local justices of the peace were the mayor, the Bishop of Lichfield, the Bishop's chancellor, the recorder, the town clerk, the outgoing mayor, and four senior alderman.[3] Local opposition that did arise, usually encouraged by local Tories, was combated through the creation of honorary freemen or 'bastard burgesses', known as faggots, often from the vast Cavendish estates. However, there was always a sizeable Tory vote in the borough, as the 1775 election demonstrates, when the Tory candidate Daniel Parker Coke was only defeated by a mere fourteen votes after strenuous efforts by the Whigs. A Tory club existed from the 1770s and formed the nucleus of what later became the True Blue Club in 1812.[4] As predominantly the only newspaper until the 1820s, the *Derby Mercury*, particularly under the Drewrys, strove to be neutral, only becoming openly Tory during the 1820s when the Whig *Derby Reporter* appeared. According to the Derbyshire-born historian Thomas Hinton Burley Oldfield (1755–1822), the Whig Cavendish/corporation hegemony was an 'evil' where a 'very large and opulent town' such as Derby could not 'maintain its independence' as it could have done if it had been 'relieved from the tyranny of a corporation, under which no town can be free'. However, despite its oligarchic aspects, the corporation was fairly repre-

18 The Rev. Charles Stead Hope, painted by T. Barber and engraved by W. Say.

sentative of the urban elite and more so than the reformed body after 1836, which initially excluded leading Tories such as the Rev. Charles Stead Hope (1762–1841), president of the True Blue Club, wealthy landowner and mayor of Derby in 1797, 1805, 1817, 1825 and 1831. It is notable that although they made some criticisms, the municipal corporation commissioners concluded that the corporation had been broadly successful in promoting the interests of Derby through economic and charitable activities. Likewise, though powerful, the Dukes of Devonshire, via their agent, were careful to take account of the views of Derby's urban elite.[5]

Dissent, natural philosophy and political reform

As in other provincial centres such as Manchester and Norwich, there was a strong association at Derby between scientific culture, dissent and political reform campaigns. Although there were important differences between nonconformist denominations, most dissenters retained a distinctive identity founded upon a strong sense of their own historical development as well as the shared experiences of campaigning against civil restrictions. As the resolutions of the Midlands Protestant dissenters, including Peter Crompton and George Benson Strutt, affirmed in 1790, 'our reverence of Britain, her government and laws is only in subordination to our reverence of God and of human nature', by which they demanded 'an equal participation in all civil privileges'.[6] The Rev. James Pilkington, minister at the Friargate Unitarian Chapel in Derby, campaigned for social and economic reforms, and successors, including Edward Higginson and Noah Jones, were equally assiduous in reform campaigns, as were dissenting ministers from other denominations such as the Rev. James Gawthorne (1775–1857), the Congregationalist minister.[7] Dissenters in Derby such as the Strutts enjoyed considerable power and influence, which increased during the eighteenth century, and most, but not all, supported limited political reforms. A comparison between membership of local government organisations and the Friargate congregation demonstrates that the town was similar to Nottingham, Coventry, Norwich, Bristol, and other places where dissenters played an important role in local government.[8] Derby dissenters were prepared to practise occasional conformity in order to hold urban office, while continuing to attend chapel and baptise children as Presbyterian or Unitarian. In the last decades of the eighteenth century, for example, the proprietor of Derby's only successful newspaper, the most powerful family of bankers, around half the corporation, and various mayors were all nonconformists. Symbolised by the Strutt's calico mill, the tallest building in Derby after All Saints' Church tower, wealthy dissenting families such as the Leapers, Cromptons, Binghams, Foxes, Strutts and Rowlands were firmly entrenched in the urban oligarchy, intermarrying with each other, holding multiple offices through the generations, and becoming major landowners in defiance of civil restrictions.[9]

Although there was support within Derbyshire for Wilkes during the 1760s, the first county parliamentary reform campaign was begun during the 1780s, stimulated by opposition to the American war.[10] Whig gentry such as Francis Noel Clark Mundy (1738–1815), a friend of Darwin, Anna Seward and John Whitehurst, opposed the war and advocated limited political change. Darwin, Priestley and their Lunar associates were members of philosophical networks which saw the American struggle as synonymous with the drive for scientific progress; the arts and sciences were felt to be 'the offspring of freedom' and the antithesis of the tyranny of empire.[11] This sympathy for the colonists was shared by some of the borough and county elite. Encouraged by Mundy, the Wyvillite campaign in Yorkshire was supported by some Derbyshire gentry at a county meeting in 1780, though the emphasis was placed on economic rather than constitutional reform. Mundy was supported by the independent Tory Daniel Parker Coke, who had just been unseated as the Derby MP, the Presbyterian banker Samuel Shore (1738–1828) and Samuel Crompton (1750–1810), another Presbyterian banker and mayor of Derby in 1782 and 1788. Opposition came from Tories such as Nathaniel Curzon, who supported the North administration. The metropolitan Society for Constitutional Information was backed by some Derbyshire Whig gentry and the Unitarian John Drewry, the *Mercury* proprietor, who sold copies of the American constitution. In an editorial, he called for an end to a war caused by 'the late dangerous administration', and 'a due correction of that inequality in the representation of the people … the true source of all our national grievances'.[12] There was another Derbyshire petition in 1783, but the movement petered out as a result of lack of Cavendish enthusiasm and the peace.[13]

The renewed reform agitation of the early 1790s, given impetus by the French Revolution, was significant because, with the foundation of the Derby Society for Constitutional Information, urban-based reformers assumed leadership from the county gentry.[14] While maintaining close links with national reformers, they also asserted a distinctive provincial identity against national and metropolitan political commentators; as Brooke Boothby contended, the reformers were 'not the least learned and reflecting men in this kingdom', and were not 'confined to courts and capitals'.[15] The political philosophy of the Derby Society was outlined in a series of publications, public addresses and editorials in the *Mercury*. Its purpose was to 'arouse the people to a peaceable pursuit of reforms of government', primarily, 'full, free and frequently elected representation', with the labouring classes at least theoretically being encouraged to join. It was organised into divisions of ten people who paid a subscription of two shillings per annum and was thus closer in organisation to the Sheffield Political Society than to the more exclusive Manchester Society. The Society distributed hundreds of copies of works by Paine sent by the London Society for Constitutional Information, helped to encourage the formation of a Nottingham political society, and made efforts to spread the campaign to other counties such as Shropshire.[16] One address of July 1792 attacked opponents for 'inflaming the public mind' and argued that government should be instituted according to Rousseau's general will. Excessive government was condemned, including repressive laws such as those concerning game and excise; corruption, pensions, the neglect of the poor, and burdensome taxation feeding a 'swollen military establishment' all came in for criticism. Another notice in 1793 supported Fox in his campaign against the war argued that he would 'rank amongst the greatest of legislators' in the future, while Burke would be condemned with 'execration and disgust'.[17]

The inspiration provided by enlightenment science is evident from the fact that Derby Philosophical Society members took the lead in the political societies, including William and Joseph Strutt, Darwin, Samuel Fox, William Brookes Johnson, Peter Crompton and John Hollis Pigot. Of the non-resident Philosophical Society members, Brooke Boothby, Robert Bage, John Storer and Robert Waring Darwin were also prominent political campaigners, the last founding a political society at Shrewsbury. This helps to explain how mutual interest in scientific and technological innovation brought the Derby philosophers to the work of Thomas Paine rather than his political publications. By 1789 William Strutt was taking a keen interest in the use of iron for construction and in March he noted to Samuel Oldknow that 'a man has got a patent for making iron bridges cheaper than of stone' and was preparing a plan for one over the Schuylkill in America. Paine was more successful as a publicist than an engineer but he had been working with the Walker family, who were friends and suppliers for the Strutts, and had married into the family of Samuel Need, Jedediah Strutt's old partner. William Strutt was then engaged in the architectural and engineering problems of constructing St Mary's Bridge over the Derwent at Derby. In 1792–93, as we shall see, he had incorporated a fire-resistant design of iron casings and pillars with hollow pots into the family's Derby mill, and Fitton and Wadsworth consider it likely that he met Paine in 1789 during the latter's tour of cotton mills, potteries and iron mills in the north and midlands.[18] Paine wrote to Thomas Walker on 24 February 1789 that 'Mr. Fox of Derby called again on me last evening respecting the bridge but I was not at home.' This was Samuel Fox, the prominent hosier, Derby Philosophical Society member and later the primary contact between the Derby and London political societies, to whom copies of Paine's works were distributed, and it seems that Strutt had asked him to make enquiries concerning the bridge.[19]

The French Revolution was initially regarded optimistically as a move towards a British-style balanced constitution. Writing to his old friend Richard Dixon in October 1792, Darwin remarked that 'the success of the French against a confederacy of kings gives me great pleasure, and I hope they will preserve their liberty, and spread the holy flame of freedom over Europe.' Joseph Strutt dined with the reformers of the Revolution Society at the London Tavern in 1792 and his brother William distributed copies of Paine's *Rights of Man* to workers at Belper, which apparently caused unrest, resulting in the family mills being flooded and some of the books being burnt. With reference to George III, Darwin provocatively and jokingly commented to Dixon that 'Mr. Pain (sic) says ... he thinks a monkey or a bear, or a goose may govern a kingdom as well, and at much less expense than any being in Christendom, whether idiot or madman out or in his royal senses; adieu dear citizen.'[20] Two young radical writers, William Ward and Charles Sambrooke Ordoyno, publicised the activities of the Political Society. However, although the secretary was Thomas Pratt, a tailor and parish clerk, they did not attract large numbers of the labouring classes and never obtained more than a hundred members.[21] Until 1791 Ward was apprenticed to John Drewry at the *Mercury*, who allowed sympathetic articles and editorials by his junior to appear. Ordoyno worked as a printer and bookseller in Derby and set up the short-lived *Derby Herald* as a radical organ, which allegedly failed because he 'loved ale more than republicanism'.[22]

The delegation to Paris

Perhaps the most notorious and ambitious of the Political Society's ventures was to dispatch a delegation on behalf of themselves and the political societies at Nottingham and Belper to the French National Convention to foster European accord. Their trip was widely known in the county; as the Rev. Robert Wilmot of Morley put it, from Derby they have 'actually sent two persons to the national convention of France to invite the French over to this country to create the same anarchy here which is there triumphant'.[23] The delegation reflected the enlightenment ideal of an international philosophical republic transcending national rivalries, shared by scientific figures in Birmingham, Manchester, Sheffield and Derby.[24] The Derby delegation also followed communications between the London Revolution Society and the French, and a notorious visit to Paris by James Watt junior and Thomas Cooper of Manchester, who while ostensibly organising business in March and April 1792, had sought to promote links between British political societies and the Jacobins. This culminated in their appearance in a triumphant procession on the Champs de Mars, and they had been accused of being agents in a conspiracy with the Jacobins by Burke in the House of Commons.[25] The two Derby delegates were Henry 'Redhead' Yorke (1771–1813) and William Brookes Johnson. Born in the West Indies to a mixed-race mother, Yorke was of striking appearance, and began his polemical career with a pro-slavery pamphlet dedicated to Bache Heathcote, a local Tory. He rapidly became one of the most radical Political Society activists, being dispatched by William Strutt to assist the Sheffield Political Society in creating its organisation.[26]

Johnson's involvement with the Derby Philosophical Society and the inspiration he obtained from progressive enlightenment science are evident from his scientific activities and publication of a *History and Present State of Animal Chemistry* (1803). Educated at Cambridge, where he became close friend of the chemist Smithson Tennant, Johnson used the Derby Philosophical Society library while studying to become a physician during the 1780s. He borrowed texts including Smith's *Wealth of Nations*, volumes of the *Manchester Memoirs*, the *Mémoires de l'academie de Dijon* and Sparman's *Voyage to the Cape of Good Hope*.[27] He used social and professional contacts gained at Derby and Cambridge to expand his national contacts, and was recommended as a Fellow of the Linnean Society by Darwin after acquiring a large herbarium from South Amercia and composing the botanical sections of Pilkington's *View of Derbyshire* (1789). Johnson also assisted Jonathan Stokes FLS, the Derbyshire physician, botanist and Lunar Society member, with his botanical works. The chemical history was not an original work but a compilation of sources for medical men and the 'enterprising experimentalist'. The third volume, for instance, included extracts from Beddoes, Davy, de la Métherie, Rush and Ferriar, and was devoted to animation and death, revealing Johnson's materialist sympathies. Inspired by his involvement with the Derby philosophers, Johnson exuded confidence in the triumphant powers of enlightenment science and the potential of intellectuals to improve society, while emphasising industrial applications of animal chemistry. Because happiness or misery depended 'ultimately on the proportion of the different parts of the substances that compose them', so, Johnson contended, the mind would be in reach of 'the powers of the analytical science' and 'some distant and future age may be indebted for an explanation not only of the functions of the body, but of the faculties of the mind, to chemical analysis'.[28]

The address was read in Derby at a meeting chaired by Peter Crompton on 22 November 1792, probably at the Bell, and it was 'unanimously' decided that it be presented to the National Convention. Yorke and Johnson claimed that the revolution ought to excite 'the attention of Englishmen' and admiration, and wished to 'pay personal homage to that enlightened and regenerated REPUBLIC'. The two were intending to travel anyway to peruse the 'political writings which daily appear in that asylum of liberty and justice' and to attend the trial of Louis Capet (Louis XVI). Yorke claimed to be 'highly gratified to be the harbinger of the glad tidings of good fellowship' between rival nations, and to be 'instrumental in bringing about a re-union in the great family of men'.[29] The declaration was addressed to the 'brethren and fellow citizens of the world' and wanted the French to be aware 'how deeply' the political societies in Derby, Nottingham and Belper were 'interested in the happiness of France, and ... the cause of general freedom'. With much high-flown rhetoric they praised the French exertions 'in favour of oppressed men' and the reclamation of ancient rights. Tyrants had been destroyed and now the 'poorest cottager of France smiles upon ... delicious plains' previously 'watered with his tears'. He had revolted against 'oppressions', while the French armies had repulsed the armies of Austria and Prussia 'impelled by tyrants' for the sake of humanity. European peace had been 'deluged in blood' at the 'command of kings and priests', and as Paine had contended, 'reason and philosophy' were now 'making hasty strides' as 'precedent and hereditary notions' went 'fast into decline'. By teaching 'mankind' that 'they are all equal in rights', the French had 'dedicated a glorious edifice to liberty' which would always prove 'the dungeons of tyrants and the asylum of the oppressed'. The political societies therefore hailed 'the halcyon days of France, of liberty, and of philosophy'.[30]

In December 1792 Yorke and Johnson arrived in a Paris seething with discontent and tumultuous faction, having been doubtless forewarned by Smithson Tennant, who was in the city until 9 August before meeting Gibbon in Switzerland. The promise of peaceful change heralded by some intellectuals was disappearing as the revolution became more violent, and in the summer of 1792 the monarchy was overthrown, although this was not yet mob rule. Between 1789 and 1791 the moderate middle classes, acting through the Constituent Assembly, had governed through a constitutional monarchy likened by some to the British model, which was partly why English reformers had been attracted to Paris. The liberal assembly created lasting institutional achievements with reforms such as the introduction of the metric system and the emancipation of Jews. The sale of church and aristocrat lands released some Old Regime wealth, but political tension increased markedly after the King's attempted flight in June 1791, which strengthened the republican position. The declarations of war with the European powers and finally England and the execution of Louis XVI helped to force the revolution along a more violent path, which culminated in the Reign of Terror and the rise of Napoleon. To defend the nation the government was forced to mobilise a unified national effort, and the monarchy was overthrown in August and September 1792 as the Girondins took over the National Convention. However, the revolution unleashed forces beyond their control and the limited reforms of the enlightened European despots, the code of gentlemanly warfare and the moderation of the Girondins were early casualties as the demagogical Jacobins took over.[31]

When Yorke and Johnson arrived they were accepted into a British community resident in Paris since the beginning of the revolution, most of whom regarded the French convulsions as the equivalent of the 'Glorious Revolution'. Between 1790 and 1792, the London Revolution Society had corresponded with the French National Assembly and provincial Jacobin clubs. By November 1792, British sympathisers were congregated at White's Hotel or the 'Hotel of the English' at 8 Passage des Petits-Pères in Paris next door to the 'Hotel of the Philadelphians', where visitors from the infant American republic tended to stay. On 18 November, the British assembled at White's for dinner to celebrate French victories and call for the abolition of hereditary titles and 'feudal distinctions' in England. Paine, who was present, was toasted, as were the 'English patriots', Priestley, Fox, Sheridan, Christie, Cooper, Tooke and Mackintosh. A congratulatory address from the English, Scottish and Irish signed by around half those present was adopted to be presented to the Convention and published in the *Moniteur* on 29 November. By 16 December a 'Society of Friends of the Rights of Man' was meeting which convened twice a week and was joined by Johnson and Yorke. Reports reached the British government via Captain George Monroe, who had remained in Paris after the withdrawal of the British embassy in August 1792. He reported that by 11 January 1793 the society, then 'dwindling into nothing', had quarrelled over an address by Paine and Merry. Johnson and Yorke, who were staying with Paine and William Choppin at White's Hotel, were present at this meeting. Choppin, a close friend of Johnson, Paine and Yorke, had known them from London and been present when the original declaration was prepared on 24 November 1792. According to Yorke, the disputed address called for the Convention to 'rescue England from slavery', but it was rejected by a majority of one after Johnson and he had opposed it. However, it was brought forward again and presented to the Convention on 22 January 1793, although Johnson and Yorke drew up a remonstrance and seceded from the society.[32]

Despite their secession, Johnson and Yorke had not argued with Paine and Choppin, although Yorke quarrelled with John Oswald, a soldier, traveller, poet and pamphleteer, who told him, in what may have been a pointed reference to his mixed-race origins, that he was 'not fit to live in a civilised society'. Oswald had been in Paris since at least 1790 and was connected with Watt and Cooper, who claimed to be acting as representatives of the Manchester Constitutional Society. At first friendly, Robespierre seems to have quarrelled with the Manchester deputation. In autumn 1792 or spring 1793, Oswald was sent by Paine to support an Irish rising. He was secretary to the British Club, or Society of Friends of the Rights of the People, until this ended in February 1793. Yorke had taken his pro-slavery pamphlet to Paris to compose a retraction and this was left with Rayment, an English economic writer and friend of James Watt junior. On leaving Paris, Yorke apparently intended to settle in France with his family, which probably means his mother and a brother later living in Manchester. Around this time, he went to Holland, where James Morgan, recently in Paris, told him that Rayment had gone to the committee of general security and denounced him as an English spy. The pamphlet was produced as evidence, it being said that his real name was 'Redhead', and the committee then issued a warrant for Yorke's arrest, seized his belongings, and interrogated several Englishmen. Yorke appears to have later accepted this account by Morgan, but it is hard to see the motives for Rayment's actions if true, and Alger was sceptical.[33]

At a Sheffield meeting of 7 April 1794, Yorke spoke about having 'materially assisted' the revolution in France, although the meaning of this is uncertain. His obituary stated that he was an officer in the French army, a member of the National Convention and 'personally acquainted with all the leading characters in the Revolution', which seems to be supported by confused newspaper reports that he had been killed in fighting. However, this seems hard to square with his opposition to the motion calling for France to intervene in England. As we have seen, Derbyshire was awash with rumours concerning Yorke and Johnson's activities. However, by the end of February 1793 the former seems to have been back in Derby, where he completed his *Reason Urged against Precedent*. Johnson remained in France, staying with Paine and Choppin, with whom he took lodgings at number 63 Faubourg Saint-Denis, nine kilometres north of Paris. They played games, conversed and relaxed in the old garden in the warmth of summer and received distinguished visitors including Brissot, Bancal, General Francisco Miranda, Thomas Christie, Joel Barlow and Mary Wollstonecraft. Wollstonecraft had come to Paris alone in 1792 to see the King on his way to trial and composed her own account of the French Revolution in 1794. Gilbert Imlay, the American explorer and businessman with whom she had an affair, was another visitor and it is likely that Paine, Imlay and Johnson, who had already been to America, would have discussed the role of the American republic in French affairs.[34]

The violence in Paris increasingly threatened the supposed neutrality of international philosophy. Johnson's friend Smithson Tennant had returned early in 1793 and they may well have met up to discuss the rising tension and hostility to the British. Tennant called upon Jean-Claude de la Métherie, the editor of the *Journal de Physique*, and found 'the door and windows closed as if the owner was absent'. The philosopher was found to be in a back room sitting in candlelight with the shutters closed during the middle of the day, and after a hurried and anxious conversation, de la Métherie urged Tennant not to return in case the visit became known to the authorities, which would be serious for both of them. Tennant was back in London by the spring.[35] By the first week of March 1793 in the company of Johnson and Choppin, Paine was working on what was probably to become his most controversial book, *The Age of Reason*, his most sustained attack upon organised religion, in which he argued that churches would decline in the face of the inexorable and inevitable advance of rationality. Presenting the deism of Hume and Gibbon in popular language, Paine reasserted the evidence for a benevolent creator while attacking the Bible as a 'book of riddles' and churches as mere human organisations for the purpose of control. His position was thus less extreme than that of Robespierre and the Jacobins, who would try and create a deity of the goddess of reason – a secular religion; however, political differences with the Jacobins caused much more immediate trouble.[36]

On 21 January, Louis XVI was executed at the instigation of the Jacobins led by Robespierre and Marat, and on 1 February 1793 war was declared on England. Paine had been prominent in opposing the royal execution, and his retreat to Saint-Denis had been an attempt to step back from the battle with the Jacobins, but Robespierre never forgave him for defending the hated monarch, and as one of Paine's closest supporters, Johnson was drawn into this quarrel. Marat accused those who argued for royal clemency of being traitors, and the leading Girondins Guadet and Brissot ordered that he be tried before a revolutionary tribunal for inciting the nation to riot and anarchy. A few days before the trial Johnson was in a state of paranoia; convinced by rumours that Marat's call to arms was a command to massacre all foreigners in France, he stood at the top of the hotel stairs and stabbed himself twice in the chest. Thinking that he was dying, Johnson then fell into the arms of his friend Choppin, offered Paine his blood-spattered watch, and gave Choppin a special message to be presented to the French Assembly that he had composed before. Fortunately, a local doctor arrived in time to staunch the flow of blood and save his life without damage to major organs. News of this extraordinary event had rapidly spread from Saint-Denis to Paris and Paine had allegedly given Johnson's letter to Brissot, who passed it to the editor of his newspaper, *Le Patriot Français*, with a description of events. The sensational account that was published on 17 April helped to make the episode the talk of Paris. It was said that Johnson had 'before dying' written

> with his trembling hands these words which we have read on a paper now in the hands of an eminent foreigner: 'I came to France to enjoy liberty, but it has been assassinated by Marat. Anarchy is even more cruel than despotism. I cannot endure the grievous spectacle of the triumph of imbecility and inhumanity over talent and virtue.'

Johnson was therefore portrayed as a strong opponent of Marat and despite his annoyance there was nothing he could do to counter this impression.[37]

When Marat was brought to trial, with Paine and Johnson as reluctant witnesses, the whole episode was dragged into the public domain again. Of the trial, Sampson Perry, an eyewitness to Johnson's suicide attempt, stated in a letter that 'on the whole it is a mysterious affair, and ought to be cleared up, some people regard it as a farce, others a tragedy'. About halfway through the proceedings, Paine was called to the stand as Jacobin supporters hissed and shouted from the public gallery. Unlike Choppin and Johnson, he had to speak through an interpreter, and when asked to explain Johnson's note, stated that he was unable to see what it had to do with the accusation against Marat. He admitted that Johnson had stabbed himself twice after hearing that Marat was going to denounce him, at which point Marat interrupted to say that 'the young man did not stab himself because I was going to denounce him, but because I wanted to denounce Thomas Paine'. Paine replied that Johnson had long experienced anxieties, and when Johnson himself took the stand he admitted that he had been ill when writing the last testimony, and was now less sure that foreigners were in danger. At the time he had read a newspaper report that Marat intended to order the massacre of all deputies who had equivocated over the fate of the King, and incorrectly believing that Paine was among the deputies, he decided to try and protect him by embracing martyrdom. His friendship for Paine led him to want to kill himself.[38]

This extraordinary trial, as well as demonstrating the devotion that Johnson had for Paine, served to hamstring the case against Marat, who made an impressive speech in his own defence and was found not guilty. Robespierre and Danton overthrew the Girondins and had Brissot and twenty others of their leaders guillotined at the beginning of June, which marks the beginning of the bloody 'Reign of Terror'. Paine had made few friends by supporting the Girondins to the end, and on 13 July Marat was assassinated by Charlotte Corday, a member of the Girondins. His body was committed to the revolutionary temple of the Panthéon with great public ceremony, only to be removed a year later. In the meantime, Paine, Johnson and Choppin were still at Saint-Denis but remained worried about how Robespierre would treat them. Realising the threat, Johnson and Choppin were urged to flee, and in November Paine used the influence that he still commanded to obtain passports for them. One night, just before four o'clock in the morning, Johnson and Choppin fled across the border to Basle, Switzerland just in time, as many of the English were seized as hostages for the Royalist capture of Toulon. On a bitterly cold November night two days after Johnson and Choppin had fled, guards arrived at the hotel with muskets and a warrant to arrest them. Paine recognised that despite his support for the Revolution he would be next, but had nowhere to flee from the Jacobins and received little support from the suspicious American minister to France, Governeur Morris. He could not cross the Atlantic for fear of being arrested by the British for treason, and on Christmas night 1793, after the landlord had been taken away, Paine was arrested on the orders of the committee of general security. He was to remain in the Luxembourg Prison until the fall of Robespierre and was trapped in France until the temporary peace in 1802. Johnson and Choppin wrote to Paine from Basle, as he informed Lady Smith, 'highly pleased with their escape from France, into which they had entered with an enthusiasm of patriotic devotion'.[39]

So ended the mission of the Derby Society for Political Information to the French Assembly, with Yorke returning to hostility and the threat of arrest in England and Johnson fleeing by night to avoid execution or imprisonment. Yet Johnson and Yorke remained British patriots who supported moderate constitutional reforms, and Yorke claimed to 'ever consider my embassy as far more honourable and useful than that of courts, and the abuse I have shared, as a flattering testimony of my praise'. He was 'highly gratified' to have been 'the harbinger of the glad tidings of good fellowship between two nations hitherto rival ... to be instrumental in bringing about a re-union in the great family of men', although by this time the two nations were at war, a conflict that was to last for over two decades.[40] It is noteworthy that although Yorke is usually presented as being a firebrand or ardent revolutionary, Yorke and Johnson, acting on behalf of the Derby, Nottingham and Belper societies, asserted a moderate line and refused to countenance support for French interference in British affairs. As the address of 16 July 1792 had emphasised, 'we are ... fully sensible that our situation is comfortable, compared with that of the people of many European kingdoms'. The Derby reformers hoped that a 'rich harvest' would be reaped from the 'seed which France has so admirably sown' and 'the peoples of each nation' be inspired to reform themselves without external interference. Similarly, Johnson's melodramatic suicide attempt seems to have been motivated by profound support for Paine in his attempts to moderate the extremists as well as fear for their lives.[41]

The loyalist reactions

In the face of news of the terror and war with the French, a loyalist reaction gathered pace which eventually split the Derbyshire reformers, forcing them to cease constitutional campaigns and direct their opposition to the war itself. Even though, as we have seen, enlightenment science provided probably the greatest inspiration for their progressive philosophy, the loyalist reaction forced the Derby philosophers to try and deny this. The first serious evidence of the loyalist threat to provincial scientific communities came when Joseph Priestley's house and property in Birmingham and four dissenting chapels were burnt down by a church-and-king mob on 14 July 1791, with the acqui-

19 Burning of Priestley's house at Fairhill, Birmingham in 1791, engraving from 1792 reproduced in S. Smiles, *Lives of the Engineers: Boulton and Watt* (1904).

escence of local magistrates.[42] In response, the Derby Philosophical Society sent a letter consoling Priestley for the attacks. This was a bold move in the circumstances, given that the Manchester Literary and Philosophical Society had decided against such action for fear of being associated with Priestley, leading to the resignation of James Watt junior and Thomas Cooper.[43] The Rev. Charles Hope was expelled from the Derby Philosophical Society after protesting that the letter had been agreed with too few members present and complaining to the newspapers, despite significantly having praised Priestley's contribution to natural philosophy. However, the fact that he was alone suggests that under Darwin the Derby philosophers were united in the stance they took.[44]

Despite attempts by the Derby philosophers to emphasise the differences between Priestley's scientific, political and theological activities, the attacks also cemented in loyalist eyes the perceived relationship between natural philosophy, dissent and political radicalism. Almost certainly composed by Darwin although signed by Richard Roe, the letter, which was published in the *Mercury*, condoled with Priestley upon the loss of his experimental apparatus, library and 'your more valuable manuscripts', congratulating him for having 'escaped the sacrilegious hands of the savages at Birmingham'. Priestley was likened to intellectual and scientific heroes such as Socrates and Galileo who had been attacked by enemies 'unable to conquer [their] arguments by reason'. They had 'halloo'd upon you the dogs of unfeeling ignorance, and of frantic fanaticism' and kindled fires 'like those of the inquisition, not to illuminate the truth, but, like the dark lantern of the assassin, to light the murderer to his prey'. However, his 'philosophical friends' in Derby hoped that he would not again risk his person 'amongst a people, whose bigotry tenders them incapable of instruction' and that he would 'leave the unfruitful fields of polemical theology' and cultivate science, 'of which you may be called the father'. It was through his international promotion of the discoveries of natural philosophy, which, 'by inducing the world to think and reason', would 'silently marshal mankind against delusion' and 'overturn the empire of super-stition'. Priestley's fame, 'already conspicuous to every civilised nation of the world', would ensure that he persevered in 'improvements of science' and rose like a phoenix from the flames of his laboratory, with 'renewed vigour, and shine with brighter coruscation'.[45] In his reply, Priestley expressed thanks but declined to discontinue his religious polemical works: 'excuse me if I still join theological to philosophical studies, and if I consider the former as greatly superior in importance to mankind to the latter'.[46]

The Derby Political Society found itself indirectly on the end of the Pitt administration crackdown in 1793 and 1794 when, in a test of the new act, the proprietors of the *Morning Chronicle*, John Lambert, James Perry and James Gray, were tried for seditious libel for publishing the original address in July 1792. During the trial, the address was alleged to have been composed by a writer who 'ranked as the first poet of the age', a reference to Darwin, which he considered necessary to refute by taking out a notice in the *Mercury*, but not, significantly, in the national newspapers. As the *Mercury* made clear, the author was actually William Ward, a 'youth, to whom nature alone has been bountiful – and whose genius has not been improved by the refinements of a liberal education'. The address was dissected in court as a 'false, wicked, scandalous' libel, which 'published and industriously dispersed' would tend to 'excite tumult and disorder, by endeavouring to raise jealousies and discon-tents in the minds of' the King's subjects. The attacks upon the constitution, corruption and the international war were singled out as especially seditious. In 1794 Joseph Strutt was called to London as a defence witness in the trial of Thomas Hardy, causing much anxiety for himself and his wife Isabella, who was 'tormented with many apprehensions' and 'frightful dreams' as he wasted 'day after day upon these cursed trials'. Nevertheless, though he considered it a 'very provoking and tiresome business', Strutt thought it a 'great cause' and was determined to help Hardy, a 'sober, honest, peaceable, quiet and pious man', who had been unjustly maligned. Strutt stated that the object of the Derby Society was to support a reform of the House of Commons and it had no other view with respect to the monarchy or the House of Peers, emphasising that it never contemplated using force and never met after the petition was rejected.[47]

Nevertheless, Tory opponents continued to emphasise the association between the Derby philosophers and political radicalism. At the consecration of the Derby militia colours in 1799, the Rev. Charles Stead Hope, mayor of Derby in 1797, was still dramatically denouncing the 'destructive principles ... adopted by the infidel, and propagated by the philosopher' for poisoning 'the humble content and simplicity of rural life' and turning the manufacturer and mechanic 'from their important labours ... to the midnight contemplation of rebellion and massacre'.[48]

In the face of the loyalist reaction, a print war and the burning of Paine effigies in Derbyshire and elsewhere, the Derby philosophers strove to assert their loyalty while continuing to call for limited parliamentary reforms and an end to war. An address to Fox in March 1793, for instance, was parodied by a local writer and this forced a sarcastic response, which was published as a broadside and again called for moderate reforms while attacking those who gloried in the war and opportunities for killing. A loyalist address to the king in December 1792 was signed by William Strutt, Darwin, Pigot, Thomas Evans, Francis Mundy and other local reformers.[49] A subscription fund to provide clothes for soldiers was signed by some reformers including Mundy and James Oakes, but opposed by a rival resolution signed by William Strutt and Peter Crompton, which was passed around the meeting but ignored by the mayor, William Snowden. This stated that subscriptions for the relief of British troops were unnecessary, unconstitutional, and might 'tend to promote the continuance of the present calamitous war'. Strutt and Crompton's resolution was also supported by some local friendly societies, which passed resolutions opposing the 'destructive and calamitous war' and petitioned the King for peace.[50]

The reformers became disillusioned with events in France, and the *Derby Mercury* changed the tone of its political coverage after the new proprietor, John Drewry, who succeeded his uncle in October 1794, emphasised that he would only include articles that furthered the 'propagation of truth, the advancement of religion, virtue and morality'. Although reformers continued to argue their case in the newspapers, the focus came to be opposition to the restrictions to civil liberty, particularly the suspension of Habeas Corpus and the Seditious Meetings and Treasonable Practices Acts passed by Pitt's coalition in 1794 and 1795, which succeeded in isolating the Foxite rump, and to the war with its huge economic cost and 'effusion of human blood'.[51] Meetings called to consider the 'security of the county' and form a yeomanry in response to wildly exaggerated reports of attempts to assassinate the King, whose carriage window had been broken on 29 October 1795 when travelling to the Lords, were supported by local gentry including some Whigs. Continuing county divisions were evident from the publication of dissenting notices beside the loyalist address by the county sheriff, which was signed by many Whigs and Tories, including Thomas Evans and Richard Arkwright. These urged that people attend the meeting and voice support for 'liberty and peace' and opposition to civil repression, calling for an end to the 'impolitic and disastrous war'. Opponents claimed to have obtained 2,290 names in two days, but Drewry was attacked for printing the notices close to the loyalist address, which, it was claimed, threatened to disturb the peace of the county. Although popular opposition to the war continued, the middle-class Derby reform leaders seem to have decided to avoid public political campaigns in the immediate future.[52]

The political philosophy of the Derby reformers

The activities of the Derby Political Society are, however, only one aspect of the avenues employed by the Derby reformers for the promotion of their progressive enlightenment ideas. As well as being contributions to distinctive genres, the poetry and prose of Erasmus Darwin, the novels of Robert Bage, and polemical political works by Brooke Boothby and the Rev. James Pilkington provide valuable information concerning the political philosophies of the group, such as their attitudes towards religion and social equality. Darwin's political philosophy, as we shall see, was part of a developmental worldview inspired by enlightenment science and articulated in his address to the Derby Philosophical Society and a series of books. Bage looked to the new American state rather than the old world for inspiration, Boothby employed analogies drawn from natural philosophy and mechanics to provide a justification for constitutional improvement, and Pilkington appealed to reason as well as the usual biblical passages to justify his call for social equality.

A corresponding Derby Philosophical Society member and one of the three members of the Lichfield Botanical Society assisting Darwin with his translation of Linnaeus, Boothby argued in response to Burke that the French, Irish and American revolutions were 'great and pregnant experiments' in the 'science of politics'.[53] While emphasising his love of monarchy and constitution, he contended that 'neither the hereditary succession of ages nor the acquiescence of millions' could sanctify abuse or change evil into good. Echoing the old Whig analogy between British constitution and Newtonian system, Boothby argued that the English system was a mix of democratic, aristocratic and monarchical systems, with government a 'centripetal force that confines these bodies within their orbits', although corruption and 'the dry rot of influence' posed constitutional threats if unchecked. Natural science exemplified by the Newtonian system demonstrated, by the actions of opposing forces, the fragility of human structures through time. It was the essence of power to 'encrease by its own force; wherever the greatest quantum is found, so that all inferior quantities will gravitate to a common centre'. Only God could fabricate physical or moral 'eternal machinery' and even time was subject to error and change, as the introduction of the Gregorian calendar demonstrated. 'Poor' human institutions 'like our watches', required to be 'periodically wound up and frequently repaired', because all contained 'in their very essence and original concoction latent principles of destruction'. With his detailed first-hand knowledge of France, Boothby likened Burke's reading of French affairs to false science and asked by what 'alchemy' the 'base materials of the French system' could be transmuted into the 'silver and gold of standard currency'. Burke had been 'so enamoured of the rising beauties of the Dauphiness' and the court that he ignored the social squalor around fostered by the 'destructive splendour of a court' and the 'invidious wealth of a lazy and luxurious priesthood'. Boothby warned that the 'destruction of inveterate tyranny' and the 'probable' establishment of a free constitution would have to be purchased at the expense of 'a few years anarchy and disorder', as the 'decomposition and combination of elements will be attended with commotion and effervescence'.[54]

Bage, on the other hand, tempered the moderate political philosophy of his novels with diffuse and digressive plot structures, self-referential irony, and humour and satire in the British comic tradition.[55] He employed satire to attack social and political abuses and aristocratic power in his novels with pompous noblemen, idealised lovers, fashionable females, dutiful servants, rakes, fops and upright women, though these last included some strong-willed and independent characters. Earlier novels were sympathetic to the Americans, parallels being drawn between domestic and political tyranny, *The Fair Syrian* (1787) beginning in America. Although *Man As He Is* (1792) explicitly ridiculed Burke, the most political novel was *Hermsprong* (1796). This features Lord Grondale, a pretentious aristocrat who employs bribery and corruption to seemingly maintain private and public authority, using pocket boroughs and government subsidies. His daughter, Caroline Campinet, tries to be obedient but comes to admire Hermsprong, the practical and intellectual American hero. After having Caroline refuse her father's command to marry the ludicrous fop Sir Philip Chestrum, Bage introduces references to Wollstonecraft and has Hermsprong and Miss Fluart, the tormenter of Grondale, argue for greater equality of the sexes, a compliment reciprocated by Wollstonecraft in a favourable review in the *Analytical Review*.[56] Following the enlightenment tradition of Gibbon, Voltaire and Hume, ridicule of the church was a constant theme of Bage's novels and he told Godwin that d'Holbach's notorious materialist *Système de la Nature* was one of his favourite books. The slimy sycophant Dr Blick, desperate to please Grondale, is contrasted with homely parson Brown and Woodock the curate. Blick, who preaches a sermon on the Birmingham riots which attacks radicalism and dissent, is a caricature of Bishop Horsley, who preached a Tory sermon on the riots which tried to blame the dissenters. Bage's friend Hutton had lost much of his property in the riots. Additionally, Bage used *Hermsprong* to attack the repressive measures of the Pitt government, and Grondale tries to have the hero arrested for reading Paine's *Rights of Man*.[57]

Like the public pronouncements of the Derby reformers, Boothby and Bage strove to emphasise their political moderation, distancing themselves from levelling imputations. In *Hermsprong*, when the philosophical hero is faced by a riotous mob of miners demanding better wages, he responds by questioning their demands: 'my friends, we cannot all be rich, there is no possible equality of property which can last a day', and even if capable of desiring this, they must 'wade through such scenes of guilt and horror' to obtain it as they would 'tremble to think of and then finish the horrid conflict by destroying each other'. In any case, why was it necessary to desire the luxury and disease of the rich? When one worker responds by accusing him of being a spy of King George and no better than his master, Hermsprong responds by knocking him down 'a little bloody', because 'to revile your King, is to weaken the concord that ought to subsist betwixt him and all his subjects, and overthrow all civil order'. He then gives him half a crown, but the imputation is clear: Hermsprong is no leveller, charity is fine, but suppression of social differentiation is dangerous.[58] Likewise, Yorke was careful to distinguish between the 'sublime principle of equality of rights' and mere levelling, which he argued had been foisted upon the reformers by reactionary opponents. The concept of levelling was 'unknown to the thoughtful men of this age' and only to be found in visionary writers such as Plato and Thomas More. Equality of rights was, however, 'the aim and end of the social union', defined as protection of lives, liberty and property, and meant equality under the law through the 'aggregate power of the community', distinctions only being made by 'merit' and 'public utility'. Levelling property was bad and commerce and industry good, but *contra* Mandeville and Smith, private vice would not automatically lead to public virtue, which required reform of the law and government action to prevent the tyranny of a minority.[59]

Tories and loyalists continued to accuse the Derby reformers of levelling tendencies despite such qualifications. Boothby tried to distance himself from the more extreme Painite pronouncements, which would destroy the entire social edifice, and worried that, as the Birmingham riots demonstrated, the 'temporary evils' accompanying the French convulsions might 'bring liberty itself into disrepute'. He emphasised the value of hereditary legislature, the 'wisdom and integrity' of the House of Peers, and the dangers of 'sacrificing' religion, laws, morals, customs and manners 'upon the altar of I know not what deaf and dumb idol'.[60] Coming from a group largely made up of professional and manufacturing reformers, the need for minimal government and taxation and an emphasis upon rights, duties and the interests of individuals was paramount, rather than calls for government subsidies for the poor or equalisation of property. As Hermsprong emphasised in Bage's novel, they should not submit to being 'fettered and cramped throughout the whole circle of thought and action', the need being as much to recover supposed ancient British freedoms as to create something new. Yorke was 'ready to prove at any time that even in the celebrated writings of Thomas Paine, there is not a political maxim which is not to be found in the works of Sydney, Harrington, Milton and Buchanan'. Similarly, Bage referred to English political writers of the seventeenth century in his novels, and Boothby cited Bacon, Holt and Somers in support of his arguments.[61] An editorial in the *Derby Mercury* in January 1793 tried to explain what the Derby Society had meant by demanding equality, claiming that it had 'never intended to convey an idea of equal distribution of property' but 'equality of rights'. All were entitled to equal voting rights, to equal treatment under the law and to enjoy 'the produce of their own industry'. Equal 'partition of lands and money' was 'as absurd and unjust in theory, as it is impossible in practice', because if achieved, the 'difference of men's characters, moral and intellectual, would cause an inequality of fortune', discouraging 'every exertion, mental or corporeal' in the arts and sciences. The idea was so absurd that 'no one but an idiot' could propagate an opinion so totally subversive of society'.[62]

Arguably it was Godwin rather than Paine who had the greatest impact upon the Derby philosophers, and by 1795 the Philosophical Society library contained copies of his *Political Justice* (1793) and works by Paley, Smith, Bentham and Gisborne. The enthusiastic reception of Godwin's ideas is evident from Elizabeth Evans's comments to Joseph Strutt, her brother-in-law, in a letter from 1793:

> Mrs. Drewry tells me you are impatient to see Godwin's Enquiry. I am impatient that you should, and therefore send you the first volume which you may be going on with, whiles I finish the second – and oh my dear Joseph read attentively, meditate, discuss, disseminate these precious opinions, these divine truths, with all the zeal which their importance demands. They have penetrated my heart, may they raise yours above prejudice, enable you to use all your efforts to ameliorate the condition of mankind. The grand desideratum in politics is the diffusion of knowledge and morals amongst the poor. – This the manufacturer has it in his power ... to promote and is culpable in the neglect of it.[63]

Intellectual adherence to Godwin's ideas was not enough, as wealthy manufacturers had a duty to lead others in realising them. Despite Godwin's emphasis upon the gradual removal of government and law and the impractical system of autonomous parish republics sending delegates to a central legislative assembly, his language and ideas were regarded as of more relevance to the reforming middle classes, anxious to challenge elite power but not to force social dislocation. Godwin also wanted greater equality in property distribution as part of an ideal society which would facilitate intellectual emancipation, although he rejected the concept of natural human rights, organised political societies and the notion of popular resistance. Universal benevolence would encourage equal justice for rich and poor, forming a utopian state featuring the moral obligation on all to employ wealth for the social good.[64]

The idea of social equality, however, had not gone away, being advanced again in Derby by James Pilkington, minister at the Friargate Unitarian chapel and author of the *View of Derbyshire* (1789) to which Darwin, Johnson and other Derby philosophers had contributed. In his *Doctrine of Equality of Rank and Condition* (1795) Pilkington contended that distinctions between legal, political and material equality were impossible, because political and economic power were concentrated together. The doctrine of equality accorded with Christianity and the 'principles of reason', philosophers and Jesus having always sought to countermand pernicious wealth and power. Jesus' language on the subject was 'neither figurative nor doubtful' and, given his divine status and authority, 'we cannot doubt' that 'those who hanker after rank and wealth, would 'either in this or ... another world be brought down below' the poor and humble. Pilkington contrasted the early Christians who created a 'community of goods' with modern society, like Gibbon and Yorke, comparing the desire for wealth and sensual gratification which 'enfeeble and debase the mind' with 'rational, manly, benevolent, and devotional exercise'.[65]

Basic needs such as 'plain, wholesome and substantial food', 'convenient houses', warm clothing and health of body, peace of mind, private friendship, moral worth and divine favour, needed to be distinguished from illusory ones. Happiness was not increased by sums squandered on magnificent houses, 'ostentatious equipages, attendance of numerous servants, gay diversions, and luxurious entertainments'. Equality would bring less likelihood of 'cruel and ruinous' wars for the spoils of wealth, the poor would become more independent, and there would be no need for 'excessive labour' for ten or twelve hours a day but only 'useful industry'. It was little wonder they were ignorant for they had little time to learn, when all should have the 'leisure for gaining knowledge'. Excessive work had led to exhaustion of strength and spirit leading to drinking and other pernicious forms of amusement, when the poor could have 'leisure for acquiring that knowledge ... requisite for improving ... morals, and refining their amusements'.[66]

Aware that equality was 'contrary to the common and received opinion of mankind', Pilkington admitted it had never yet been achieved, with the possible exception of the Quakers, and had anyone 'ever seen in our jails or our workhouses any persons of their denomination?' Though he accepted that it would be difficult to realise, Pilkington countered with the utilitarian argument that equality 'ought to be admitted or rejected in proportion to its tendency to advance or diminish the happiness of mankind'. How this was to be achieved was left an open question. However, Pilkington suggested the rich might 'with divine help' contract their expenses and transfer some 'worldly substance' annually until reduced to a state of equality, the equivalent of a share of 'an exact division of the outward blessings of life' made throughout the nation or world. This should depend upon family size and individual position, difficulties arising not through 'want of knowledge' but inclination. Rather forlornly, Pilkington hoped that the example of improvements in the 'useful arts and sciences' and discoveries made by 'persons in the middle station of life', the advancement of general happiness, and the joy of distribution among fellow men rather than material reward would encourage 'ingenuity and industry'.

Pilkington's book seems to have begun as a sermon and would have been perhaps unremarkable as that; however, in the political climate of the mid-1790s it was perceived as politically controversial and split his Derby Unitarian congregation. Publication of his book by Joseph Johnson, the reformist London publisher, identified it with other authors in his stable, including Priestley, Maria Edgeworth, Anna Barbauld, Godwin, Cowper, Wollstonecraft, Joel Barlow and Darwin.[67] Mindful of the French violence, Pilkington emphasised that attempts to introduce greater equality could be made 'without injustice, confusions, and mischiefs in human society'. While the rich would voluntarily surrender their 'distinctions and superfluities', the professions would be pursued with the 'same regularity and success as they are at present'. Civil government would continue and the lower orders 'not be excited to violence and insurrection'. However, change could only come from the rich as an 'equalisation of property, if produced by any other means' was 'neither consistent' with Christian precepts nor with 'the principles of reason and benevolence'. If those in lower and indigent conditions were to foment revolution, 'it would become the duty' of magistrates 'to oppose and defeat their designs'.[68] Despite Pilkington's closing qualifications and the vagueness concerning how his plan was to be executed, his language was unambiguous and was too much for some of his congregation, and he resigned in 1796. However, a meeting called to consider the matter unanimously resolved that the 'persecution of punishment for speculative opinions would be inconsistent with the principles of the friends of truth and free enquiry' and that the 'objections urged do not appear sufficient for an acquiescence in Mr. Pilkington's resignation'. The expression 'friends of truth and free enquiry' was virtually a quotation from the address of the Derby Political Society and implies the backing of the Strutts, Peter Crompton and other wealthy and influential congregation members. However, others seem to have chosen to leave and eventually joined the Anglican Church, including other members of the Crompton family.[69]

Conclusion

This chapter has argued that the Derby Society for Political Information was not created in a vacuum but owed its existence to the activities of the Derby philosophers, inspired by their progressive enlightenment philosophy. This is evident from the role of leading Derby philosophers such as Erasmus Darwin, Joseph and William Strutt, Peter Crompton and Samuel Fox in creating and supporting the Political Society, both practically and intellectually. Although there were differences between the Derby reformers, as we have seen, they all supported a broad, moderate, progressive programme of reforms within a constitutional framework. Inspired by progressive enlightenment science and acting through the Society for Political Information, they were able to attract support or at least considerable sympathy from members of the Derby corporation and others in the local political elite. It was only as the French Revolution turned to anarchy and violence that the organised public reform campaign was curtailed. The most spectacular failure was the delegation to the French Assembly, which despite idealistic intentions attracted considerable notoriety and rapidly became caught up in Parisian political tensions, almost resulting in the death or imprisonment of Johnson and Yorke.

Notes

1 C. Hogarth, 'The Derbyshire parliamentary elections of 1832', *Derbyshire Archaeological Journal*, 89 (1969), 68–71.

2 J. V. Beckett, *The East Midlands from AD 1000* (London: Longman, 1988), 355; S. Glover, *History and Directory of the Borough of Derby* (Derby, 1843), 7, 13; W. Woolley, *History of Derbyshire*, ed. C. Glover and P. Riden (Chesterfield: Derbyshire Record Society, 1981), 41; A. W. Davison, *Derby: Its Rise and Progress* (London: Bemrose, 1906), 254.

3 Woolley, *History of Derbyshire*, 41; Glover, *History and Directory* (1843), 5–7; Davison, *Derby*, 254.

4 R. Simpson, *A Collection of Fragments Illustrative of the History and Antiquities of Derby*, 2 vols. (Derby, 1826), I, 205; *The History and Proceedings of the Derbyshire Loyal True Blue Club* (Derby, 1829); *Derby Mercury* (5 August 1813); *List of the Honorary Burgesses, of the Borough of Derby* (Derby, 1776), broadside, DLSL.

5 T. H. B. Oldfield, *The Representative History of Great Britain and Ireland*, 6 vols. (London, 1816), III, 278; Report of the Municipal Corporation Commissioners on the Borough of Derby, *Parliamentary Papers*, 25 (1835), 1857–58; *Derby Mercury* (23 December 1835); *Derby Reporter* (31 December 1835, 7 January 1836); M. Craven, *Derby: An Illustrated History* (Derby: Breedon, 1988), 131–4.

6 *Derby Mercury* (28 January 1790); J. Seed, 'Unitarianism, political economy and the antinomies of liberal culture in Manchester, 1830–50', *Social History*, 7 (1982), 1–25; J. Seed, 'Theologies of power: Unitarianism and the social relations of religious discourse, 1800–50', in R. J. Morris (ed.), *Class, Power and Social Structure*

in *British Nineteenth-Century Towns* (Leicester: Leicester University Press, 1986), 108–56; B. Smith (ed.), *Truth, Liberty, Religion: Essays Celebrating Two Hundred Years of Manchester College* (Oxford: Manchester College, 1986); M. Watts, *The Dissenters*, 2 vols. (Oxford: Clarendon Press, 1995); R. Watts, *Gender, Power and the Unitarians in England, 1760–1860* (London: Longman, 1998); P. Wood (ed.), *Science and Dissent in England, 1688–1945* (Aldershot: Ashgate, 2004); J. Seed, 'History and narrative identity: religious dissent and the politics of memory in eighteenth-century England', *Journal of British Studies*, 44 (2005), 46–63.

7 J. Pilkington, *The Doctrine of Equality of Rank and Condition examined and supported on the authority of the New Testament* (London, 1795); E. Higginson, *The Doctrines and Duties of Unitarians: A Sermon Preached before the Association of Unitarian Dissenters* (Lincoln, 1820), 14–17.

8 Watts, *Dissenters*, I, 482–5.

9 Friargate Chapel, Derby, congregation compiled from birth and baptismal registers (M690 RG 4/5, M695 RG 4/2033, M695 RG4/2034, M691 RG4/499), and minutes and accounts from the period (1312 D/A1, A2), *c*.1700–1860, Derbyshire Record Office; R. S. Fitton and A. P. Wadsworth, *The Strutts and the Arkwrights, 1758–1830: A Study of the Early Factory System* (Manchester: Manchester University Press, 1958); E. Fearn, 'Reform Movements in Derby and Derbyshire, 1790–1832' (MA dissertation, 1964); G. Turbutt, *A History of Derbyshire*, 4 vols. (Cardiff: Merton Priory Press, 1999), III, 1218–88, IV, 1559–1622; S. Glover, *The History, and Gazetteer, and Directory of the County of Derby*, 2 vols. (Derby, 1829, 1833), II, 573–4, 578–9, 581–4, 594–7; *Modern Mayors of Derby*, I (Derby: Derbyshire Advertiser, 1909), 4.

10 *Considerations upon the Intended Petition; offered to the Freeholders of the County of Derby* (Derby, 1769), broadside, DLSL.

11 A. Thackray, 'Scientific networks in the age of the American revolution', *Nature*, 262 (1976), 20–4, p. 21.

12 *Derby Mercury* (2 May, 25 July 1782).

13 Fearn, 'Reform Movements in Derby and Derbyshire', 25–39.

14 E. Fearn, 'Derbyshire reform societies, 1791–93', *Derbyshire Archaeological Journal* (1968), 47–59; A. Goodwin, *The Friends of Liberty: The English Democratic Movement in the Age of the French Revolution* (London: Hutchinson, 1979), 136–267; H. T. Dickinson (ed.), *Britain and the French Revolution, 1789–1815* (Basingstoke: Macmillan, 1989).

15 B. Boothby, *A Letter to the Right Honourable Edmund Burke* (London, 1791), 68–70.

16 Derby Society for Constitutional Information, *Rules of the Derby Society for Political Information* (Derby, 1792); *Ibid.*, *Address to the Friends of Free Enquiry and the General Good* (Derby, 16 July 1792); Fearn, 'Reform Movements in Derby and Derbyshire', 'Derbyshire reform societies'; J. Money, *Experience and Identity: Birmingham and the West Midlands, 1760–1800* (Manchester: Manchester University Press, 1977), 224–5; Goodwin, *Friends of Liberty*, 230–1.

17 *Derby Mercury* (7, 14 March 1793).

18 M. D. Conway, *The Life of Thomas Paine*, ed. H. B. Bonner (London: Watts, 1909), 108; Fitton and Wadsworth, *The Strutts and the Arkwrights*, 199.

19 E. Strutt (attrib.), 'Memoir of William Strutt, esq., FRS', *Derby Mercury* (12 January 1831), 19; Conway, *Thomas Paine*, 98–109; Fitton and Wadsworth, *The Strutts and the Arkwrights*, 198–9; D. F. Hawke, *Paine* (New York: Harper & Row, 1974), 162–87; E. L. Kemp, 'Thomas Paine and his pontifical matters', *Transactions of the Newcomen Society*, 49 (1977–78), 21–40; J. G. James, 'Thomas Paine's iron bridge work, 1785–1803', *Transactions of the Newcomen Society*, 59 (1987–88), 189–221.

20 Letter from J. Strutt to I. Douglas (4 November 1792), Galton papers, Birmingham Central Library, Archives and Heritage, MS3101/C/E/4/8; E. Darwin, letter to R. Dixon (25 October 1792), in King-Hele (ed.), *The Collected Letters of Erasmus Darwin* (Cambridge: Cambridge University Press, 2007), 409–11; Money, *Experience and Identity*, 225; *Manchester Mercury* (13 November 1792), referred to in Goodwin, *Friends of Liberty*, 231.

21 *British Universal Directory* (1791), 887.

22 Fearn, 'Derbyshire reform societies', 47–59.

23 C. Kerry, *Smalley in the County of Derby: Its History and Legends* (London: Bemrose, 1905), 42–3.

24 H. Yorke, *Reason Urged against Precedent, in a Letter to the People of Derby* (Derby, 1793), 41–7; E. Robinson, 'The English "philosophes" and the French Revolution', *History Today*, 6 (1956), 116–21.

25 E. Robinson, 'An English Jacobin: James Watt, Junior, 1769–1848', *Cambridge Historical Journal* (1955), 349–55; Goodwin, *Friends of Liberty*, 200–3.

26 H. Yorke, *Thoughts on Civil Government* (London, 1794); after a term in jail Yorke later renounced radicalism, marrying the prison governor's daughter and becoming a successful publisher and writer. Obituary of Henry Yorke, *Gentleman's Magazine* (1813), i, 283–4; E. Fearn, 'Henry Yorke, radical traitor', *Yorkshire Archaeological Journal*, 42 (1967–70), 187–93.

27 R. P. Sturges, 'The membership of the Derby Philosophical Society, 1783–1802', *Midland History*, 4 (1978), 212–29, p. 223.

28 W. B. Johnson, *The History and Present State of Animal Chemistry*, 3 vols. (London, 1803), introduction.

29 Yorke, *Reason Urged against Precedent*, 46–7.

30 Yorke, *Reason Urged against Precedent*, 41–7.

31 Conway, *Thomas Paine*, 142–9; A. Cobban, *History of Modern France*, 2 vols. (London: Penguin, 1961), I, 149–259; Hawke, *Paine*, 250–64; E. Hobsbawm, *The Age of Revolution* (London: Abacus, 1977), 73–100; S. Schama, *Citizens: A Chronicle of the French Revolution* (London: Allen Lane, 1989).

32 Goodwin, *Friends of Liberty*, 124–35, 249–55; J. G. Alger, 'The British colony in Paris, 1792–93', *English Historical Review*, 13 (1898), 672–94, pp. 672–5; R. Eagles, *Francophilia in English Society, 1748–1815* (Basingstoke: Macmillan, 2000).

33 H. Yorke, *Letter to Bache Heathcote Esq., on the Fatal Consequences of Abolishing the Slave Trade* (Derby, 1792); Alger, 'British colony in Paris', 679.

34 M. Wollstonecraft, *Historical and Moral View of the Origin and Progress of the French Revolution* (London, 1794); Alger, 'British colony in Paris', 679; Conway, *Thomas Paine*, 180–4.

35 D. McDonald, 'Smithson Tennant, FRS, 1761–1815', *Notes and Records of the Royal Society of London*, 17 (1962), 77–94, p. 82.

36 T. Paine, *The Age of Reason* ([1794]; Secaucus NJ: Citadel Press, 1974), with a biographical introduction by P. S. Foner, 30–9; Conway, *Thomas Paine*, 195–7, 228–45.

37 Alger, 'British colony in Paris', 679–80; Conway, *Thomas Paine*, 175–80; C. Conner, *John Paul Marat: Scientist and Revolutionary* (Amherst: Humanity Books, 1999).

38 Alger, 'British colony in Paris', 680; Conway, *Thomas Paine*, 175–80.

39 Alger, 'British colony in Paris', 680; Paine, *Age of Reason*, biographical introduction, 32–3; Conway, *Thomas Paine*, 187–228, p. 196.

40 Yorke, *Reason Urged against Precedent*, 47.

41 Derby Society for Constitutional Information, *Address to the Friends of Free Enquiry*.

42 R. B. Rose, 'The Priestley riots of 1791', *Past and Present*, 18 (1960), 78–92; Money, *Experience and Identity*, 220–9; Goodwin, *Friends of Liberty*, 180–2.

43 Robinson, 'An English Jacobin', 349–55.

44 Untitled broadside, 10 October 1791, DLSL.

45 'An address to Dr. Priestly agreed upon at a meeting of the Philosophical Society', *Derby Mercury* (29 September 1791); King-Hele (ed.), *Letters of Erasmus Darwin*, 386–7.

46 *Derby Mercury* (29 September 1791).

47 T. B. Howell (ed.), *Complete Collection of State Trials* (London, 1816–28), 12 (1793), 34 George III, 954–70, 24 (1794), 1099; I. Strutt, letter to J. Strutt (1 November 1794), J. Strutt, letters to I. Strutt (29, 31 October, 2 November 1794), Galton papers, Birmingham Central Library, Archives and Heritage, MS3101/C/E/4/8. *Derby Mercury* (19 December 1793, 13 November 1794).

48 C. S. Hope, *An Address to the Derby Volunteers ... at the Consecration of their Colours* (Derby, 1799), 7.

49 *Derby Mercury* (10 January, 25 April, 6 June 1793).

50 *Derby Mercury* (19, 26 December 1793).

51 *Derby Mercury* (26 February 1795, 13, 21, 28 November, 12 December 1793); H. S. Watson, *The Reign of George III* (Oxford: Clarendon Press, 1960), 359–61; Goodwin, *Friends of Liberty*, 386–91.

52 *Derby Mercury* (3 December 1795).

53 D. King-Hele, *Erasmus Darwin: A Life of Unequalled Achievement* (London: Giles de la Mare, 1999), 179–82; J. Zonneveld, *Sir Brooke Boothby* (The Hague: De Nieuwe Haagsche, 2003).

54 Boothby, *Letter to Burke*, 6, 8, 20, 15, 31, 34, 37, 77–9, 41–2, 32–3, 102; B. Boothby, *Observations on the Appeal from the New to the Old Whigs, and on Mr. Paine's Rights of Man* (London, 1792).

55 P. Faulkner, *Robert Bage* (Boston: Twayne, 1979); G. Kelly, *The English Jacobin Novel, 1780–1805* (Oxford: Clarendon Press, 1976); M. Butler, *Jane Austen and the War of Ideas* (London, 1975); J. H. Sutherland, 'Robert Bage: novelist of ideas', *Philological Quarterly*, 36 (1957), 211–20; P. Faulkner, 'Man as He is Not', *Durham University Journal*, 57 (1965), 137–47.

56 R. Bage, *Hermsprong, or Man As He Is Not*, first edition, 3 vols. (London, 1796), II, 168, III, 94, 230; *Analytical Review*, 24 (1796), 398–403, 608–9.

57 Bage, *Hermsprong*, I, 108, III, 174; Kelly, *English Jacobin Novel*, 49, 52–3, 59; W.

Godwin, letter to Mary Wollstonecraft, 15 June 1797 in D. Wardle (ed.), *The Letters of William Godwin* (London: Lawrence, 1967), 101; R. Bage, *Man As He Is*, 4 vols. (London, 1792), IV, 272.

58 Bage, *Hermsprong*, III, 196–8.

59 H. Yorke, *These are the Times that Try Men's Souls! A Letter Addressed to John Frost, a Prisoner in Newgate* (Derby, 1793), 436.

60 Boothby, *Letter to Burke*, 68–70; Boothby, *Observations on the Appeal*, 22–4.

61 Yorke, *These are the Times*, 20.

62 *Derby Mercury* (10 January 1793).

63 *Catalogue of the Library of the Derby Philosophical Society* (Derby, 1795); Elizabeth Evans, letter to Joseph Strutt, 24 October 1793, 'Memoir of William Strutt', Derbyshire Record Office, D2943/M, fols. 10/1, 12.

64 W. Godwin, *An Enquiry Concerning Political Justice*, ed. I. Kramnick ([1793]; London: Penguin, 1976), 174–7, 282–92, 701–35.

65 Pilkington, *Doctrine of Equality*, 6–8.

66 Pilkington, *Doctrine of Equality*, 55–6.

67 G. P. Tyson, *Joseph Johnson: A Liberal Publisher* (Iowa City: University of Iowa Press, 1979), 84; Watts, *Gender, Power and the Unitarians*, 91.

68 Pilkington, *Doctrine of Equality*, 60–1.

69 Derby Friargate Chapel minutes and accounts, 1697–1819, meeting of Chapel subscribers (27 March 1796), DLSL, BA 288, DER A.22207; J. Birks, *Memorials of the Friargate Chapel, Derby* (Derby, c.1890), 10.

6

Enlightenment and improvement: the campaign for the enclosure of common lands during the 1790s

Introduction

Although the concept of European enlightenment remains controversial, there is now some agreement that it embodied counter and contradictory elements in addition to progressivism. While national strands have been identified, perhaps two central trends in enlightenment thought are often identified: the emphasis upon individual rights and the search for universal rational principles modelled upon the laws of natural philosophy, especially those exemplified by Newtonianism.[1] The geography of enlightenment has been enriched by studies of individual nations, the American colonies and provincial centres, extending it far beyond French *philosophes* and the courts and capitals of 'enlightened' despots. In the German states and Scotland, for example, towns such as Hamburg, Munich, Dresden, Aberdeen and Glasgow were almost as important enlightenment centres as Paris, London or the imperial capitals. In England, while the metropolis was easily, of course, the largest town, urban centres such as Manchester, Birmingham, Bath and the larger county towns were also major cultural and intellectual foci. Inspired by Habermas's now familiar formulation, the rise of a distinctive public sphere in the European enlightenment has been recognised with related physical and intellectual manifestations. Although this had certain individual characteristics according to local and national factors, there is general agreement that the public sphere entailed striving towards a new public arena for rational debate founded upon autonomous private individuals and the rationalisation of the urban environment. As we have already seen at Derby, the former was manifest in the expansion of print culture and the reading public, and the development of sites of cultural and intellectual interplay such as salons, coffee-houses, antiquarian, literary and scientific societies, Masonic lodges and other forums, and including counter-enlightenment spaces. The latter was manifest in the trend towards improved, rationalised, clean, illuminated, socially differentiated and governable public spaces that fostered the civic and leisurely interaction of middling and gentlemanly urbane citizens.[2] The enlightenment is now seen to have embraced and involved formal and informal associations, including freemasonry, the largest pan-European secular association, and a host of literary, philosophical, debating and improvement societies across Europe. In a comparative study of Paris, London and Hamburg, for instance, Munck has argued that enlightenment ideas had an important impact beyond the intelligentsia and metropolitan elite, and were diffused around provincial and non-elite groups.[3]

The role of Georgian urban development in facilitating changes in society and culture has been well recognised, with Clark arguing that 'urban improvement generated further momentum for the growth of public sociability in the Augustan era' and 'helped to create the physical context and space for the enactment of new forms of fashionable socialising'. However, the role of enlightenment ideas in facilitating English provincial urban improvement has received less attention despite the evidence of the development of London, Edinburgh, Manchester, Bath and other centres.[4] Enlightenment philosophers celebrated the application of reason to the progress of society including urban improvement.[5] A major inspiration was, of course, Bacon, whose *New Atlantis* depicted a highly urbanised society engaged in philosophical studies and conducting experiments to facilitate agricultural and social improvements, while the language and concept of improvement spread from agriculture to the urban context in the seventeenth century. The role of natural philosophers such as Hooke and Wren in the development of the metropolitan baroque and neo-classical townscape was an important example of urban improvement. Likewise, the role and inspiration of philosophers in Scotland – who were 'an essential characteristic of the intellectual, social, and cultural life' of the country – in urban development is well accepted.[6] Enlightenment rationality employed the language of neo-classicism to provide a grammar of creative and intellectual endeavour that smoothed the fissures of jagged modernity, an underpinning utilitarian aesthetic, a means of managing probability, and a resource of knowledge that could be applied to urban space and institutions. Enlightenment philosophy encouraged the development of townscapes that were supposed to embody the careful application of the principles of order, balance, harmony, symmetry and regularity at the microcosm of small spaces and individual buildings and the macrocosm of the townscape.[7]

Much previous analysis of enclosure has focused on rural rather than urban areas despite the fact that the enclosure of common lands surrounding towns to pay for urban improvements was quite a common occurrence and engendered considerable controversy. As a result, broad economic or social explanations for enclosure have been favoured that are not necessarily applicable in urban contexts where various intellectual, economic and political arguments were articulated in the public sphere and much depended upon local social and political factors.[8]

This chapter examines controversies concerning plans to enclose common lands around Derby to pay for improvements and aims to demonstrate the relationship between these and British enlightenment culture. It explores the inspiration and origin for the proposals, the paper war that ensued, and the individuals and groups who opposed and promoted them, before making some comparisons with the enclosure campaign in nearby Nottingham during the 1780s. The chapter argues that provincial *savants* such as Erasmus Darwin and William Strutt were inspired by enlightenment ideas to promote urban improvement and enclosure, although opponents were motivated by other, equally and dialectically integral, aspects of enlightenment thought, especially the development of the concept of individual rights. These cross-currents of enlightenment intellectual discourse had a greater impact upon local enclosure debates than economic arguments, which were employed by both sides without decisive impact. They opened up a sophisticated, witty and increasingly democratic public sphere of debate which transcended conventional religious and political allegiances, requiring established oligarchies to appeal to social groups beyond the gentry, middling sort and wealthier burgesses.[9]

Industry, improvement and enclosure

Urban growth and population expansion helped to bring the issue of urban improvement to greater prominence in Derby and Nottingham after 1760. Both towns benefited, as we have seen, from the opening up of road, river and canal communications in the later seventeenth and eighteenth centuries, especially the growth of the turnpike network and commerce on the Trent and Derwent rivers utilising the rapidly growing port of Gainsborough, Lincolnshire.[10] By the mid-eighteenth century, textile industries dominated and Lombe's Derby silk mill established a model of water-powered manufactory that was utilised in other silk and cotton mills in the region and beyond. Copper and iron working, china manufactory, and mineral and spa working were other important industries, and Derby and neighbouring towns benefited from their proximity to coal and lead mining districts, and lead mining had an important technological impact upon the textile industries.[11] The population of Derby grew increasingly rapidly after 1770, driven by its increasing importance as a manufacturing town, reaching 8,563 in 1791 and 10,828 in 1801, expanding at a rate of about 10% per decade. By comparison, at Nottingham, the rate of growth was even greater, driven by immigration and a high birth rate, expanding from 17,711 in 1779 to 28,861 by 1801, that is by 11,000 or 60%.[12]

Improvements had taken place since the later seventeenth century funded by parishes, corporations and subscribers such as the provision of water-supply systems like that installed by George Sorocold, which helped to fund street lighting. Parishes were responsible for paving, lighting, watching and, in co-operation with corporations, fire fighting.[13] The need for urban improvements potentially united all the political and religious communities and social classes of Derby, as they were thought to benefit trade and public culture; however, the precise form that they took, the responsibility to carry them out, and how they were to be paid for were always controversial issues. Some of the measures, however, only benefited the wealthier inhabitants, as only the main streets tended to be paved, cleaned and illuminated, and piped water supplies were limited to subscribers, everyone else being dependent on rivers and wells. The poorer sorts were beginning to be concentrated in the courts and alleys of lower-lying districts such as along Markeaton Brook. Hence, proportionately, the ancient rights accorded to common and pasturage probably assumed greater importance to poorer burgesses and inhabitants, given their exclusion from improvements available to subscribers.

Under Queen Mary after the Henrician dissolution, the corporation was granted various estates that had belonged to religious foundations, including some eighty-six houses and about 216 acres of land, tithes, mills and other privileges. These included the Siddals, the Old Meadows, Bradshaw Hayes, the Castlefields estate, the Holmes and the 48-acre Nun's Green. Despite the sale or disposal of the Castlefields estate to the Borough family during the 1730s, partial enclosure and encroachments, and the loss of part of Nun's Green after 1768, the commons were of considerable economic value, and it was estimated by Hutton in 1791 that they raised a total of £2,000 for the corporation per year. The burgesses, who he thought numbered some 900 in 1791, had various privileges on the land unavailable to freeholders, poor burgesses being entitled to right of common on the Siddals, the Old Meadows and other lands after removal of hay by midsummer. The Holmes, which adjoined the River Derwent, was planted with trees and formed a pleasant walk, also serving as common pasture and a place for bleaching cloth. The burgesses were sometimes prepared to take measures to preserve these rights, ranging from rioting and burning fences to legal action, although tenants and other inhabitants were generally able to increase their stake in the common lands.[14]

The Nun's Green was used for a variety of purposes, including the borough kennels, a bowling green, a cockfighting arena, beast and horse fairs, shooting, militia practice and gravel pits. Traders including brick makers, carpenters and bakers stored materials and worked on the Green, and a corn mill was powered by the brook. Some trees were planted for public walks but the Green acquired something of a reputation as the haunt of thieves and vagabonds and the site of various illicit activities such as duelling, romantic liaisons and prostitution.[15] Common lands such as the Green were usually situated close to the edge or beyond borough boundaries, both spatially and culturally, and frequently acted as tolerated sites for illicit social and cultural activities. Enclosure therefore often represented an important act of policing, reflecting changes in moral values and extending elite control over social and urban spaces.

The first major sale of common land to fund improvements occurred when an act of 1768 released part of Nun's Green to be sold in plots. The money raised was supposed to pay for the clearing and improvement of the remainder under the auspices of trustees consisting of urban gentry, professionals and tradesmen. Although this enabled the sale of common land and the enrichment of entrepreneurs and builders who constructed the elegant houses of Friargate such as the architect and builder Joseph Pickford, the measures excited little opposition. This was probably because most of the commons remained untouched, the money was used for improvement of the rest of the Green, and the trustees were chosen to reflect the spectrum of religious and political affiliations.[16] However, the success of the 1768 act and the major impact that it had on the townscape, providing entrepreneurial profits and elegant housing for the middling sort and gentry, showed how important enclosure could be in facilitating improvement. Further general improvement measures were floated in Derby and Nottingham during the 1770s, but political tensions discouraged their enactment. Although the Whig oligarchies made stronger efforts to retain control, there were significant Tory challenges concerning the conduct of borough and county elections and complaints of bribery and corruption. At Nottingham, the junior council, consisting predominantly of Tories, was finally recognised after much pressure, and at Derby the Tory candidate, Parker Coke, and his supporters were able overturn the return of the Whig, John Gisborne, as borough MP, by challenging the use of bastard burgesses and bribery to secure his election.[17] Periodic interventions by corporation and parish and private subscriptions such as those used to rebuild the bridges in Derby were still the preferred means of effecting improvements. But as populations grew, by the 1780s, open spaces in both towns started to disappear, and in 1781 Sylas Neville wrote that Derby was in some of the streets 'not paved at all, others but badly paved with small pebbles and all of them narrow, except for a few'.[18]

The Derby Improvement Committee proposals

The case for improvement seems to have been supported by most of the gentry and middling sort, and a mixture of professionals, manufacturers and traders in Derby during the 1780s. Committees were appointed at town meetings to consider what improvements to undertake and how they were to be funded without putting fresh burdens upon poorer inhabitants. A Derby committee formed in 1789 sought information concerning urban improvement schemes elsewhere and quickly decided that 'the town is ill-paved, ill-lighted, and dirty, and amendment is necessary' and that 'the most oeconomical means' of answering the resolution had to be sought.[19] They resolved that it was necessary to sell the remainder of the Nun's Green by parliamentary act to pay for improvements, and that an improvement tax should also be levied based upon the house tax of dwellings valued at £5 or above per annum, money being offered by the Duke of Devonshire and wealthy dissenting businessmen to defray the expenses on condition that the act be unopposed.[20]

The Derby proposals provoked major opposition, although they had the support of a majority of the corporation. Enclosure supporters solicited the support of inhabitants to circumvent burgess opposition by collecting thousands of signatures in referendums, while opponents claimed to have secured equally large numbers of people hostile to the measures. After an attempt to negotiate a compromise with opponents had failed, the proponents of enclosure in Derby went ahead, with the support of the corporation, in obtaining another improvement act in 1792.[21] This authorised the sale of the remainder of Nun's Green and the creation of an improvement commission to oversee paving, lighting, seweraging and street clearing measures. Funds were also to be raised from a rate based upon the house tax already in existence, which after the third year could not be above nine pence in the pound of the annual property value. Only those with houses of annual rent or value of £20 or above were entitled to be commissioners and they had to meet annually in April to settle the accounts, publishing notice in the press and on boards.[22]

The Derby philosophers and improvement

In both Derby and Nottingham support for enclosure of common lands to pay for urban improvement cut across religious and political affiliations and social classes and was led by gentry, senior corporation figures, clergy, professionals, textile manufacturers, merchants and intellectuals. The issues transcended religious and political affiliations, forging unusual alliances, partly because of differences in the socio-economic character of the two towns. Well-established corporation figures supported enclosure although only a majority of the governing elite in Derby. At Nottingham, these included urban gentry such as John Sherwin, John Fellows and Thomas Smith, the Fellows and Smith families being prominent Presbyterians. Some major textile manufacturers were also prominent enclosure supporters, including members of the Strutt and Swift families at Derby. Professional supporters of enclosure in Derby were inspired by the enlightenment ideas articulated by local intellectuals such as Erasmus Darwin and William Strutt, which reflected and encouraged their social and economic aspirations. Urban improvement was regarded as part of a broader progressive enlightenment reform agenda encompassing religious and political reforms and charitable ventures. Prominent Derby medical men who supported enclosure included Peter Crompton and Darwin. Medical men were significant, as we have seen, as educated gentlemen of some social standing and prominent figures in British enlightenment culture, so support for enclosure from eminent physicians such as Darwin was significant. Natural philosophers, intellectuals and professionals established networks in the region centring upon associations such as subscription libraries and the Derby Philosophical Society, which fostered progressive enlightenment thought through use of the library and convivial discussion.[23]

Darwin intervened directly to support the campaign for enclosure in a broadside that left little doubt of his views and articulated the case for improvement made by many provincial enlightenment progressives, who argued that it would help to regenerate the urban environment. He and Strutt also encouraged their friend the natural philosopher, mechanic and Lunar Society member, Richard Lovell Edgeworth, to contribute another broadside supporting enclosure in Derby. As Darwin emphasised, animal and vegetable remains left in the streets and stagnating water from defective drainage were 'known to afford one of the usual causes of putrid and infectious fevers' according to miasmatic medical theory. By cleaning, lighting, draining and paving the streets, the health and morals of the poor would be improved, business would be increased, and so would the pleasure of all Derby's inhabitants. Improvements would encourage more from the country to spend time in town 'who do not chuse the hurry and expense of London'. Thus an improved town would become 'a kind of Metropolis to that and the neighbouring counties, as York, Shrewsbury, Lincoln', enticing 'a conflux of wealthy and fashionable' residents and visitors to the benefit of 'all trades and professions'. Property would increase in value and the cost of improvement would be rapidly repaid. Darwin expected that most of the plots would be sold for gardens, which would 'contribute to the health' and 'pleasure' of the purchasers, and ten-foot fashionable walks were planned for either side of the brook, which itself was to be improved by flood prevention schemes. The poor would not be burdened with new taxes, and improvement money would not be 'carried out of the country like the taxes levied by the government, but [be] distributed amongst the poor labourers', which was 'certain to return immediately for food and raiment to the baker and the clothier'. On this principle, public works such as 'canals, roads, bridges' were 'doubly advantageous' both as local improvements and 'because by employing the labouring people they thin the workhouses without carrying the money out of the country'. With its endorsement of the social, economic, cultural and governmental benefits of improvement, Darwin's language is strikingly close to similar proposals such as the plan for the improvement of Edinburgh in 1752 and that proposed by John Gwynn in his *London and Westminster Improved* (1766). The latter emphasised the many benefits to be derived from more rational metropolitan street development, and the former argued, like Hume in his essays, that an improved city would attract in the gentry and 'naturally become the centre of trade and commerce, of learning and the arts, of politeness, and of refinement of every kind' which would 'diffuse themselves through the nation, and universally promote the spirit of industry and improvement'.[24]

Another broadside, by 'Amicus', supported Darwin's arguments in powerful enlightenment rhetoric, emphasising the importance of universal rational progress and the fundamental brotherhood and equality of individuals. It was characteristic that John Bull 'most abominably hates innovations of every kind', whether new roads, canals, enclosure or even 'every patent for ingenious discovery in mechanics, by which expense and labour are saved'. However, where philosophy had 'open'd the mind to reason, and conviction', the writer was 'impressed with this important truth – that men are brothers in every rank, and also in every distant clime'. But 'John's opposition only goes to a certain height, and then falls to the ground.' It was a 'heavenly disposition to enjoy the blessings of liberty and the benefit arising from improvements in commerce, agriculture, and arts', because 'by every elegant accommodation of life, the value of property' would be 'annually increased', with cash being extended 'thro' every rank of men'. For Darwin and 'Amicus' social, political and intellectual progress were related to economic success, the spread of wealth and improvement of the quality of life. Never could 'opposition to the public good' for any 'selfish gratification' be allowed to force a 'return of ages wherein dirt, darkness, and barbarism reigned triumphant', instead of 'general convenience, elegance and happiness'.[25]

Besides taking part in the paper campaign, Darwin took an active role in the early business of the improvement commission, attending meetings with his son, Erasmus the attorney, between April 1790 and July 1792. They were replaced in June 1794 but only seem to have stopped attending because the commission had become established and everyday business required a quorum of just five commissioners.[26] Darwin wrote to his friend Josiah Wedgwood at Etruria in Staffordshire on Strutt's behalf, to obtain 3,000 paving bricks 'for a trial of paving the foot-paths of the streets of Derby with'. The intention was to observe the impact of these bricks, and if successful, to pave the whole of the town with them, and Darwin thought that he 'should get it done better and sooner' by applying to Wedgwood 'than to any other person'. His letter to Wedgwood demonstrates that, with Strutt, he went to considerable trouble to select exactly the best and most economical type of bricks. They had to be in the 'shape of common brick', which were thought to be cheaper than 'squares or quarries', and should be as light as possible to reduce carriage costs. Wedgwood was requested to provide information concerning the paving of the streets in Newcastle-under-Lyme and whether the bricks had mortar under them, the joints filled with mortar or a combination of both. Strutt and Darwin hoped that he would be able to provide an estimate for paving both footpaths on either side of the Derby streets, thought to require some 100,000 to 200,000 bricks.[27]

The association between enlightenment progressivism, urban improvement, dissent and industrial wealth was utilised by Tory opponents of enclosure in Derby, who recognised the role of the philosophical community in promoting it. One broadside identified the enclosure promoters as 'these philosophers' who wasted money on 'ingenious' plans. Another complained that 'these philosophers' after having wasted a thousand pounds would then need to 'have another ingenious plan to think of'. According to another who pointedly rejected their celebration of industry and the social application of industrial methods, Strutt and his friends had made themselves 'ridiculous' by 'wishing to appear what they really are not'. Having had their 'mental powers ... confined to the improvements and regulations of a manufactory' and acquired a 'slender knowledge of mankind', they had 'come forward with the importance and sublimity of philosophers' pretending to 'lay down rules and regulation to govern a new world'. 'Philosophy and legislation' constituted 'no part of a *stock in trade*' nor were they 'usually to be met with in a warehouse'. There were also sarcastic references to the public support of Dr Priestley by the Derby *savants*, and his *'true adherence'*, *'loyalty'* and *'attachment'* to the established church and constitution.[28] Another ballad attacked the 'established church' for taking in these 'Presbyterians, Jesuit like' and their domination of the town hall, while another called for unity against 'this presbytree', the Whig aldermen and corporation, a 'vile nest, who strives us to enslave', who should 'ne'er presume to sit in chairs, not honoured be with town affairs, but stay at home and say their prayers'.[29]

Darwin's influential broadside was attacked directly in another by 'Tully', which sought to sever the link between improvement and enclosure and redirect the kind of 'medical' arguments that he employed against the scheme. While Tully claimed to entertain 'the highest respect and deference for Doctor Darwin and his opinions', he considered that these had been perverted by 'his attachment to individuals' who supported the sale, it was hinted, for political reasons. Tully argued that as a 'free circulation of wholesome air' was universally acknowledged to 'highly conduce to the health of every town', so the retention of the Nun's Green would allow the circulation of such wholesome air to be maintained. Enclosure and the erection of houses and other buildings would interrupt the free circulation of air and deprive the inhabitants of that which is so essentially necessary to the preservation of health. Though, as he sarcastically put it, he might not reason like a philosopher, Tully contended that most inhabitants would prefer to walk on the Green 'in its present airy state' rather than in small pent-up gardens attached to 'each small dirty house proposed to be built'. Darwin's argument that the course of the brook could be straightened to lessen the chance of flooding was ridiculed on the grounds that 'the learned doctor unfortunately forgot that he was forcing his reason against the stream'. Similarly, Darwin's arguments were parodied with the claim that the lead works promised for the brook would be justified 'in the next hand-bill Doctor Darwin may publish' in which he will 'learnedly descant upon the salutiferous effects the effluvia of lead will have upon the animal economy', while the desire to visit the works would provide 'a means of bringing them into the air, and ... tend greatly to their health'. If the doctor's arguments were all built upon 'such slender foundations', and his 'opinions on great and important occasions are no better founded', then 'how dreadfully' were the 'wise men of this town misled'.[30]

Opposition to improvement

Prominent figures within the Derby elite opposed enclosure and were able to capitalise on burgess hostility, although the majority of the corporation were supporters. Opposition to enclosure was led by the Derby Tory MP for Nottingham, Daniel Parker Coke, the Tory attorney John Harrison, and a group of remaining trustees from the 1768 act. The enclosures were supported by an influential group that included James Pickford's widow Mary, the artist Joseph Wright, the instrument maker John Stenson, the chemist Henry Browne, the clockmaker John Whitehurst (II), the mechanic and tutor Thomas Swanwick and the mineralogist Henry Browne junior. The main arguments against the sale of the Green were that the town and burgesses would lose ancient rights and privileges as well as suffering financially, while those who had purchased the parcels of land 'at very high prices' had done so 'under the full expectation' that the Green would 'forever remain an open common'. Like their Nottingham counterparts, the Derby opponents emphasised the benefits that the Green brought and the additional burden that would be caused by an improvement rate based upon the house tax rather than the poor rate on all types of property above the value of £5 per annum. The opposition argued that a poor-rate style tax was fairer, as it did not force the poor to pay, whereas the house tax excluded assembly rooms, play-houses, coach houses, 'all mills, malt houses, warehouses' and many other buildings 'of a large amount that will be improved by ... the paving and lighting'. This was dismissed by the proponents as a tax upon industry which would threaten the competitiveness of Derby manufacturers against those from other towns, and 'a tax upon the poor inhabitants' whose wages would have to be lowered to pay the extra tax.[31]

Enclosure controversies and public culture

The controversies concerning enclosure in Nottingham and Derby, like the paper war resulting from the Nottingham election in 1802 and 1803, reveal the sophisticated nature of the Georgian public sphere.[32] Allegiances were to some extent contingent, transcending religious and political affiliations, as one ballad at Derby put it: 'my song concerns not church or king, neither Whig nor Tory'.[33] The enclosure of common lands was, of course, controversial in other towns, with positions on the subject being equally dependent upon local factors. At Newcastle-upon-Tyne, for example, proposals by the corporation to enclose the town moor in 1769 were opposed by freemen who feared the loss of their rights, yet at Bath it was the freemen who wanted the common lands to be developed and the common council members who opposed them. At Leicester in 1754, a plan by the Tory corporation for enclosure in the south field common land was vehemently opposed by local Whigs backed by county gentry, who exploited the discontent felt among freemen, with both sides enrolling hundreds of additional freemen for the borough election.[34] At Derby, although Tories tended to side with Parker Coke, and Whigs with the corporation and enclosure group, many either remained neutral or supported different forms of taxation. The regular attendees at the early meetings of the improvement committee included the Anglican clergymen the Rev. Charles Hope, his son the Rev. Charles Stead Hope, the Anglican brewer Thomas Lowe, the staunch Tory alderman John Hope, the Presbyterians Peter Crompton, William Strutt and Lamech Swift, and Darwin the deist.[35] Among those who declared opposition or neutrality to the sale were reformers such as Charles Ordoyno, James Pilkington the Unitarian minister, and John Hollis Pigot, James Oakes, Henry Browne and Thomas Swanwick, all members of local philosophical circles. As we have seen, Ordoyno joined Strutt and Darwin in the Derby Society for Political Information to campaign for parliamentary reform. Parker Coke was a member of Drewry's Book Society with most of the *savants*, and Darwin, Strutt, Coke and Harrison were members of the Derby canal committee in 1791.[36] Similarly at Nottingham, the corporation and most prominent enclosure opponents acted with supporters such as the physicians John Storer and Snowden White, the Unitarian minister the Rev. George Walker, and the classicist and scholar the Rev. Gilbert Wakefield on charitable committees such as those for the General Hospital and the Sunday school.[37]

The vehemence of the enclosure controversies in both towns did not prevent major co-operation on other matters, signifying the consensus underpinning religious and political differences within elites and the ability to separate disputes concerning specific subjects from other allegiances. At Nottingham, despite being one of the leading supporters of enclosure, Walker continued to be effectively the intellectual and spiritual leader of the corporation. At Derby, fundamental elite consensus is apparent from the acceptance of the improvement commission by erstwhile enclosure opponents, with Parker Coke 'as a friend to good order and government', offering to assist the promoters of the act in 'enforcing obedience to it' in June 1792. By 1796, Coke, Harrison and other opponents were joining the Strutts, Cromptons and other supporters at improvement commission meetings.[38]

Town government was subjected to a degree of democratic, political and economic scrutiny seldom witnessed previously as a result of the enclosure controversies. Arguments were conducted in newspapers, hand bills, pamphlets and broadsides in an effort to influence public opinion.[39] Enclosure opponents and trustees of the 1768 act were forced in the paper war to publicly explain how money raised had been spent, while proponents had to justify every stage of their plan and suggested modes of taxation. The poor state of the Green and the town streets was emphasised by proponents of the sale, who characterised it as the haunt of vagabonds and prostitutes. One ballad complained of the mud, dirt and stones, women of easy virtue, and broken bones which only benefited the surgeon or the cobbler, all to preserve:

> A Den for Thieves,
> And brutal Knaves,
> Who court by night on benches
> A sorry Tribe,
> From every Side,
> Stroll there with sorry wenches.[40]

Another suggested that after the sale the green would be 'covered with gardens and plantations, and pretty snug houses', while characterising the opponents as opposing every improvement so that 'it shall be our glory hereafter to have it published to the world, that OBSTINACY, SELFISHNESS, and FOLLY, have made DERBY, the *dirtiest* town in England.'[41] An opponent of enclosure at Derby demanded that the land-tax money be fully and publicly accounted for, and supporters called for the trustees of the 1768 act to produce all their accounts as 'public property'. In both towns searches were made for alternative sources of revenue that could be used for urban improvement. These included various forms of taxation, tolls, revenue arising from the canal at Nottingham, and the county militia-fine money from the Seven Years War at Derby. Questions were even raised about the Rev. Hope's collection of money from the dancing assemblies, which, 'as the whole money has arisen from the contributions of the public, common honesty requires that it should be applied for the public benefit'.[42]

The public support given to political reform by some enclosure proponents was contrasted by both reforming and Tory opponents with their apparent trampling upon ancient rights. The language of rights, responsibilities and public duties was employed by both parties, resulting in a debate about the meaning of these. At Nottingham during the contest between Collishaw and Heywood, a flag was produced by the burgesses ornamented with a blue border on which was written: 'NO STOOPS AND RAILS – NO ENCLOSURE – NO PAVING THE STREETS WITH THE BURGESSES' PROPERTY – NO BRIBERY NOR CORRUPTION – COLISHAW AND LIBERTY FOREVER.'[43] The opponents of enclosure at Derby emphasised the power of the Cavendish family and corporation influence, maintained by the creation of honorary 'bastard' burgesses and a disregard for the rights and privileges granted to the burgesses centuries past. As 'Veritas' put it, 'arbitrary power, under the specious colouring of Whiggism and liberty principles' was 'secretly raising its iron hand to subjugate the deluded inhabitants of the town of Derby', particularly 'the middle class of inhabitants'. This theme was expressed with much vehemence in Ordoyno's contributions, in which the sale was 'a violent and bare-faced attack upon the poor remains of [the citizens of Derby's] ancient rights and privileges'.[44] The ballad of 'The Sorrowful Lamentation, Last Dying Speech and Confession of Old Nun's Green' parodied the popular souvenirs hawked around Georgian public executions. It noted the value of the Green 'to the poor people of this town', 'a great and good gift, by John of Gaunt', and featured a woodcut of Nun's Green personified being led to the gallows. Rhetoric from the national controversies concerning parliamentary reform was appropriated, with opponents claiming that the sale of the green was an infringement of ancient rights and freedoms and associating the proponents with Paine's *Rights of Man*, the French revolutionaries and 'specious' Whig and 'liberty principles'. Tully thundered in Burkean language against the hypocrisy of 'men who declaim with the greatest violence in support of liberty'. Yet no one would doubt that the inhabitants of Derby enjoyed the possession of the commons 'by prescription', just as the highly 'venerated ... unwritten maxims and customs' of 'our ancestors' were respected in the constitution'.[45]

Both proponents and opponents of enclosure claimed to be differentiating the 'public good' from 'private interest' and to be representing 'the sense of the town'.[46] Both sides at Derby and Nottingham offered to subject their plans to public ballots, with lists of names being published and collections of signatures solicited, the awkwardness of the language employed revealing the novelty of the procedures. These were major exercises in democracy, emphasising the judgement of autonomous individuals and going beyond conventional meetings of inhabitants designed to solicit opinion led by gentry and corporation figures. Nottingham enclosure supporters resolved to gain public approval by publishing their proceedings and organising a vote of the town's inhabitants. Inhabitants were reassured that no 'particular mode or modes of effectuating the improvements desired' were assumed at this stage, it being stated in the 'preamble to the application for individual concurrence', that agreement carried with it 'no obligation to adopt any one mode of revenue in preference to another', this to be broached 'when it can be laid before them with more information'.[47] The petitions at Derby carried at least 3,000 names, almost a third of the population. The requirement for a public mandate, including support from immigrant individuals and women, encouraged the expansion of the electorate beyond the burgesses towards all the town inhabitants.

Women were encouraged to take part in the enclosure debates and therefore to play a greater role in the public sphere. They already had a public role as supporters of charitable institutions, but urban improvement encouraged more to contribute to town affairs.[48] Of the 89 people who subscribed to rebuild St Peter's Bridge and the 478 who subscribed to St Mary's Bridge in Derby, 9% and 7% respectively were women. Of these, about one-third were gentry and in the case of St Mary's Bridge, 18% of the women were widows, who could act in their own right in public ventures. Likewise, of the 159 individuals who had subscribed to a fund for procuring cheap coals for the urban poor by January 1793, 25, or 16% were women.[49]

Proponents and opponents of enclosure made determined attempts to encourage female participation. Proponents contrasted bad women such as the 'sorry wenches' who cavorted on the Green with the maids of the town fetching their shopping, who dropped their eggs on the muddy uneven pavements and had to pull their smock 'quite up to the knee' bedraggled with dirt. One broadside directed at 'the ladies of Derby' and ostensibly penned 'by one of their own sex' argued that women were confined 'within doors many months of the year' by the poor state of the pavements. They had to sit waiting for good weather so that they might 'share the delight of gossiping and shopping', to prevent scuffed shoes, sprained ankles and filthy gowns. It was insinuated that just as the Chinese crippled their women's feet with tiny shoes, the opponents of the sale were deliberately forcing women to remain at home to 'imprison us in the same manner'. The author of the broadside was made to throw her tea over her brother's waistcoat and pull off 'his artificial tail' for his opposition. She remarks that as 'women have rights as well as men; and the act of Parliament says, the Nun's Green belongs to every inhabitant of Derby', so they had a right to voice their opinions. Women were called to arms with a final flourish:

> LADIES! now exert yourselves to forward the sale of this useless common …
> – Matrons, command your sons! – Wives, teaze your husbands! – Sisters, coax your brothers! – Maids … – make all promise and swear … that … they will vote at the public meeting … for the sale of the Nuns'-Green.

Enclosure opponents went further than urging women to influence their men folk, actively canvassing them for signatures on the petition against the sale. The result was that 276 women expressed their views against the sale or the form of taxation, while 56 women were neutral to the argument. Those against included Mary Pickford (1740–1812), widow of Joseph the builder and architect, who had run a worsted mill on Nun's Green after her husband's death with a partner and then on her own.[50] Women accounted for almost 21% of the 1,609 signatures on the petition. Of these some 20% of those against the sale were women, but almost 28% of those who expressed themselves to be neutral were women, the higher figure perhaps reflecting continuing pressure on them not to become publicly involved in politics.

Conclusion

Economic and political differences between Derby and Nottingham help to explain why the enclosure proposals succeeded at the former and failed in the latter. Elections at Nottingham were notoriously tumultuous affairs which could pose problems for the Whig elite, while the relatively large number of burgesses made them a powerful force to antagonise. Thus, although supporters in both Nottingham and Derby tried to appeal to a wider constituency beyond the burgesses for support including women and immigrants, this was a more effective tactic in the Derby context. The enclosure of Nun's Green to pay for improvements had already begun during the 1760s, setting a precedent. Enclosure in Derby also had the support of the majority of the corporation and the *Derby Mercury*, whereas George Burbage's Tory *Nottingham Journal* was more neutral. The *Mercury* proprietor, Presbyterian printer John Drewry, carried editorials in favour of enclosure and detailed the transformations in trade and fashionable status resulting from improvement in Lincoln, Peterborough and other towns.[51] Derby enclosure proponents conducted a more effective propaganda war than their counterparts in Nottingham, who were wrong-footed to some extent by their opponents and their vitriolic hand bills. Although the Derby campaigners also met fierce opposition in print, they had probably learnt from the failure of the Nottingham measures and produced a series of witty, powerful and effective broadsides composed by Strutt, Darwin and friends such as Edgeworth. Although the Nottingham and Derby *savants* were supported by some gentry, manufacturers and traders in their campaigns for enclosure, differences between the economies of the two towns meant that this support was more decisive in Derby than Nottingham. Derby manufacturing and business families such as the Strutts and Cromptons were wealthier than their Nottingham counterparts, who were mostly hosiers and traders.

Although economic arguments were employed by both sides, enclosure and improvement were encouraged and justified by the progressive enlightenment philosophy articulated by local intellectuals such as Darwin and Strutt and friends such as Edgeworth and Wedgwood. Fostered by the intellectual conviviality of local associations such as the Derby Philosophical Society, they supported enclosure on the basis that urban improvement was a major manifestation and facilitator of economic, social and intellectual progress. Opponents were also motivated by major enlightenment concerns in contending that enclosure would destroy precious ancient rights and were able to make emotional appeals for the protection of these from local oligarchies. These dialectical intellectual counter-currents help to explain, in a much more satisfactory way than economic selfishness alone, why opinion during the enclosure controversies transcended political and religious affiliations. The debates concerning enclosure in Derby, Nottingham and elsewhere reveal how sophisticated the provincial English enlightenment public sphere had become as an arena of public discourse on the eve of the French wars.

Notes

1 K. H. Baker and R. M. Reill (eds.), *What's Left of Enlightenment? A Postmodern Question* (Stanford: Stanford University Press, 2001).
2 M. Ogborn, *Spaces of Modernity: London's Geographies, 1680–1780* (London: Guilford Press, 1998); J. Van Horn Melton, *The Rise of the Public in Enlightenment Europe* (Cambridge: Cambridge University Press, 2001); T. Munck, *The Enlightenment: A Comparative Social History, 1721–94* (London: Arnold, 2000).
3 Munck, *Enlightenment*.
4 P. Clark, *British Clubs and Societies: The Origins of an Associational World* (Oxford: Clarendon Press, 2000), 166; D. Spadafora, *The Idea of Progress in Eighteenth-Century Britain* (New Haven: Yale University Press, 1990); S and B. Webb, *English Local Government, IV: Statutory Bodies for Special Purposes* (London: Longmans, Green & Co., 1922); E. L. Jones and M. E. Falkus, 'Urban improvement and the English economy in the seventeenth and eighteenth centuries', in P. Borsay (ed.), *The Eighteenth-Century Town, 1688–1800* (London: Longman, 1990), 116–58; P. Borsay, *The English Urban Renaissance: Culture and Society in the Provincial Town, 1660–1770* (Oxford: Clarendon Press, 1989); R. Sweet, *The English Town, 1680–1840* (London: Pearson, 1999).
5 Ogborn, *Spaces of Modernity*.
6 R. L. Emerson, 'Science and the origins and concerns of the Scottish enlightenment', *History of Science*, 26 (1988), 333–66; A. J. Youngson, *The Making of Classical Edinburgh, 1750–1840* (Edinburgh: Edinburgh University Press, 1966); P. Reed, 'Form and content: a study of Georgian Edinburgh', in T. A. Markus (ed.), *Order and Space in Society: Architectural Form and its Context in the Scottish Enlightenment* (Edinburgh: Mainstream, 1982), 115–45; C. Philo, 'Edinburgh, enlightenment, and the geographies of unreason', in D. Livingstone and C. Withers (eds.), *Geography and Enlightenment* (Chicago: Chicago University Press, 1999), 372–98.
7 For instance, H. Home, Lord Kames, *Elements of Criticism*, seventh edition, 2 vols. (Edinburgh, 1788), II, 432–3, 455; W. Hogarth, *The Analysis of Beauty* (London, 1753), ed. R. Paulson (New Haven: Yale University Press, 1997), 25–48; E. Burke, *A Philosophical Enquiry into the Origin of our Ideas of the Sublime and Beautiful*, ed. and introd. Adam Philips ([1756]; Oxford: Oxford University Press, 1996).
8 R. Williams, *The Country and the City* (Oxford: Oxford University Press, 1975), 60–107; J. A. Yelling, *Common Field and Enclosure in England, 1450–1850* (London: Macmillan, 1977); J. M. Neeson, *Commons: Common Right, Enclosure and Social Change in England, 1700–1820* (Cambridge: Cambridge University Press, 1993).
9 W. G. Hoskins, *The Making of the English Landscape* (London: Penguin, 1970),

279–89; R. Sweet, 'Freemen and independence in English borough politics, c.1770–1830', *Past and Present*, 161 (1998), 84–115.

10 J. D. Chambers, *Nottinghamshire in the Eighteenth Century* (London: Frank Cass, 1966); M. Thomis, *Politics and Society in Nottingham, 1785–1835* (Oxford: Basil Blackwell, 1969); E. Biagini, 'The complex pattern of opposition to the Nottingham enclosure', *East Midland Geographer*, 19 (1996), 20–37; J. V. Beckett *et al.* (eds.), *A Centenary History of Nottingham* (Manchester: Manchester University Press, 1997); P. Elliott, 'The politics of urban improvement in Georgian Nottingham: the enclosure dispute of the 1780s', *Transactions of the Thoroton Society*, 111 (2007), 87–102; W. Woolley, *History of Derbyshire*, ed. C. Glover and P. Riden (Chesterfield: Derbyshire Record Society, 1981); W. Hutton, *The History of Derby: from the Remote Ages of Antiquity* (London, 1791); R. Simpson, *A Collection of Fragments Illustrative of the History and Antiquities of Derby*, 2 vols. (Derby, 1826); S. Glover, *The History, and Gazetteer, and Directory of the County of Derby*, 2 vols. (Derby, 1829, 1833); G. Turbutt, *History of Derbyshire*, 4 vols. (Cardiff: Merton Priory Press, 1999).

11 J. V. Beckett, *The East Midlands since AD 1000* (London: Longman, 1988); J. V. Beckett and J. E. Heath, 'When was the industrial revolution in the East Midlands?', *Midland History*, 13 (1988), 77–94.

12 Chambers, *Nottinghamshire*, 83.

13 M. E. Falkus, 'Lighting in the dark ages of English economic history: town streets before the industrial revolution', in D. C. Coleman and A. H. John (eds.), *Trade, Government and Economy in Pre-Industrial England* (London: Weidenfeld and Nicolson, 1976), 248–73; M. Craven, *Derby: An Illustrated History* (Derby: Breedon, 1988), 113, 115; *Derby Mercury* (11, 18 November 1774, 12 November 1762, 29 April 1774); A. W. Davison, *Derby: Its Rise and Progress* (London: Bemrose, 1906), 204–6.

14 Woolley, *History of Derbyshire*, 42; Hutton, *History of Derby*; Davison, *Derby*, 247–55.

15 Nun's Green material, parcel 202 and broadsides box 27, DLSL; Woolley, *History of Derbyshire*, 30, 33, 42; Simpson, *History and Antiquities*, I, 273–6; Craven, *Derby: Illustrated History*, 92.

16 An Act for Selling Part of a Green Called Nun's Green, in the Borough of Derby (8 Geo. III, 1768).

17 *The Proceedings of the Committee appointed to try the Merits of the Derby Election* … (1776).

18 *The Diary of Sylas Neville, 1767–1788*, ed. B. Cozens-Hardy (London: Oxford University Press, 1950), 277.

19 'Report of the committee at a general meeting 23 November 1789 to prepare a plan for paving and lighting the streets of Derby', DLSL, Nun's Green parcel 202; *Derby Mercury* (26 November 1789).

20 W. Strutt jun., 'Report of the Committee … to Prepare a Plan for More Effectually Paving and Lighting the Streets', DLSL; *Derby Mercury* (24 February 1791).

21 Minute Book of the Paving and Lighting Commissioners, 1789–1825, DLSL, BA 625/8 acc. 16048 (8, 29 March, 16 April 1791).

22 *Derby Mercury* (24 October 1792).

23 'To the author of an address to the inhabitants of Derby, who signs himself an observing inhabitant', broadside (27 January 1792), DLSL.

24 Youngson, *Classical Edinburgh*, 4–12; John Gwynn, *London and Westminster Improved* (London, 1766); D. Hume, *Essays and Treatises on Several Subjects*, 2 vols. (Edinburgh, 1825); Ogborn, *Spaces of Modernity*, 98–104.

25 Amicus, 'To the candid and impartial public, on the intended improvement of the town of Derby', broadside (6 February, 1792), DLSL. Amicus claimed to be a supporter of 'that bold asserter and champion of liberty and freedom, Mr. Fox'.

26 Minute Book of the Paving and Lighting Commissioners, DLSL.

27 Darwin, letter to Josiah Wedgwood, 7 October 1791, in D. King-Hele (ed.), *The Collected Letters of Erasmus Darwin* (Cambridge: Cambridge University Press, 2007), 395–6; letter from Josiah Wedgwood to Erasmus Darwin (29 March 1791), observations on Newcastle brick paving, letter addressed to Dr. Darwin and memorandum of particulars respecting the contracts for paving at Liverpool, DLSL, Nun's Green parcel 202, 1/19, 1/20.

28 DLSL, Nun's Green box: 'To the inhabitants of the town of Derby', undated handbill; 'To the author of an address to the inhabitants of Derby who signs himself an observing inhabitant', undated broadside; 'To the author of a letter who signs himself a lover of social peace', broadside signed by 'Tully' (1792).

29 'Paving and lighting: a new song, to the tune of chivy chase' and 'A birch rod for the Presbyterians', undated broadside, Nun's Green box, DLSL.

30 'Letter to Samuel Crompton, esq., by Tully', broadside box 27 (24 February, 1793), DLSL.

31 DLSL: Undated draft petition against the bill, petition asking that the act of 1792 be altered and amended, Nuns Green parcel 202, 1/1 and 1/2; 'To the inhabitants in general of the town of Derby', handbill (6 April 1791); 'To the honourable the commons of Great Britain in parliament assembled', petition against the sale, signed by Edward Chamberlain (5 April 1791); 'An inhabitant of Derby and a Nottingham freeholder': 'To Daniel Parker Coke, esq', broadside (5 January, 1792).

32 *The Paper War at the Nottingham Election of 1803* (Nottingham, 1803); Thomis, *Politics and Society*, 146–9, 164–6; Sweet, *English Town*, 132–3; Elliott, 'Politics of urban improvement'.

33 'The Nun's Green rangers, or the triple alliance', broadside, DLSL.

34 Sweet, *English Town*, 135; R. W. Greaves, *The Corporation of Leicester, 1689–1836*, second edition (Leicester: Leicester University Press, 1970), 101–2.

35 Minute Book of the Paving and Lighting Commissioners, DLSL.

36 *Derby Mercury* (2 November 1791).

37 Elliott, 'Politics of urban improvement'.

38 *Derby Mercury* (21 June 1792); Minute Book of the Paving and Lighting Commissioners (11 June 1796), DLSL.

39 *Nottingham Journal* (4 November 1786).

40 'To the Author of a New Song, Set to the Old Tune of Chevy Chase', undated broadside ballad, Nun's Green box, DLSL.

41 Dialogue against paving and lighting between Jerry Snip, the tailor, and Crispin Snob, the cobbler, pamphlet, Nun's Green box, DLSL.

42 'To the Inhabitants of Derby, Paving and Lighting', broadside (31 January 1792), Nun's Green box, DLSL.

43 *Nottingham Journal* (1 September 1787).

44 DLSL, Nun's Green box: Veritas, 'Modern liberty', undated broadside; 'To the Inhabitants of the Town of Derby', broadside (25 January 1792), probably by Charles Ordoyno.

45 DLSL, Nun's Green box: 'The Sorrowful Lamentation, Last Dying Speech and Confession of Old Nun's Green', undated broadside; Veritas, 'Modern liberty'; 'A Remonstrance to the Public by the Two Tullys' (Derby, 21 February 1792), 4; Tully junior, 'A Letter to Major Trowell' (1792), 12–13.

46 'To Every Individual Inhabitant in the Town of Derby', broadside (5 April 1791), Nun's Green box, DLSL.

47 *Nottingham Journal* (4, 11 November 1786).

48 *Derby Mercury* (11, 18 February, 1790).

49 *Derby Mercury* (10 January, 1793).

50 Petition calling for amendment of the act, 1792, Nun's Green parcel 202, DLSL; E. Saunders, *Joseph Pickford of Derby: A Georgian Architect* (Stroud: Alan Sutton, 1993), 38.

51 *Derby Mercury* (27 October, 24 November 1791, 12, 26 January 1792).

7

The Derby Literary and Philosophical Society

Introduction

Literary and scientific associations such as the Derby Philosophical Society, as we have seen, existed in most British Georgian towns, although industrial centres such as Liverpool, Birmingham and Manchester have received most attention. This chapter exploits a variety of manuscript and printed sources to examine a related scientific society, the Derby Literary and Philosophical Society, which, although created by some members of the Derby Philosophical Society, aimed to be more active than the earlier society in promoting a public platform for science. Although it only existed for six years, the Literary and Philosophical Society demonstrates the transition between more private Georgian associations and the institutions of Victorian civic science. The journal book provides details of membership, accounts of some meetings, copies of research papers and other material including printed circulars and broadsides, revealing a relatively democratic association in which members held equal shares and the rules of conduct during meetings were carefully defined. The chapter argues that the Derby philosophers were encouraged by the success of civic charitable and political activities to make a determined effort to forge a public platform for science, promoting scientific education as a means of fostering social improvement. They would probably have done this using the established Philosophical Society, but it appears that some members of the latter perceived it primarily as a private society and library rather than a public institution. Encouraged by the success of the Derbyshire General Infirmary (1810), renewed opposition to the war and demands for reform, the Derby philosophers used the Literary and Philosophical Society to organise public lecture programmes, collect scientific instruments and encourage scientific research.

The chapter begins by exploring the activities of the Derby Literary and Philosophical Society in terms of social improvement, education and civic science, comparing it with the Philosophical Society and employing prosopography to examine the occupational status and religio-political characteristics of members. Informed by case studies of chemistry and galvanism, the chemist Charles Sylvester, and the composition of John Farey's *Agriculture and Minerals of Derbyshire* (1811–17), it argues that the Derby philosophers and other provincial *savants* were able to overcome provincial marginality. Inspired by love of science, hostility to international conflict and belief in intellectual progress, they exploited national and international commercial, political and religious networks to foster a national membership in a way never achieved by the Philosophical Society. The chapter shows that despite the political reaction against enlightenment natural philosophy from the 1790s, progressive scientific ideals continued to inspire the activities of provincial *savants* during the Napoleonic wars.

Organisation

Although the Derby Literary and Philosophical Society was founded in 1808 modestly as a 'Mutual Improvement Society' modelled on Scottish university associations familiar to medical men, members had wider objectives than private betterment. In his opening address to the society, Richard Forester noted that 'in an increasing town' such as Derby, the association would be able to 'gradually diffuse', through encouraging 'observation and attentive reading', an 'active spirit of investigation and research', hitherto lacking in 'mere book-clubs'. Like the Derby Philosophical Society and the Manchester Literary and Philosophical Society, both residential and non-residential members were enrolled in order to attract broad support, as the adoption of the title of literary and philosophical society in 1812 signified. Membership was more varied than that of the Derby Philosophical Society and included cotton merchants, maltsters, brewers, Anglican and Dissenting clergy and tutors. Despite discouragement of medical papers, as in the Philosophical Society, medical men still made up the largest group, and as at Manchester, there was a close relationship between the local infirmary and scientific culture. They included the physicians Forester, Thomas Bent, Thomas Haden and his son the surgeon, Charles Thomas Haden, the surgeons John Wright, Richard Bennet Godwin, George Wallis, and visiting medical men such as Hanson Beasant, assistant surgeon to the 37th Regiment of Foot. Of these, Forester, Bent, C. T. Haden, Wright and Godwin worked at the Derbyshire Infirmary. Bent was physician from 1811, a fellow of the Linnean Society and prominent supporter of Derby scientific culture who succeeded Forester as Derby Philosophical Society president, while Wallis became senior physician and clinical lecturer at Bristol Infirmary (1828–55).[1]

The Literary and Philosophical Society members tried to promote original scientific and intellectual work and forge a sustained public platform for science. They did this in five ways, which we will look at in turn: through holding regular meetings where papers were read and discussed, by equipping a laboratory for investigative and educational purposes, through advocacy of the commercial and industrial utility of science, by promoting public lectures, and by trying to use the society to construct national and international scientific networks.

Regulations were typical of many Georgian societies and reminiscent of professional medical, legal and ecclesiastical proceedings familiar to many members. Meetings were originally held monthly on the third Saturday, with an annual gathering in September at the King's Head on the Market Place, when the vice-president and secretary were chosen and changes to the rules considered. The conduct of meetings was carefully structured and papers had to be read twice in succeeding weeks without interruption before questions or observations could be put, 'conducted with temper, in the spirit of improvement, to elicit truth, and not for victory'. A series of penalties from fines to expulsion was specified for failure to give papers, non-attendance or delays in paying membership fees, while only the president or senior figure present could make additional comments and a hammer is listed as one of the earliest acquisitions.[2]

The emphasis upon intellectual improvement and the social and industrial utility of science contrasts with the privacy of Philosophical Society meetings and is evident from the choice of subjects for papers. Laboratory equipment was assembled for experimental and educational purposes, utilising the proceeds from public lectures to pay for acquisitions, such as some expensive galvanic apparatus procured for the chemist Charles Sylvester. The value of a share in the society fluctuated according to the state of finances: in December 1809 it was £2 5s. 6d., when the apparatus was worth £63 15s. and there were fourteen residential members, and in 1811 it had risen to £3 4s. 8d.[3] Rather than just celebrating intellectual confraternity as Darwin had done in 1784, Forester's address provided detailed instructions concerning the composition of scientific papers and emphasised the need to avoid figurative expressions common in 'ornamental walks of literature', using the 'simplicity and exactness of science'. Franklin's essays were recommended as models rather than the works of the geologist James Hutton, who 'wearies and disgusts the student by his verbose and very defective style'.[4] Although most of the titles of the sixty-two papers read to the society between 1808 and 1816 are unknown, the rest suggest subjects ranged across the sciences, literature, economics, political economy and medicine, despite its official preclusion. Papers on social and political questions included Joseph Strutt on the 'Relative advantages and disadvantages of the English and Scottish universities', Higginson on 'The employment of children in the silk, woollen and cotton manufactories', and Mousley on 'Historical view of the liberty of the subject'.[5]

Although the Derby Literary and Philosophical Society ceased to exist in 1816, primarily because of the post-war recession and the deaths of Richard

Roe and Thomas Swanwick, two of its most active organisers, it played a major role in establishing science in Derby public culture. There was no official merger with the Derby Philosophical Society, but Literary and Philosophical Society members who had not already done so joined the older society, which assumed a more public platform under its energetic new secretary, William George Spencer. The Philosophical Society made a greater effort to publicise its activities and attract new members by reducing membership fees, advertising in the *Derby Mercury* and publishing a new edition of its regulations and library catalogue.[6] For the next seven or eight years, the Strutts, Forester, Higginson and their associates turned their attention to other forms of campaigning, notably against taxation, for the abolition of the Test and Corporation Acts and for parliamentary reforms. As we shall see, however, the public platform for science was sustained by continuing lectures on scientific subjects and the foundation of other institutions.[7]

Scientific education

The Derby philosophers considered that scientific education would foster socio-political progress and serve important moral and economic objectives. They advocated a form of experiential and empirical learning that emphasised the importance of individual experience and use of scientific and mathematical objects, equipment and observation in addition to reading textbooks. Darwin's *Plan for the Conduct of Female Education* (1797) argued that in addition to making visits to manufactories, women should be taught mathematics and natural philosophy using demonstration experiments, while William Strutt provided assistance to the Edgeworths in the composition of their *Essays on Practical Education* (1811). Like the philosophical communities of Manchester, Liverpool and Nottingham, as we shall see, the Derby *savants* were major supporters of educational institutions locally and nationally, including the Derby Lancasterian School (1812), Manchester College, York, University College, London, and the Derby Mechanics' Institute (1825). Derby Literary and Philosophical Society members including Higginson, Sylvester, Roe and Swanwick supported themselves partly by teaching scientific subjects. As we have seen, Roe taught scientific subjects with the help of 'experimental illustrations', and his patrons included local philosophical families such as the Darwins, Cromptons, Strutts and Foxes.[8]

Thomas Swanwick senior was a surveyor, mathematician, inventor and natural philosopher who kept a commercial academy at various locations for thirty years, being succeeded by his son, the surveyor Thomas junior (1792–1852) in 1814. Swanwick senior taught both sexes subjects including writing, arithmetic, book-keeping and geometry as well as aspects of natural philosophy, carried out land surveying and also published meteorological tables. He assisted Darwin, who described him as 'a very ingenious philosopher', and Strutt in scientific experiments such as those using the galvanic pile described in the *Temple of Nature*. Encouraged by Darwin, Swanwick also invented an improved seed drill based upon Jethro Tull's device, and a governor or regulator for mills (with the assistance of James Fox), while contributing regular meteorological reports to the *Derby Mercury*, the *Gentleman's Magazine* and other publications. Finally, William George Spencer, another of Mozley's teachers, also became a major figure in local science and a philosophical teacher. Although he never seems to have joined the Literary and Philosophical Society, some of its activities appear to have encouraged him to take an important role in public scientific culture.[9]

The progressive educational philosophy of the Derby philosophers was partly summarised by Joseph Strutt in a paper on the comparative advantages of English and Scottish universities.[10] As we saw, in response to civil restrictions and the perceived limitations of classical education, dissenters had founded their own educational institutions, employed domestic tutors and tended to emphasise personal development and 'modern' practical and utilitarian subjects, such as practical mathematics and natural philosophy. Strutt accepted that the classics had value as the basis of 'refinement and elegance in literature and morals', but too much time was 'sacrificed' in learning them and too little given to the 'general improvement of the understanding', the sciences, and the formation of moral character. This was evident from the 'temptations to idleness and dissipation' among young men, especially at large public schools and colleges, where many were assembled and 'debauchery often begins at one and increases at the other'. Too much time was sacrificed in the '*minute* acquisition' of classics, of 'little use in the general affairs of the world', which contributed little 'to the individual happiness of the professor'. Scottish universities were therefore preferred because of the breadth of their education, which included political economy, natural philosophy, medicine and ethics. Encouragement of disciplined private study and self-reflection, which the Derby Literary and Philosophical Society was endeavouring to foster, was superior even to good lectures. Education could 'restrain ... vicious inclinations', steer youths away from falling in with 'idle and dissolute characters', and lay a foundation for the advance of human intellect and character.

The encouragement that Unitarian ministers in Derby gave to literary and scientific education further demonstrates the importance of the associations between nonconformity and scientific culture. Edward Higginson, minister at the Friargate Chapel from 1811, conducted another small school. According to Thomas Mozley, a pupil, who came from a family of Tory printers, with its limited classical education, Higginson's school was the only one really for 'gentlemen in Derby' because of the state of the grammar school. However, Higginson also took day pupils generally drawn from the wealthy shopkeeping and trading families in the town. The tone of the school was 'good' and 'genial', with Higginson having little truck with social snobbery, the boys mixing on fairly equal terms and playing in a field that had been loaned by a local surgeon.[11] Other pupils included Edward Strutt and the Swedenborgian silk throwster Thomas Madeley (1804–72), mayor of Derby 1853–54.[12] In Mozley's biased but entertaining view, Higginson was 'a very clever, witty, and well-informed man' who loved his disabled wife but was also 'very fond of good houses where there were either politics to be discussed or pretty women to be agreeable to'. Higginson composed some articles for the *Derby Mercury* on religious and political matters, though his 'prevailing character was cynicism'. However, Mozley had to admit that he was 'much liked by the Strutts, by his congregation, by Tommy Moore [the Romantic poet], by all women, by his children, and, what was more than all, by his wife'. He was 'always amusing' and 'if he did not teach much professionally, his scholars had the benefit of many a humorous turn and many a racy expression' so that 'he had the whip-hand of us all, as indeed he had of the whole town'.[13]

Public lectures

The Derby Literary and Philosophical Society public lecture programme was intended to promote the 'advantages' of natural philosophy to a wider audience and pay for experimental apparatus. Newspaper notices and printed leaflets emphasised that the 'character of pure science' as well as of literature was 'very communicable' and the promoters appealed to their 'fellow townsmen' to co-operate in the 'extension of literary and scientific knowledge'. It was argued that in addition to its utilitarian value, scientific education would provide 'a fund of profitable entertainment ... rendered attractive by novelty' and attended by 'none of those unhappy effects' which they thought accompanied other forms of amusement. Citing the 'ingenious' Thomas Beddoes' contention that 'a tribute of wholesome relaxation may be levied upon the experimental sciences and mechanic arts', it was claimed that philosophical societies had 'by popular encouragement risen to great respectability' in science and literature, and been effective 'in influencing the taste and ennobling the pursuits of their respective neighbourhoods'. Personal improvement extended 'the little circle' of sociable individuals and, as knowledge was a 'peculiar species of good' becoming 'more abundant by communication', so those who distributed these blessings were benefactors 'to [the] race'.[14]

The first set of Literary and Philosophical Society lectures in 1809 were provided by Sylvester and featured chemical apparatus and 'a very powerful galvanic battery' intended to illustrate Davy's latest discoveries 'relative to the decomposition of the alkalies'. At 10s. 6d. for the course, one ticket to admit two children, and non-subscribers 2s. 6d. per lecture, the cost was, however, prohibitive to the working classes, although the lectures were supported in advertisements and newspaper editorials. Drewry expressed the hope that they would be 'well attended', as Derby 'has an interest in giving encouragement to the undertaking'.[15] The lectures appear to have been successful, although more work was required than expected, and Sylvester was offered a share of the profits, annually varying between 10 guineas and £15, because of the 'very serious sacrifice' of his time. He also had to dispose of two galvanic troughs, 'the expense having exceeded … expectations'.[16] Others who either gave lectures, provided equipment or executed experimental demonstrations included Forester, Higginson, Oakes, Swanwick, Duesbury, Charles Haden and last but not least Roe's maid, who was paid a guinea 'for her trouble' during the first set of lectures.[17] The number of lectures was increased in 1812, the first year of the new programme to be occupied by philosophical and experimental chemistry; the second embraced 'the application of chemistry to arts, manufactures, and agriculture' using the programme detailed in Sylvester's *Elementary Treatise on Chemistry* (1809). The third examined 'mechanics, hydrostatics and pneumatics', the fourth 'mineralogy and the subjects connected with it', the fifth, 'electricity, galvanism and magnetism', and the sixth, 'optics and astronomy'. In conjunction with these a shorter course of lectures by Higginson was projected 'on subjects connected with general literature – to begin with the literary history of our own country'. It was contended that it was 'scarcely necessary' to justify the choice of subjects as knowledge of them was 'essential to the ordinary arrangements of domestic life' or 'important to the commercial interests' of Derby and its locality. The remainder were 'intimately connected with the valuable mineral treasures' of Derbyshire, and all solicited attention 'as subjects of improving research to the curious and ever active mind of man'. Without 'affecting to direct the pursuit of their fellow townsmen', the committee wanted to emphasise the importance of scientific education and the tendency of their plan to 'furnish valuable preparatory information' for those intending on a university education. In hoping that those who had spare time from 'business or severe studies' or those 'whose rank exempts them from commercial cares or professional labours' would benefit from attendance, they made it clear that the middle classes were the intended audience.

By promoting natural philosophy as a regular public activity, the lectures tried to raise the civic status of science. Although civic science tends to be regarded as a later nineteenth-century phenomenon, the support obtained from varied political and religious groups is striking. Prominent Whigs and members of the corporation who joined the Derby Literary and Philosophical Society included James Oakes (mayor of Derby in 1820), the Rev. Henry Lowe (mayor in 1812 and 1821) and Charles Matthew Lowe (alderman and mayor in 1831). Similarly, the Unitarian bookseller and publisher John Drewry (mayor in 1806, 1814 and 1823) promoted the activities of the Literary and Philosophical Society strongly in his

Derby Mercury.[18] The fact that Tories such as Thomas Haden and the attorney William Eaton Mousley remained on board, despite the political campaigning, suggests that scientific culture retained an aura of 'neutrality'.[19]

Industrial utility

By the early 1800s, Derby's commercial and industrial importance was well established. Literary and Philosophical Society members and lecturers such as Stancliffe repeatedly emphasised the benefits of scientific study for local manufacture and industry, and the society succeeded in attracting more individuals with industrial interests than the Derby Philosophical Society had done. These were reflected in papers on practical engineering and technology given to the Literary and Philosophical Society, such as George Henry Strutt 'On wheel carriages', Lowe on 'Suggestions on improving the structure of pumps' and Thomas Rawlinson 'On a fire escape'.[20]

20 Blue John Cavern from *Black's Tourist Guide to Derbyshire* (1879).

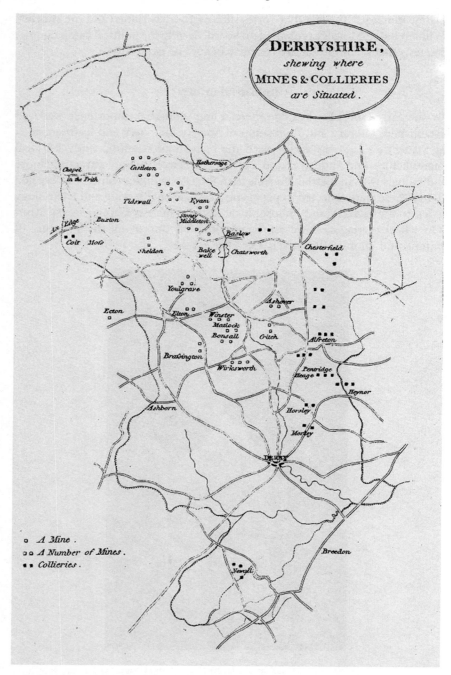

21 Derbyshire mines and collieries from J. Mawe, *Mineralogy of Derbyshire* (1802).

Chemistry and geology were two subjects promoted as likely to benefit industry, and Sylvester's galvanic and chemical research must have attracted members with industrial interests including Forester, James Oakes, the attorney Nathaniel Edwards, William Duesbury III and Archibald Cochrane, ninth earl of Dundonald. Oakes was a partner in the Ridings Ironworks at Somercotes with Thomas Saxelbye, Edwards and Forester, which was managed by the leading iron founder and metallurgist David Mushet.[21] As we shall see, they served with William Strutt on the design committee for the Derbyshire Infirmary, which featured an innovative iron and glass dome, fireproof ceilings and iron window frames supplied by the Riddings company. This may explain why Cochrane became a corresponding member, with his interests in iron manufacture, the production of synthetic soda, the application of coal tar, the development of gas lighting, the application of chemistry to agriculture and the production of alum for silk and calico printers.[22]

Duesbury, whose family had established the Derby Crown China manufactory, attended John Stancliffe's lectures on chemistry and the arts in 1806 and joined the Literary and Philosophical Society presumably because of his artistic, chemical and manufacturing interests. After gradually relinquishing control of the china business around 1811, he managed 'an establishment for the making of colours at Bonsall, near Matlock, where he was the first to introduce the use of baryta as a substitute for white lead as paint'. He emigrated to the United States in 1826 after 'falling into difficulties' in a commercial panic, and settled at Lowell where he worked as a textile chemist in the mills, painted watercolours and designed the Universalist Church, and his son, a physician, followed him to America. In this he was encouraged by members of the Boott family, who had strong Derby connections and helped to establish cotton manufactories on the Strutt model at Lowell.[23]

National scientific networks

The role of the Boott and Haden families supporting Duesbury and helping to establish US textile manufacture on the Derbyshire model, illustrates the importance of late-Georgian national and international philosophical networks. Although in Darwin's lifetime the Derby Philosophical Society succeeded in attracting a regional membership, the Literary and Philosophical Society had a much wider – if less occupationally and geographically concentrated – spread of honorary members who exploited scholarly, business, religious and political networks to counter the growing metropolitan domination of the sciences. Sylvester enrolled new members on his lecture tours and the society gained the prestige of association with other scientific societies. New members that Sylvester encouraged to join included medical men such as William Goodlad and William Wilson, surgeons of Bury and Manchester respectively, the Rev.

John Holland, Unitarian minister and tutor at Bolton, and William Stone-house of Manchester and Brazenose College, Oxford.[24] Sylvester's Liverpool and Manchester lectures produced entrepreneurs such as Joseph Sandars and prominent natural philosophers such as Thomas Stewart Traill, Peter Clare, William Henry, Richard Dalton and Robert Bakewell. Sandars, a Liverpool corn merchant and industrialist, was later co-initiator of the Liverpool and Manchester Railway. Clare, a Manchester clockmaker, engineer and friend of John Dalton, lectured on natural philosophy for decades and served as secretary of the Manchester Literary and Philosophical Society. Traill was a founder member and first secretary of the Liverpool Literary and Philosophical Society (1812) and prominent supporter of other Liverpool scientific institutions, including the Royal Institution, Botanic Gardens and Mechanics' Institute, before becoming a professor at Edinburgh University. He contributed a paper on the migration of swallows to the Derby Literary and Philosophical Society sent in by Sylvester and published in *Nicholson's Journal*.[25]

Besides constructing national and international scientific networks, the Derby *savants* were able to exploit local knowledge and social position to transcend and exploit their provinciality, just as they utilised such networks, as we shall see, in political campaigning. Assistance was provided to others trying to establish their careers such as Farey, while Sylvester, the botanist Francis Boott (1792–1863) and other natural philosophers used provincial science to try to develop national careers.

Torrens, Rudwick and others have demonstrated how communications between British and Continental geologists continued during the war through publication and correspondence despite travel obstacles, and this is confirmed by the activities of the Derby *savants*.[26] As we have seen, the varied landscape, topography and geology of Derbyshire had long been celebrated in the 'Wonders of the Peak' tradition; the development of the mining and textile industries and the mineral trade provided additional tourist interest, blurring the distinction between natural and artificial wonders. Cavern tours, the exhaustion of accessible lead veins and the belief in the industrial utility of science fostered practical geology. British philosophers and mineralogists sent specimens to the Continent and Continental counterparts came to conduct observations, such as Alexandre Brongniart, who met the Bakewell mineralogist White Watson in 1790 and was accompanied on a tour of the Peak after he had been shown his geological sections.[27] Such visits were reciprocated. The mineralogist John Mawe crossed the Channel during the Peace of Amiens to further his trade and visited Paris, where he met de la Métherie and was informed that his work was 'in great fame'.[28] Collections of fossils, minerals and geological works were made by the Derby *savants* and utilised by visitng philosophers. Geologists were invited to join the Derby Literary and Philosophical Society as corresponding members, such as Bakewell and the French

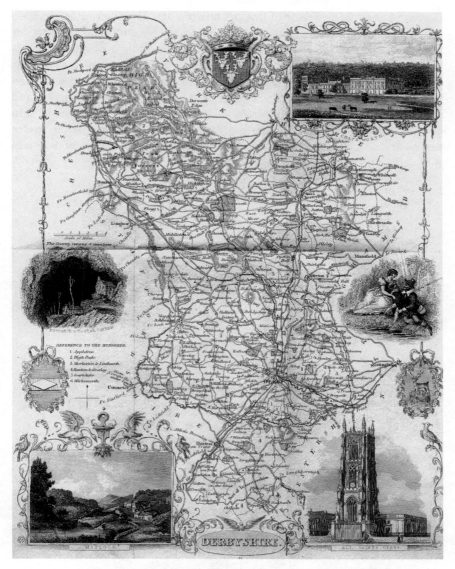

22 Map of Derbyshire from W. Adam, *Gem of the Peak*, second edition (1840).

mineralogist Jean-André-Henri Lucas, who presented a copy of his book on mineralogy to the Philosophical Society library.

As an editor of a major and comprehensive French dictionary of natural history and keeper of the mineralogical collection at the Muséum National d'Histoire Naturelle, Lucas was a colleague of Georges Cuvier and Jean-

Baptiste Lamarck, professor of invertebrate biology. His *Tableau Méthodique* provided a description of the museum mineral collection and support for René-Just Haüy's *Traité de Minéralogie* (1801), including a system of classification based classes, orders, families and species. Lucas made a particular study of Italian volcanoes and the vulcan minerals they spawned and tried to encourage the international exchange of specimens, and his work featured specially commissioned translations of mineral species from French into German, Italian, Spanish, English and Latin. Although not one of the twelve professors, as a philosopher from a nation against whom a major war was

23 Peverill Castle from *Black's Tourist Guide to Derbyshire* (1879).

being conducted, Lucas's membership of the Derby Literary and Philosophical Society and international activities was highly symbolic. It underscored the enlightenment ideal of philosophical progress, neutrality and fraternity espoused by the Derby philosophers and recalled the objectives behind the Political Society delegation to Paris.[29]

A friend and correspondent of John Lindley, Joseph Hooker and Charles Darwin, Francis Boott became a major figure in British and American botany. His brother Kirk Boott junior founded the mills at Lowell modelled on the Strutt manufactories, and after being educated at Harvard, Francis travelled to Derby in 1811 where he met members of the Derby Literary and Philosophical Society and probably attended the courses of public lectures. During his three-year residence among the Derby philosophers, according to his friend Asa Gray, Boott 'formed both the scientific and social attachments which determined the aims ... of his whole after life'. In 1820, he married Mary Hardcastle (1794/5–1871), daughter of Derby botanist and schoolmistress Lucy Hardcastle (1771–c.1835) and author of *An Introduction to the ... Linnaean System of Botany* (1830) and botanical sections of Stephen Glover's *History of Derbyshire*. Lucy's education had been supervised by Darwin, whom Charles Darwin and Hooker suspected to be her natural father, and she helped her son-in-law develop his botanical studies.[30]

Like Boott, Sylvester's career illustrates how provincial manufactures and scientific culture could help to launch a national career. It also shows the importance of connections between science, industry and manufacture. Having begun as a plate-metal worker in Sheffield, Sylvester pursued an itinerant scientific lecturing career, moving to Derby in 1808 after meeting William Strutt during a lecture course that he was giving; Strutt employed him as tutor for his son Edward.[31] To support his large family, he worked as private tutor and businessman, exploiting patents that utilised the galvanic properties of zinc as a coating agent and malleable material for the manufacture of domestic and industrial goods.[32] According to Brewer, Sylvester was an 'ingenious experimental chemist', who 'greatly established' by many years residence the 'philosophical character of Derby', and his pupil Edward Strutt considered him 'one of the most remarkable men I ever knew', despite having 'had scarcely any education' prior to sixteen and being 'entirely' self-educated. Though 'somewhat careless about his exterior', like Higginson, Sylvester was 'a man of very refined mind', full of 'sympathy and kindness' and a 'very agreeable converser' with a 'most keen appreciation of humour'. Only his deism was controversial, thanks to his 'utter inability to understand why religious persons (who were more orthodox than himself)' were 'unwilling to have this subject openly and fully discussed'.[33]

With the support of the Derby philosophers, Sylvester continued his lecturing career and branched out into other businesses such as carbonated mineral water manufacture and manufacturing zinc products in partnership with the local plumber, glazier and Philosophical Society member George Hood. He continued his galvanic and chemical experiments and became established as a scientific writer publishing a series of articles in *Nicholson's Journal*, Thomson's *Annals of Philosophy* and Abraham Rees's *Cyclopaedia* and *An Elementary Treatise on Chemistry* (1809). As we shall see, he assisted William Strutt with the development of the Derbyshire Infirmary, using it to investigate experimentally the effects of moisture upon the specific gravity of gases and the behaviour of gases in different conditions. The Literary and Philosophical Society helped to stimulate and provide audiences for Sylvester's work as tutor and lecturer and in turn he played a major role in supporting the society as principal scientific lecturer, presenting and interpreting the latest scientific developments. He encouraged William Henry to become a corresponding member, who submitted a paper on 'Additional experiments on the muriatic and oxymuriatic acids' to the Literary and Philosophical Society, which was read and copied into the journal book.[34] Sylvester also encouraged Derby scientific enthusiasts such as Duesbury and the attorney Charles James Flack (d.1837) to conduct chemical and galvanic experiments. Flack equipped a domestic laboratory that included a microscope, electrical machine, air-pump, furnace and barometers, and collected a scientific library that included Thomson's *Annals of Philosophy*, Darwin's works, botanical texts and other scientific books.[35] With Strutt's support, Sylvester went into business manufacturing and installing domestic and institutional heating systems, which he developed after he had left Derby for London in 1821, and the concern was continued by his son John (1798–1852).

The opportunities available in provincial scientific culture, and particularly the networks and support available, encouraged Sylvester to develop a system of chemical symbols for his pupils. The *Elementary Treatise* utilised a series of 'symbolic characters' for individual substances, which Sylvester considered an improvement, judging 'from the success' which had attended his educational plan in teaching 'those who have been placed under my care'. However, after discussion with Thomas Thomson, the author of the influential *System of Chemistry* (1804) and supporter of Dalton's atomism and Prout's law, in a later paper Sylvester propounded a method for expressing chemical processes by algebraic characters. Lavoisier had utilised symbols to denote both constitution and quantity, though not an equals sign. However, in Sylvester's system the sign for equality was used to express the equality of the elements before and after decomposition. The weight of atoms was to be represented by letters, and compounds were to be expressed as algebraic products; thus he suggested that carbonic oxide was to be 'co' and nitrous gas 'ao^2'. Parenthesis was used to distinguish acids from bases, and where more than one atom of acid occurred, an exponent was used, thus ao^5 was nitric acid whereas (ao^5) was nitrate potash.

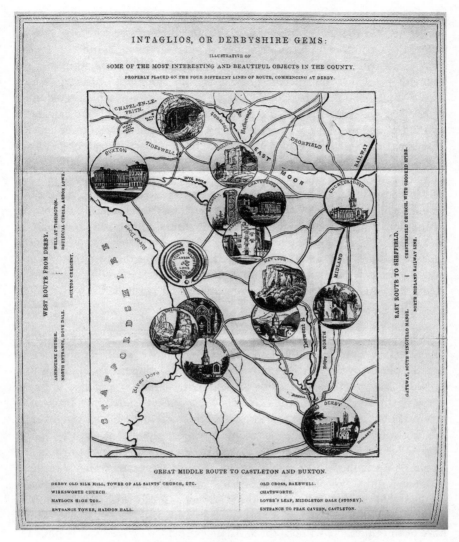

24 Intaglios or Derbyshire gems, from W. Adam, *Gem of the Peak*,
second edition (1840).

The decomposition of nitrate of ammonia by heat to obtain nitrous oxide, for example, was expressed as $(ao^5)(ah^5) = (ao)^2(ho)^3$, two atoms of nitrous oxide and three atoms of water.[36]

It was conventional when chemical changes took place among gaseous bodies to state their proportions by volume, which was more 'easily managed' if the specific gravity of gases was known. Sylvester noted a 'remarkable connection' between the specific gravity of a gas and the weight of its atom, which he said had been pointed out to Dalton, Henry and Thomson 'long before the account of it was published' by Prout in Thomson's *Annals of Philosophy*. Thus when hydrogen was made unity, the weight of the atom of a gas was either equal to the specific gravity, or some multiple of the same by a whole number. So if the ratio of the weight of any two gases which acted upon each other was known, then the ratio of their volumes could be calculated by multiplying the ratio of their weight by the inverted ratio of their specific gravities. In a series of examples, Sylvester tried to demonstrate the efficacy and economy of the algebraic method by combining it with stoichiometry, or a demonstration of the efficacy of mathematical relations.[37] This type of method was intended to facilitate predictions concerning the chemical compositions resulting from reactions (even where some factors were initially unknown) by manipulating algebraic representations of atoms, compounds and volumes just as in a mathematical equation. Sylvester's system, which was promoted in his lecture tours and publications, helped to encourage the adoption of Latin and algebraic symbolism in Britain and ensure the favourable reception of Jacob Berzelius's system of chemical equations and symbols.[38]

Geology

The interface between national and provincial science is further apparent in the role played by the Derby philosophers in stimulating geological study and providing information concerning geology, natural history and agriculture to Farey in the composition of his *General View of Derbyshire*. This 'broke new ground in British geology' and was 'an historic point in geological publication', the first volume alone making 'perhaps ... a more significant contribution to knowledge than any of the other county reviews conjured into print by the Board of Agriculture.'[39] Probably Farey's greatest importance was his role as William Smith's expositor and promoter, providing a 'crucial link' between Banks and Smith and encouraging the president of the Royal Society to support Smith's work at a time when it remained unpublished.[40] Farey's work would have been impossible without the assistance provided by Derbyshire philosophers, naturalists, mineralogists, mine owners, lead smelters and landowners, and in return he publicised and authenticated their work on a national and international stage. As George Greenough and the members of the new London Geological Society acknowledged in 1808, geological progress required the attention of philosophers and 'minute observation and faithful record' from local miners, quarrymen, surveyors, engineers, colliers, iron masters and others who had the 'opportunities of making geological observations.'[41] This symbiosis between provincial and metropolitan natural philosophy represented, to some extent, by Farey's work, ensured its reputation. John Claudius

Loudon, for instance, who utilised it extensively, regarded it as 'one of the most interesting and valuable of the county reports' and an 'example of extraordinary industry, research and excellent general views' to be 'read with great profit by every class of readers.'[42] Farey's enlightenment vision of the importance and interrelationship of natural philosophy with geology, mineralogy, horticulture, agriculture and arboriculture, industry and manufacture, was enthusiastically supported by the Derby *savants* and their associates.

As Torrens has made clear, there was an important relationship between fossil and mineral dealers, urban scientific culture and the development of geological theory evident in the relationship between Derbyshire philosophers, mineralogists and geologists. Those who offered assistance to Farey included White Watson, Richard Brown senior (1736–1816) and junior (b.1765), John Mawe (1766–1829), William Martin (1767–1810) and Joseph Hall. The Browns were marble workers and ornament manufacturers and probably members of the Derby Philosophical Society who worked with Derbyshire fluorspars and marbles, sold mineralogical productions to Darwin, Wedgwood, Faujas de Saint-Fond and other philosophers, and collaborated with Mawe in a 'Derbyshire Spa Warehouse' opened in London in 1794.[43] The scale of this mineralogical trade is evident from Richard Phillips's description of a visit to Hall's Derby manufactory in 1827, where 'Amythistine and other spars, white and variegated marble, alabaster, etc.' were 'formed in a series of workshops, aided by a steam engine, into vases, columns, obelisks, etc.' and employing 'numerous hands' in 'polishing, sawing, fashioning'. These materials were converted into tasteful 'dogs, horses, sheep, cows, etc. for chimney ornaments', and Hall 'imitated the best vases, and ... structures of Egypt'.[44]

Farey came to Derbyshire originally at the invitation of Banks, who hoped to strike new lead veins on his Overton estate by gaining an understanding

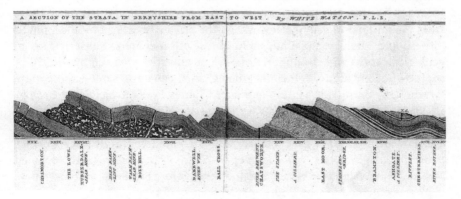

25 Geological section from W. Watson, *A Delineation of the Strata of Derbyshire* (1811).

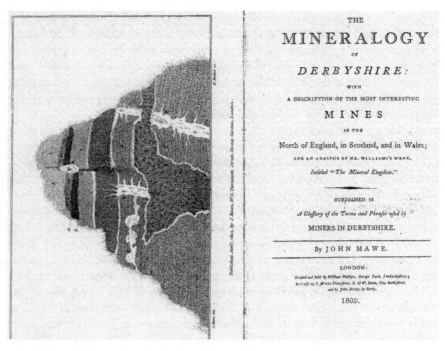

26 Frontispiece from J. Mawe, *Mineralogy of Derbyshire* (1802).

of the structural geology of the Peak.[45] Based upon a new survey inspired by Smith's methods, the *General View* contained accounts of topographical and geological features, including rock types, caverns, strata, soils, agricultural methods, minerals, collieries, mining processes and 'a theory of faults and denuded strata, applicable to mineral surveying and mining'.[46] Further evidence for the geological interests of the Derby *savants* at this time comes from the number of geological works in the private libraries of Philosophical Society and Literary and Philosophical Society members. Subscriptions to geological works and the number of geological works acquired by the Philosophical Society library, which included Cuvier, Faujas de Saint-Fond, Smith, Greenough, Humboldt, Hutton and later Charles Lyell's *Principles of Geology* (1830–3), tell a similar story. As we have seen, geologists such as Bakewell joined the Literary and Philosophical Society, and others such as Mawe and Edward Mammatt gave lectures and presented copies of their works.[47] Subscribers to Watson's *Delineation of the Strata of Derbyshire* (1811) included Oakes, John Wright, Sylvester and the three Strutt brothers, and a separate copy was obtained for the Philosophical Society library.

In return for assistance and information from Francis Agard, Dewhurst Bilsborrow, Oakes, George, William and Joseph Strutt and Sylvester, including the use of geological collections and herbariums, Farey publicised the activities of the Derby philosophers. He described how 'a philosophical society has been some years established' in Derby, which had 'done considerable good, in diffusing scientific and useful information', particularly through the activities of those members who 'deliver lectures on the subject of their particular studies or pursuits'.[48] Sylvester, who later published an inquiry into the feasibility of steam railways, supplied information on Derbyshire canals and railways. Swanwick, secretary of the Literary and Philosophical Society, 'whilst suffering under the illness, which a few months afterwards, terminated his useful life', furnished eleven years' worth of meteorological abstracts and averages which formed the basis of the analysis of Derbyshire precipitation and climate. Farey took considerable interest in the agricultural, horticultural and arboricultural improvements undertaken by the Strutt family on their Derbyshire estates. The system of sewerage and vegetable gardening at Belper was detailed, and a long chapter on arboriculture examined tree-planting schemes and methods employed by the Strutts, who had 'very laudably' applied themselves to planting some 100,000 larch, Scotch pine and other trees on their Belper estates. They also oversaw the 'pruning and management' of extensive plantations already existing and kept 'accurate and systematic accounts of the expense and time of planting', pruning, thinning, the value of cut timber, and 'the measure and value of those trees'.[49]

Science and politics

As we have seen, inspired by enlightenment progressivism, the Derby philosophers promoted improvement and religious and political campaigns, exciting local hostility and exacerbating social tensions. Darwin also obtained a significant national and international audience for progressive developmental philosophy through his publications. The close relationship between natural philosophy and political ideas evident at Derby meant that growth of opposition to the war after 1805 and the establishment of the Infirmary helped to encourage the Derby philosophers to establish the Derby Literary and Philosophical Society. It remained unclear how long the war would continue and Napoleon's 'Continental system' (1806) and the reactionary 'Orders in Council' from the British government frustrated many manufacturers by obstructing European markets. Trafalgar (1805) and the inability to defeat the French comprehensively on land suggested stalemate, with Britain secure on the oceans and against invasion. Campaigns in the Iberian Peninsula from 1808 provided signs of hope, but problems of government corruption and mismanagement were brought into relief by scandals concerning commissions and sinecures. After decades of trying, the parliamentary anti-slavery campaign, long supported by provincial reformers and dissenters, enjoyed some success in 1807 with the abolition of the slave trade. These circumstances encouraged provincial reformers led by Whigs, dissenters and manufacturers whose businesses were suffering from taxation, British government restrictions and the Continental system. Like the anti-slavery movement, peace campaigns were able to command support from some conservative Whigs, Tories and Anglicans who saw international peace as morally and commercially beneficial and were concerned that complete obstruction might foster revolution.[50]

In these circumstances, the Derby philosophers sought issues commanding broad sympathy among urban manufacturers such as moral and economic opposition to the war. Leadership came from individuals in local government, business and public culture, including protégés of Darwin such as Forester and William and Joseph Strutt. The Strutts and other local nonconformists played a leading role in reform campaigns before the war and in the agitation against the Test and Corporation Acts. They followed Priestley in regarding the promotion of public scientific culture as an important means by which dissenting social and political objectives could be achieved, but were more adept than Priestley had been at appealing beyond dissenters. The broad progressive and paternalistic philosophy of the Derby *savants* favoured gradual political reforms over the uncertainty of revolution. As William Strutt expressed it, 'all permanent alterations of public opinion' were 'made slowly and by degrees', whereas the 'red hot democrats', even if their abstract principles were 'perfectly correct', seemed to want 'revolution more than a gradual reform', which alienated public opinion.[51] As his brother put it in an address to the Derby Literary and Philosophical Society, rather than 'the world ... becoming worse' every day, the 'wonderful events' which had taken place 'within the last thirty years' and those 'which *will take place*', would 'ameliorate the condition of the great mass' of people globally. Of the 'rising generation', some would 'rule and some obey' but all would 'have influence in the management of its affairs', and it was therefore essential that the 'right principles should be instilled into their minds' and that they acted 'their parts well under any state'.[52]

As the Strutt brothers recognised and the anti-slavery movement demonstrated, reform campaigns could not be pressed too far without alienating moderate Whigs and Tories, and the greatest strain on elite consensus occurred when the peace campaign organised by the Derby philosophers threatened their scientific and charitable activities. This campaign has been described as 'the most ambitious attempt to organise antiwar opinion during the whole war period' and 'geographically, the most extensive protest since the largely Whig inspired effort of 1797'.[53] Meetings were held in support of Wardle's motion for an inquiry into the disposal of military honours and, capitalising upon the economic problems caused by Napoleon's continental system and fear of Luddism, the peace campaign aimed to mobilise national discontent against the war. Encouraged by national reformers such as Sir Francis Burdett

4

INTRODUCTORY LECTURE,

BY THE

REV. EDW. HIGGINSON.

Monday, March 8th, 1813.

COMPARATIVE view of the human mind in an uncultivated and an improved state. Means of intellectual advancement. Speech. Written language. Printing. Literary & scientific institutions. Objects of the Derby Literary and Philosophical Society. Lectures. Speculation on the probable future improvement of the intellectual and moral faculties of man.

~~~~~~~~~

CHEMICAL LECTURES,

By MR. SYLVESTER.

LECTURE II.

Friday, March 12th, 1813.

PROPERTIES of MATTER. ATTRACTION. REPULSION. States of bodies. SOLID. LIQUID. ELASTIC, OR AERIFORM. CALORIC. NATURE OF. QUANTITY OF. Specific CALORIC. TEMPERATURE. Effects of CALORIC. EXPANSION. Thermometers. Motion of CALORIC. BY RADIATION. BY CONDUCTORS.

5

LECTURE III.

Monday, March 15th.

Analysis of the ATMOSPHERE. OXYGEN. NITROGEN. STEAM. EVAPORATION. RAIN. DEW. Constitution of bodies. Advantages of the MNEMONIC SYSTEM as applicable to Chemistry. COMPOSITION. DECOMPOSITION. Combinations of OXYGEN. WITH HYDROGEN. NITROGEN. SULPHUR, PHOSPHORUS & CARBON; forming WATER, the NITRIC, SULPHURIC, PHOSPHORIC, & CARBONIC ACIDS.

~~~~~~~~~

LECTURE IV.

Wednesday, March 17th.

ACIDS with unknown bases. Compound inflammable bodies. CARBURETTED HYDROGEN. OLEFIANT GAS. SULPHURETTED HYDROGEN. PHOSPHURETTED HYDROGEN. AMMONIA.

~~~~~~~~~

LECTURE V.

Friday, March 19th.

ALKALIS and EARTHS,—with their inflammable bases. Combinations-with SULPHUR, PHOSPHORUS, AND ACIDS. METALS. General Properties. Combinations-with SULPHUR AND PHOSPHORUS. Alloys. Salts.

27 Pages from a Derby Literary and Philosophical Society
lecture programme (1813).

and Major John Cartwright, hundreds of leaflets were distributed nationally exploiting business, dissenting and philosophical networks and east midland counties peace meetings were held at Loughborough and Derby chaired by the Strutt brothers.[54] Although broad distaste for war united religious and political opponents, radicals such as James Robertshaw, leader of the Derby Hampden Club, continued to call for wider reforms.

As we shall see, charitable ventures such as the Derbyshire General Infirmary secured support from those with varied religious and political affiliations manifest in the activities of political societies. Despite the Tory leader the Rev. Charles Stead Hope's denunciation of philosophers, for example, he supported the Infirmary, as did Tory printer Henry Mozley, and other Tories complained that local loyalists 'mixed too readily' with Whigs and reformers, and that old distinctions, clearer in the 1790s, had now been 'nearly effaced'. It was hoped that the new True Blue Club, which replaced an earlier more informal Tory society led by Daniel Parker Coke, would help to re-animate divisions by encouraging Tory candidates to oppose the local Whig hegemony and rewarding poor supporters with charity.[55]

The extent to which the Infirmary was identified with Strutt and the Derby philosophers was exploited by Tory opponents, who pointed to their open political campaigning and opposition to the war, which they tried to portray as unpatriotic. According to Maria Edgeworth, whose family, as we have seen, remained close to the Strutts, many in Derby in 1813 were standing by and hoping that the Infirmary would fail and William Strutt would be humiliated. Some of the opposition was from individuals who had not forgotten

OLD ASSEMBLY ROOM, DERBY.

Dr. STANCLIFFE,

RESPECTFULLY informs his Subscribers and the Public, that his LECTURES on CHEMISTRY, WILL BE CONTINUED AS FOLLOWS :

LECT. 3. FRIDAY Oct. 3.
On the Animal, Mineral, & Vegetable Alcalies. Their various valuable uses in different Branches of Manufacture: When the Galvanic Aura is made to pass through Vegetable Infusions it is shewn that an acid, Alcali, or neutral Salt may be formed at pleasure. The Paradox explained. Decomposition of Nitre, and other methods of obtaining the Alcali Pot-Ash. Useful Subtuitutes & Hints regarding a more easy and cheap method of obtaining Alcalies. Flaming Nitre, its very unique properties, not sufficiently attended to.

LECT. 4. MONDAY Oct. 6.
Earths. Characters, Combinations, & Habitudes of this Class of Bodies. Glass. Staining of Glass. Curious Experiments on Glass by Electricity, shewing that the most elegant figures may be formed and fixed on Glass, never before exhibited or published being the Features of a New ART. On the Arts of Pottery, Porcelain, Brick-making, and on natural and factitious Cements. This Lecture embraces a view of the most useful Qualities of Lime, Clay, Sand, Magnesia, the ponderous Earth Silex, & Strontian. The uses of Marble, Alabaster, and the varied Spars.

LECT. 5. FRIDAY Oct. 10.
Metals. Extraction of these from their ores. Native Combinations, and their employment in the different Arts. Their tendency to union. Phœnomena of metallic oxydation. Metallic Salts. Carburets, Sulphurets, Pyrites, Galena, &c. Alloys, Plating, Silvering, Tinning, Gilding, Metallic Colours.

LECT. 6. MONDAY Oct. 13.
Metals continued. Metallic Poisons. Tests of their presence. Amidores. Zaffre. Smalt. Azure Blue. The Magnet of the Glass Maker. Pearl White. Glass of Antimony. Antimonial Wine. Mufficot. Minium. Litharge. Glass & Sugar of Lead—readier means of obtaining this & the acetous acid, so important to the Painter and Paper Stainer. On Shot, Putty or Tutty for polishing, and the Solution of Tin for the Scarlet Dye. On similoring and Bronzing.

LECT. 7. FRIDAY Oct. 17.
Metals continued. Blue Vitriol, Verd-gris, Verditer, Brass, Preparations of Platina, Gold, & Silver. Assaying, Cupelling, and Smelting. Separation of Silver from Copper. Aqua regia and Aqua regina. Nitrate of Iron, Mr. Keir's singular discovery respecting its Properties. Attempts to illustrate and account for them. Arbor Dianæ. Galvanic Elucidations.

LECT. 8. MONDAY Oct. 20.
On IRON, and the variety of Processes to which this most important of all the Metals gives Birth. Its Combinations, and their properties, illustrated by experiments. Cast Iron. Steel. Plumbago. Conversion of Cast into Forged Iron, and the last into Steel. Casting, hardening, case hardening, tempering, and distempering of Iron. Some Errors of Count Rumford on this subject exposed. The Lectures will conclude with a view of various processes connected with the products of the Vegetable and Animal Kingdoms.

Ticket for the Six Lectures transferable 16s.
Further Particulars of Dr. Stancliffe, at Mrs. Wright's, Mercer, Irongate.

28 John Stancliffe, chemistry lectures, advertisement from the *Derby Mercury* (2 October 1806).

the role of Strutt and some of the other philosophers in the Sale of Nun's Green.[56] The Infirmary had cost much more than expected thanks to the technology and ambitiousness of the plan, and when a fund-raising balloon launch organised by Forester and the committee failed disastrously, Tory critics saw their opportunity and launched a print war with handbills and broadsides reminiscent of the Nun's Green controversy. This exploited the social and economic discontent exacerbated by the continuing war and evident in Luddism, while, as we have seen, the radical, French and disorderly reputation of balloon launches was exploited.[57]

On the day of the balloon launch, a large proportion of the population of Derby came out to watch, the *Mercury* observing that it had 'seldom seen so large a concourse of people collected together in this town'. The site on the Siddals was situated well beyond the town between the Derby Canal and the Derwent, and had been carefully selected so that the event could be easily controlled and to prevent individuals who had not paid from entering. Throughout the morning 'the streets … and the various roads leading to the Siddals were crowded with passengers of all kinds in every species of convenience from the titled equipage to the handsome farmer's daughter on foot with her clean white gown and party coloured ornaments'. However, despite the 'well concerted' arrangements of the committee and although the process of filling from the reaction of sulphuric acid and zinc began well, the balloon failed to inflate 'in consequence of its being full of small holes which let out the gas as fast as it was generated'. Although the committee had tried to cover all the holes with linen, many 'rents and fissures' were observed and the process was abandoned, the balloon falling to the ground.

Eleven Years' Abstracts and Averages of Meteorological Observations made at Derby, Lat. 52° 58' N, Lon. 1° 32' W: about 170 feet above the level of the Sea: by the late Mr. Thomas Swanwick, of the Commercial Academy, who died in March 1814.

Years.	Thermometer.							Barometer.							Pluviometer.
	Hottest Days.	Degrees Farh. on those Days.	Wind ditto.	Coldest Nights.	Degrees Farh. on those Nights.	Wind ditto.	Annual means, Degrees and Tenths.	Highest observations.	Inches, and Decimals, then.	Wind ditto.	Lowest observations.	Inches, and Decimals, then.	Wind ditto.	Totals of Rain, in Inches and Decimals.
1803	July 19	80	W	Dec. 8	18	N W	46·8	Nov. 30	30·40	N	Nov. 11	28·50	S W	—
1804	June 25	83	N E	Dec. 16	14	N	44·5	Feb. 21	30·50	N W	Jan. 28	28·90	S E	—
1805	August 11	76	S	Dec. 13	22	N W	47·6	Nov. 15	30·79	N	Feb. 5	28·81	N	—
1806	June 10	82	S	Jan. 30	25	W	44·2	May 19	30·61	S	Jan. 7	28·72	W	—
1807	July 10	82	W	Jan. 17	17	N W	45·4	Mar. 1	30·75	N E	Nov. 20	28·85	N W	—
1808	July 13	90	S W	Jan. 22	12	N W	43·7	Feb. 25	30·98	N	Dec. 2	28·71	W	—
1809	July 27	78	S W	Jan. 19	17	E	45·5	Mar. 8	30·40	W	Dec. 18	28·15	S E	23·51
1810	July 7	76	S W	Feb. 21	15	S W	47·8	Feb. 21	30·36	S W	Nov. 10	28·85	E	29·59
1811	May 13	78	S	Jan. 30	10	E	48·0	Mar. 12	30·38	N E	Oct. 26	28·51	S W	24·36
1812	July 9	78	S E	Dec. 9	19	N W	43·7	Dec. 7	30·48	N E	Oct. 19	23·34	S	23·79
1813	July 30	80	S	Jan 29	17	S W	46·3	Nov. 4	30·51	N	Oct. 17	28·93	N	20·34
Averages	July 8	80	S S W	Jan. 11	16	N W	45·8	Jan. 30	30·51	N	Dec. 1	28·66	...	24·32

29 T. Swanwick, meteorological table from J. Farey, *Agriculture and Minerals of Derbyshire*, III (1817).

On this becoming apparent, despite the reassurances of the committee that Sadler would be brought back to Derby for another try at no additional cost, on leaving the Siddals members of the crowd became agitated and 'excited a disposition to tear in pieces the balloon'. Although this was 'properly quelled by the dragoons' who were present, this excited the anger of the crowd further, and the troops were attacked as they escorted the balloon from the Siddals to the town hall, as they passed over the canal bridge and down the Siddals Lane from behind the hedges 'with the most violent showers of stones'. Several were wounded as they were ordered by the magistrat to clear the streets, but no one was seriously hurt, which was attributed to 'the good temper and address displayed by Major Hankin and the Scots Greys, who bore the intemperance of the mob for a considerable time'. The soldiers 'put to flight the disorderly' using the flat of their swords, although two or three dragoons received 'severe wounds on their heads' and had their 'cap plates beaten in', and another had a finger dislocated by a stone. Tension had been somewhat relieved by the roars of laughter that greeted Samuel Dawson, a barber of St Peter's Street, who dressed as Joseph Strutt complete with spectacles and was carried in a chair by four men, bowing gracefully to all while assuring the rioters that the balloon would most certainly ascend in due time.[58]

The failure of the balloon launch and the ignominious failure of a subsequent attempt to employ Sadler for another attempt, which ended in a quarrel over money, was exploited by Tory critics to attack the Derby philosophers in series of largely anonymous letters and poems to the *Derby Mercury*. The newspaper attacks used the failure of the balloon launch to attack the design of the Infirmary, the Derby philosophers and their progressive enlightenment conception of science and social progress promoted by the activities of the Literary and Philosophical Society. As we saw, balloons had become associated with natural philosophy and political radicalism inspired by the French Revolution during the 1790s, and had been utilised in cartoons attacking Priestley, Darwin and other philosophers for exploiting the hot air of sedition. One poem complained that philosophical culture in Derby had declined, referring pointedly to Forester's father Richard French, Darwin and Wright:

> Thou wretched town, how fallen thy pride
> To arts and science once allied,
> Where WRIGHT upheld thy arts sublime
> And spread thy fame from clime to clime
> Where DARWIN sung, melodious sage!
> And science charm'd thro' every page;
> Thy FRENCH, by perfect taste refin'd,
> Belov'd, ador'd, a matchless mind;
> All gone! All sunk! And now no more ...
> Instead what remained of science had died:
> By dark assassins struck, alas!
> By hosts of F—s and Philo Gas.[59]

Another poem, by 'Philosophiae verae Amator', published in the newspaper proclaimed that while Pope had warned how a little knowledge was a dangerous thing and scholars had

> thought the theme extremely wise:
> Against the truth the many shut their eyes,
> But now, ten thousand witnesses subscribe,
> A Derby Soph's the greatest fool alive.[60]

Another theme taken up by local critics of the philosophers was the plausible evidence that the intention had been to conduct experiments using the balloon. Anonymous letters by 'Philo Gas' and a 'Constant Reader' argued that 'these schemers' ought to fund their 'philosophical experiments out of their own pockets', a theme taken up in an anonymous farce entitled 'The Derby Wakes', which saw the 'celebrated Derby amateurs of philosophy' trying to reach the moon by air balloon and Luna taking revenge through 'Harlequin Mercury' (Philo Gas).[61] Instead of being neutral as befitted a charity, the Infirmary had become a vehicle for the philosophers, who were condemned by 'Philo Gas' for propounding the doctrines of 'liberty, liberality (the comprehensive cant word of the present day) and universal philanthropy'. The cost of the Infirmary had become a tax from aggrandising 'self-created juntas of speculative theorists', whose 'ruinous taste', it was 'well known' to most subscribers, had resulted in an unnecessarily extravagant institution with little money left for patient care.[62]

The status of the Derbyshire Infirmary as charitable institution, which enabled it to appeal to many Tories and Anglicans as well as Whigs and dissenters, and the support of the local Whig elite helped to ensure the failure of the critical campaign. Forester, the Strutts and some of their philosophical friends were relatively wealthy businessmen and professionals, and William Strutt was chairman of the improvement commission and, with his brother Joseph, one of the major employers of the town and a capital burgess in the corporation. As their friend Maria Edgeworth put it, after the construction of the hospital, most people in Derby were now 'so convinced of Mr. Strutt's merit by his success, that he has the power to do what he pleases and nothing is thought well done but what he directs'.[63] The fact that Tory critics of the Infirmary had to remain anonymous in the press and in their broadsides contrasts with the scores of broadsides and hundreds of signatures attracted by the campaign against the Nun's Green sale. Supporters of the Infirmary and the Derby philosophers were also able to mount a counter-campaign in the newspapers with 'Older and Wiser' and 'A Constant Reader' emphasising that 'party spirit' was at the root of that 'class of scribblers' who attacked them. They were a mere 'clan' who had formed themselves into the True Blue Club, which was 'celebrated' for showing reverence to King and constitution. In their own view the hospital was a 'magnificent' creation which had excited general admiration being 'unrivalled for simplicity of design, convenience of arrangement' and the special adaptations that had been made for 'the benevolent purpose for which it was erected'.[64]

Conclusion

The period between the late eighteenth and early nineteenth centuries has sometimes been regarded as a second scientific revolution when, as Golinksi has emphasised, major changes occurred in 'conceptual content and practice … in institutional settings that were themselves being transformed'. In Britain, 'the enlightenment tradition of genteel individualism continued to flourish'; however, the 'formation of a wide range of new educational and research institutions' similarly 'testified to the impact of rapid technological and social change'.[65] Although London institutions such as the Royal Institution were, of course, important in this process, it is necessary to examine the formation of late-Georgian provincial philosophical associations in towns as diverse as Hull, Stafford, Nottingham, Liverpool, Newcastle, Sheffield, Birmingham, Warrington, and, as we have seen, Derby. These were part of a trend towards the institutionalisation of scientific dissemination and education which promoted science in public lectures, discussion groups, botanical gardens and other forums, while nurturing, as we have seen, national and sometimes international scientific networks. Stimulated by the success of the Derbyshire Infirmary and renewed political campaigning, the activities of the Derby Literary and Philosophical Society demonstrate how one group of provincial philosophers tried to place science at the heart of the urban community, exploiting and extending scholarly, religious and political networks across the midlands, northern England and further afield. Through its promotion of a public and civic platform for science, the Literary and Philosophical Society contrasts with the more private and inward-looking character of many Georgian literary and philosophical associations, such as the Derby Philosophical Society, and anticipates the civic science of the 1820s and 1830s.

Notes

1 Thomas Bent, manuscript case book, DLSL; *Derby Mercury* (22 May 1850); Journal Book of the Derby Literary and Philosophical Society (hereafter Journal Book), Derbyshire Record Office, D5047, 20 January 1810, 7 February, 17 March 1810; J. L. Hobbs, 'The Boott and Haden families and the founding of Lowell', *Derbyshire Archaeological Journal*, 66 (1946), 59–74, p. 72; John A. Venn, *Alumni Cantabrigienses*, 6 vols. (Cambridge: Cambridge University Press, 1954), VI, 330; Journal Book, 20 January 1810, 7 February, 17 March 1810; *Derby and Chesterfield Reporter* (5 May 1825).

2 Journal Book, 18 September 1813, 21 September 1811.

3 Journal Book, 27 November 1808; 18 February, 10 March 1809; 16 December 1809, 11 September 1811.

4 R. Forester, opening address, rules and regulations, Journal Book.

5 T. Rawlinson on the 'Derbyshire Neck', Journal Book, 18 November 1815.

6 Journal Book, 31 March, 7 April 1811, 17 April 1813, 19 February, 19 March, 16 April, 17 September, 15 October, 19 November 1814; 16 and 17 September, 18 November 1815; 17 February, 16 March, 20 April 1816; *Derby Mercury* (7 April 1814, 13 May 1815).

7 *Derby Mercury* (13 September 1810, 24 February, 17 March, 7 April 1814, 13 May 1815, 25 January, 22 February 1816).

8 *Derby Mercury* (7 September 1786, 31 May 1787, 9 October 1794).

9 E. Darwin, *The Temple of Nature; Or, the Origin of Society* (London, 1803), additional note 12; letters from E. Darwin to Georgiana, Duchess of Devonshire (November 1800), William Strutt (6 August 1801) and Samuel More (13 October 1799), D. King-Hele (ed.), *The Collected Letters of Erasmus Darwin* (Cambridge: Cambridge University Press, 2007), 533–4, 556–60, 574–5; *Derby Mercury* (8 December 1791); *Gentleman's Magazine* (1814), i, 241; *Derby Mercury* (25 July 1782, 10 July 1783, 8, 15 January, 5, 12 February 1784, 4 January 1787, 14 January 1796); P. Elliott, 'Improvement always and everywhere: William George Spencer (1790–1866) and mathematical, geographical and scientific education in nineteenth-century England', *History of Education*, 33 (2004), 391–417.

10 J. Strutt, 'On the comparative advantages of English and Scottish universities';

Journal Book, 17 December 1808, Strutt papers, Fitzwilliam Museum, Cambridge, MS 48 – 1947.

11 T. Mozley, *Reminiscences, Chiefly of Towns, Villages and Schools* (London, 1882), I, 245–50.

12 M. Craven, *Derbeians of Distinction* (Derby: Breedon, 1998), 140–1.

13 Mozley, *Reminiscences, Chiefly of Towns*, I, 243–5.

14 Journal Book, 27 November, 7 December 1808, 14 February 1812.

15 *Derby Mercury* (22 December 1808).

16 Journal Book, 29 January 1809.

17 Journal Book, 15 September 1810, 16 February, 21 September 1811, 19 September 1812, 17 April 1813.

18 M. Craven, *Derby: An Illustrated History* (Derby: Breedon, 1988), 163.

19 *Modern Mayors of Derby … from 1835 to 1909* (Derby: Derbyshire Advertiser, 1909), I, 9.

20 Journal Book, 16 December 1815, 20 January 1816.

21 R. Johnson, *A History of Alfreton* (Alfreton, 1969), 127–30; J. Farey, *General View of the Agriculture and Minerals of Derbyshire*, 3 vols. (1811–17), I, 399–400; R. M. Healey (ed.), *The Diary of George Mushet, 1805–1813* (Chesterfield: Derbyshire Archaeological Society, 1982).

22 A. Campbell, 'Archibald Cochrane, Ninth Earl of Dundonald (1748–1831)', *Oxford Dictionary of National Biography*.

23 *Derby Mercury* (21 August 1806); A. Gray, 'Francis Boott M.D.', *American Journal of Science and Arts*, second series, 37 (1864), 288–92; Anonymous, 'Francis Boott, M.D.', *Proceedings of the American Academy of Arts and Sciences*, 6 (1865), 305–8; Hobbs, 'The Boott and Haden families'; D. King-Hele, *Erasmus Darwin: A Life of Unequalled Achievement* (London: Giles de la Mare, 1999), 105, 384.

24 Journal Book, minutes, 16 March, 20 April, 21 September, 21 December 1811, 18 January 1812, 29 January, 19 February, 16 April, 17 September 1814; F. Baker, *The Rise and Progress of Nonconformity in Bolton* (Bolton, 1854); R. K. Webb, 'John Holland (1766–1826)', *Oxford Dictionary of National Biography*; John Foster, *Alumni Oxonienses*, 8 vols. (Oxford, 1891), IV, 1360.

25 Journal Book, 20 October, 17 November 1810, 16 March 1811; A. E. Musson and E. Robinson, *Science and Technology in the Industrial Revolution* (Manchester: Manchester University Press, 1969), 109–10; T. S. Traill, 'On the migration of swallows', *Nicholson's Journal*, 15 (1811), 213–14; *Gore's Liverpool Directory* (1825), 95–6, 98, 268; I. Inkster, 'Studies in the Social History of Science in England during the Industrial Revolution' (PhD thesis, University of Sheffield, 1977), 290, 640; A. Thackray, 'Natural knowledge in cultural context: the Manchester model', *American Historical Review*, 79 (1974), 672–709; G. Kitteringham, 'Science in provincial society: the case of Liverpool in the early nineteenth century', *Annals of Science*, 39 (1982), 329–48; A. Wilson, 'The cultural identity of Liverpool, 1790–1850: the learned societies', *Transactions of the Historic Society of Lancashire and Cheshire*, 147 (1997), 58–73; J. Stobart, 'Culture versus commerce: societies and spaces for elites in eighteenth-century Liverpool', *Journal of Historical Geography*, 28 (2002), 471–85.

26 H. Torrens, *The Practice of British Geology, 1750–1850* (Aldershot: Ashgate, 2002);

M. Rudwick, *Bursting the Limits of Time: The Reconstruction of Geohistory in the Age of Revolution* (Chicago: Chicago University Press, 2007).

27 S. Daniels, *Fields of Vision: Landscape Imagery and National Identity in England and the United States* (Cambridge: Polity Press, 1993), 58–60.

28 H. Torrens, 'Patronage and problems: Banks and the earth sciences', in *Practice of British Geology*, paper V, 269–71; J. Mawe, *Travels in the Interior of Brazil* (London, 1812).

29 *Nouveau Dictionnaire de Histoire Naturelle*, 36 vols. (Paris, 1816–19); J.-A.-H. Lucas, *Tableau méthodique des espèces minérales*, 2 vols. (Paris, 1806–13); E. C. Spary, *Utopia's Garden: French Natural History from Old Regime to Revolution* (Chicago: Chicago University Press, 2000).

30 Gray, 'Francis Boott M.D.'; Anonymous, 'Francis Boott, M.D.'; Hobbs, 'The Boott and Haden families'; King-Hele, *Erasmus Darwin*, 105, 384.

31 I. Inkster, *Scientific Culture and Urbanisation in Industrialising Britain* (Aldershot: Ashgate, 1997), paper IV, 99–131; *Derby Mercury* (7 February 1807).

32 B. Woodcroft, *Index of Patentees of Inventions* (London, 1851), 554.

33 *Brewer's Directory of Derby* (Derby, 1823), 29; *Derby Reporter* (24 January 1828); E. Strutt, 'Private memoir', Strutt manuscript collection, Derbyshire Record Office, D2912, 40.

34 Journal Book, 20 March 1813; W. Henry, 'Account of series of experiments undertaken with a view of decomposing the muriatic acid', *Philosophical Transactions*, 90 (1800), 188–203; 'Additional experiments on the muriatic and oxymuriatic acids', Journal Book; W. C. Henry, *A Biographical Account of the Late Dr. Henry* (Manchester, 1837); W. V. Farrar, K. R. Farrar and E. L. Scott, 'The Henrys of Manchester', parts 2, 3 and 4, *Ambix*, 21 (1974), 179–228, *Ambix*, 22 (1975), 186–204.

35 *Derby Mercury* (26 September 1837).

36 C. Sylvester, *An Elementary Treatise on Chemistry* (Manchester, 1809); C. Sylvester, 'On a method of expressing chemical compounds by algebraic characters', Thomson's *Annals of Philosophy*, 2 (1821), 212–16.

37 Sylvester, 'On a method of expressing chemical compounds', 212–16.

38 W. H. Brock, *History of Chemistry* (London: Fontana, 1992), 118–21, 154–5.

39 H. B. Carter, *Sir Joseph Banks, 1743–1820* (London: British Museum, 1988), 398; Farey, *General View*, I, introduction by T. Ford and H. Torrens, reprinted in Torrens, *Practice of British Geology*, paper VI.

40 Ford and Torrens, introduction to Farey, *General View*, I; Rudwick, *Bursting the Limits of Time*, 436–45, 463–8, 494–9.

41 Geological Society, *Geological Inquiries* (1808), 2, quoted in Rudwick, *Bursting the Limits of Time*, 466.

42 Farey, *General View*, I, v–xiv, xvii–xxv, II, 219–340, III, 685–7; J. C. Loudon, *Encyclopaedia of Agriculture*, second edition (London, 1843), 1152–54.

43 T. D. Ford, 'White Watson 1760–1835 and his geological tablets', *Mercian Geologist*, 13 (1995), 157–64; H. Torrens, 'John Mawe (1766–1829) and a note on his travels in Brazil', *Bulletin of the Peak District Mining Museum*, 11 (1992), 267–71; Darwin, letter to Thomas Beddoes (*c.*1787) in King-Hele (ed.), *Letters of Erasmus Darwin*, 174–5.

44 R. Phillips, *A Personal Tour through the United Kingdom* (London, 1828), no. 2, Derbyshire and Nottinghamshire, 124.

45 Carter, *Sir Joseph Banks*, 342–7, 398; R. Porter, *The Making of Geology: Earth Science in Britain, 1660–1815* (Cambridge: Cambridge University Press, 1977), 168–70; P. Bowler, *History of the Environmental Sciences* (London: Fontana, 1992), 215–17.

46 Ford and Torrens, introduction to Farey, *General View*, I; J. Farey, 'An account of the great Derbyshire denudation', *Philosophical Transactions*, 101 (1811), 242–56; W. S. Mitchell, 'Biographical notice of John Farey, geologist', *The Geological Magazine*, 10 (1873), 25–7.

47 *Catalogue of the … Library of the Late Rev. Sir William Ulithorne Wray* (Derby, 1808); *Catalogue of the … General and Law Library of Mr. Edwards, Solicitor* (Derby, 1826); Farey, *General View*, I, preface, v, viii–ix, xi; *Catalogue of the Library of the Derby Philosophical Society* (Derby, 1835), 62–3.

48 Farey, *General View*, III, 655.

49 Farey, *General View*, II, 219–340 (239), III, 655.

50 E. Fearn, 'Reform Movements in Derby and Derbyshire, 1790–1832' (MA thesis, Manchester University, 1964); D. Read, *The English Provinces* (London, 1964), 65–77; N. Millar, 'John Cartwright and radical parliamentary reform, 1808–1819', *English Historical Review*, 83 (1968), 705–28; J. E. Cookson, *The Friends of Peace: Anti-War Liberalism in England, 1793–1815* (Cambridge: Cambridge University Press, 1982).

51 William Strutt, letter to Edward Strutt, 2 June 1818, Strutt correspondence, DLSL, D125.

52 J. Strutt, 'On the Relative Advantages and Disadvantages of the English and Scottish Universities', Strutt papers, Fitzwilliam Museum, Cambridge, MS 48 – 1947; *Derby Mercury* (16 March 1815, 15 February, 1816).

53 M. I. Thomis, *Politics and Society in Nottingham, 1785–1835* (Oxford: Basil Blackwell, 1969), 77–99, 189–94; *Derby Mercury* (3 September, 12, 29 October 1812); *Leicester Journal* (4 September 1812); M. I. Thomis, *The Luddites: Machine Breaking in Regency England* (Newton Abbot: David & Charles, 1970); Cookson, *Friends of Peace*, 238–54.

54 F. N. C. Mundy, letter to P. Williams, 21 April 1809, in 'Letters of a Derbyshire squire and poet in the early nineteenth century', ed. W. G. Clark-Maxwell, *Derbyshire Archaeological Journal*, 53 (1932), 12–13; Letters from Sir Francis Burdett and John Cartwright to Joseph Strutt (14 November 1812, 29 December 1813), Galton papers, Birmingham Central Library, Archives and Heritage, MS3101/C/E/5/23–4.

55 *Derby Mercury* (5 August 1813).

56 M. Edgeworth, letter to Honora Edgeworth, 26 April 1813, in A. Hare (ed.), *The Life and Letters of Maria Edgeworth* (London, 1894).

57 P. Elliott, 'Medical institutions, scientific culture and urban improvement in late-Georgian England: the politics of the Derbyshire General Infirmary', in J. Reinarz (ed.), *Medicine and Society in the Midlands, 1750–1950* (Birmingham: Midland History, 2007), 27–46.

58 *Derby Mercury* (16 September 1813); A. Wallis, 'Some reminiscences of old Derby', undated typescript, DLSL, c.1880.

59 *Derby Mercury* (4 November 1813).

60 *Derby Mercury* (21 October 1813).

61 *Derby Mercury* (30 September, 21 October 1813); 'Derby Wakes for the Benefit of the General Infirmary', Nun's Green broadsides, DLSL, box 27.

62 *Derby Mercury* (21 October, 30 September 1813).

63 M. Edgeworth, letter to H. Edgeworth, 26 April 1813, in Hare (ed.), *Life and Letters of Maria Edgeworth*.

64 M. Edgeworth, letter to H. Edgeworth, 21 October, 4 November 1813, in Hare (ed.), *Life and Letters of Maria Edgeworth*.

65 J. Golinksi, *Making Natural Knowledge: Constructivism and the History of Science* (Cambridge: Cambridge University Press, 1998), 67.

8

The Derbyshire General Infirmary[1]

Introduction

Though there have been various studies of hospital architecture, few have examined in detail the application of industrial technology to medical institutions in the enlightenment and early nineteenth-century periods. This chapter offers a case study of the Derbyshire General Infirmary (1810), where, principally under the inspiration of William Strutt, a deliberate attempt was made to incorporate fire-resistant building techniques with technology developed for textile mills into a medical institution. In fact, the Derbyshire Infirmary provides one of the most interesting British examples of the direct application of novel industrial technology and organisation to medical institutions for clinical and moral purposes defined and enforced by the donors, and encapsulates many of the changes in enlightenment medicine and institutions. Taking some inspiration from Michel Foucault's theories of discipline and power, the Weberian and Frankfurt School concept of instrumental rationality, and recent work on hospital design by historians of medicine, this chapter explores the objectives of promoters of the Infirmary and examines the impact of the hospital. Focusing upon the applications of industrial technology and employing medical statistics derived from hospital records, the chapter assesses the clinical impact of the Infirmary as a medical institution for an industralising society. It also reaffirms the importance of philosophical culture in British provincial urban society and the centrality of medical men, while demonstrating the role of natural philosophy and technology in the government and control of the urban environment, evident in the discourse of 'town improvement' and the medicalisation of urban space. Finally, the chapter assesses the impact of the Infirmary upon the design of other hospitals and public and private buildings.[2]

Architecture, technology and control

Rather than being principally concerned with religious or state aggrandisement, enlightenment architecture became more concerned with the problems of population, health and urban living. The socio-economic changes of the eighteenth century, evident in the impact of population growth and the industrialisation of Georgian Britain, presented new challenges of government to urban elites including those in Derby. These were most acutely felt in the special difficulties of prisons and hospitals, widely recognised to harbour disease, where it became useful to develop economic systems of control without the need for brutal restraint and violence. There was greater concern for the health and physical vitality of urban populations and what Foucault interpreted as a shift from the narrow context of charitable aid to a more general form of 'medical police'.[3] One result was the development of a more community-oriented medicine evident from the increasing importance of doctors in society and the growth of dispensaries which aimed to retain the economic and medical advantages of hospitals while supporting family-centred communities. This politicisation of space was also evident in the growing role of doctors, who enjoyed an increasingly politically privileged position as the guardians of social hygiene, being invested with powers to manage space.[4]

Another response to the problems of urban government and health were changes in medical architecture which involved the development of hospitals from unspecialised structures akin to country mansions or town halls, into more clinically conditioned establishments. Hospital architecture and design reflected contemporary socio-political realities and different medical philosophies. The problem faced by enlightenment philosophers, philanthropists, architects and governing committees was optimising and retaining control of labouring populations while avoiding overcrowding, undue contact and physical proximity, setting the able-bodied poor to work and transforming them into a 'useful labour force'.[5] With moral and religious concerns, this helps to explain why considerable attention was paid to hospital design and management by medical men and philanthropists such as the Quaker John Howard (1726–90) and the Gloucestershire magistrate Sir George Onesiphorus Paul (1746–1820). There was a revival of ancient Hippocratic concerns with an emphasis on healthy, well-watered and airy locations, where the sick could be isolated from productive labouring populations.

Improvement of hospitals also required reorganisation of internal space determined by medical and practical requirements such as the need to keep fever patients separate in small wards, and included providing technological answers to the problems of heating and air circulation.[6] Philanthropists such as Howard and Paul drew attention to the poor conditions that characterised some institutions, which, it was suggested, harboured diseases which were thought to be spread by aerial contagion and frequently associated with the spread of vice. Developments in chemistry including the management and manipulation of quantifiable and definable 'airs' suggested that natural philosophy and medical theory could help to reduce disease, and gave renewed impetus to Hippocratic concerns about the external factors of disease causation. According to miasmic theory, expounded in Britain by physicians such as William Cullen (1710–90), a poisonous 'miasma' emanated from sources of corruption such as rotting meat or diseased bodies, and this either originated or exacerbated diseases prevalent on ships or in public institutions such as hospitals and prisons.[7] Prisons and hospitals shared some of the same problems in that both required separation of their subjects, either as punishment or for clinical reasons, but in so doing, both were thought to make the control of disease problematic because of the difficulties of heating and ventilating separate rooms or cells. Attempting to reduce the prevalence of disease in ships and public buildings, Stephen Hales knocked ventilator holes into prison walls and invented a giant fan system to supply fresh air and remove the stale. He was convinced that placing ventilators at Newgate Prison in 1752 significantly decreased mortality and sickness rates.[8] Likewise, after touring hundreds of prisons and hospitals throughout Europe, Howard made recommendations concerning location, design, structure and appearance, suggesting that such institutions should ideally be in open country. Furthermore, the Scottish physicians James Lind (1716–94) and Sir John Pringle (1707–82) used their experiences working for the British army and navy respectively to investigate the relationship between putrefaction, contagion and fever, both suggesting that the effluvia of typhus sufferers could be diluted or weakened through the practice of cleanliness and particularly the provision of efficient ventilation systems. John Haygarth (1740–1827) of Chester took this further by advocating the creation of separate fever wards which would allow separate treatment and help to prevent epidemics within crowded towns.[9] Experiments with hospital design in enlightenment France were summarised in Jacques Tenon's influential *Mémoires sur les hôpitaux de Paris* (1788). Hughes de Maret, of the Lyons Hospital, experimented with meat hung in wards and demonstrated that foul air fell to the floor rather than rising, suggesting that oval wards with rounded edges with extractors at low levels would reduce the problem.[10]

Major inspiration came from the adoption of designs, technology and organisation modelled on innovations in British industry.[11] Foucault argued that the problems of power, technology and spacial utilisation in enlightenment public institutions were encapsulated in the Panopticon, the famous idealised prison system designed by the Bentham brothers, which impacted upon the design of public institutions but was never actually constructed. Bentham advocated a re-codification of the law and various rational social and political reforms modelled on the laws of natural philosophy and guided by the famous utilitarian principle of social felicity. Minimal government was optimum, allowing free operation of commerce, individual achievement and gradual but inexorable political reform.[12] The requirements of institutional

30 Derbyshire General Infirmary, from S. Glover, *The History, and Gazetteer, and Directory of the County of Derby*, 2 vols. (1829, 1833), II.

control encouraged the use of iron and glass structures in the Panopticon, which potentially maximised strength, security, light, transparency and central control. The columns of the iron structure could be exploited for water and the integrated thermo-ventilation system which operated through inter-linked ventilator tubes passing through the cells also facilitated control for overlookers. Technology was placed, as Evans has argued 'in the service of a moral order increasing efficiency, economy and especially power and the brothers recommended that their designs could be extended to manufacto-ries, mad-houses, hospitals and schools'.[13]

Design of the Derbyshire Infirmary

The design the Derbyshire Infirmary is most significant because, like the Panopticon, it united some of the greatest preoccupations of European late-enlightenment architecture, the application of ideas from natural philosophy and medicine to the reform of public institutions, the maintenance of power and control through the reorganisation of space, and the development and application of iron and fire-resistant industrial building structures beyond industry. The design reflected many enlightenment philanthropic concerns wedded to the systems of power and labour control developed by British manufacturers and businessmen. The Infirmary innovations were detailed in Sylvester's *Philosophy of Domestic Economy* (1819), and Strutt had collaborated with Sylvester in the design and application of industrial technology at the Infirmary.

Prior to 1810, medical relief for the poor and labouring classes in Derby, as at other Georgian towns, took a number of forms. If they could afford it, most sought treatment from apothecaries and surgeons with the help of friendly societies, of which there were nine by 1803 with an average of fifty members each.[14] For the unemployed or ill, the parish provided some outdoor relief, including occasional medical treatment, although there were no infirmaries attached to the workhouses.[15] A dispensary was founded by Erasmus Darwin during the 1780s which he hoped would be the 'foundation stone of a future infirmary', but this had ceased to exist by the early 1800s.[16] Derbyshire was also served by two wards in the Nottingham General Infirmary from 1786, but Derby remained the last county town in the region without a general hospital.

As we have seen, inspired by progressive enlightenment philosophy, the Derby philosophers advocated urban improvements for their social and economic benefits, striving to obtain support from the corporation, improvement commissioners and local Tories. The Rev. Thomas Gisborne, who was, as we have seen, a founder member of the Derby Philosophical Society, offered £5,000 towards a new hospital as the executor of a will. Like Darwin, Gisborne knew the Manchester physician Thomas Percival and they were both also honorary members of the Manchester Literary and Philosophical Society. After Darwin's death, the principal medical inspiration for the Infirmary came from the physician Richard Forester. Forester was joined on the design committee appointed in 1804 by wealthy subscribers including George Benson Strutt, Thomas Saxelbye a Derby chemist, mine owner and lead smelter, Thomas Cox (1770–1842) a Tory lead merchant, Nathaniel Edwards a Derby attorney, and Dewhurst Bilsborrow (1776–?), a Cambridge-educated physician.[17]

In many respects the Derbyshire Infirmary design was quite conventional and followed the pattern of hospital building already well established in the scores of voluntary hospitals established over the previous eighty years. However, inspired by architecture, organisation and technology – in a county where nearly all of the most important technological developments in the industrialisation of the textile industry had taken place – the Infirmary was one of the first British hospitals to employ an iron and glass structure for the roof (the dome), iron pillars and beams, iron-framed windows and a fireproof ceiling (over the baths). Although economy and utility were the principal motivating factors in the design, clinical considerations were emphasised by committee medical members. Originally, it was stipulated that the building be 'plain and simple and of stone', with a fever ward of twelve beds which was to have a separate entrance, but the final building was much grander and more imposing. The whole was to have eighty beds with two day rooms (one for each sex).[18] However, although this produced 'a great number of plans', many were considered by the committee to be 'extremely defective', failing to satisfactorily incorporate the features originally specified. Some betrayed a want of special knowledge, others were too large and expensive, and generally the designs were considered to pay scant regard either to the principle of 'greatest economy in the construction', or to the 'convenience' of the medical staff, or to the advantages of the patients. These problems reflected the fact that generally accepted standardised and specialised hospital designs were still in dispute during a period of competing medical paradigms and rapid technological and socio-economic change. The committee therefore 'reluctantly' proposed a design of their own, utilising 'the greatest oeconomy' and 'least possible quantity of walling and roof', which was realised by the Derby builder and architect Samuel Brown (b.1756), who produced drawings and a model. The design specified that the wards were to be small with separate and odourless water closets, so that the medical attendant could 'separate acute from chronic diseases and the former from each other as may kill the nature of their complaints'. There were to be no 'long and sometimes gloomy passages' which had 'been so generally adopted' in hospitals and large convalescent day rooms. A simple means was to be found 'of completely and perpetually ventilating every ward with fresh uninhaled air at the same time increasing the temperature by a small expense, to any given degree'. This was because 'medical men' considered that 'in consequence of [a] certain state of the air, which more or less generally pervades hospitals, and which itself has a tendency to produce disease, if the ventilation could be copious while at the same time the warmth could be regulated at pleasure, many lives would be preserved'.[19]

Strutt was able to utilise his considerable knowledge and experience gained designing bridges and mills, especially his development of iron-framed fire-resistant buildings.[20] Although occasionally used in antiquity, the use of iron as a major structural material appears to have been partly stimulated in Derbyshire, as in Europe and North America, by the demands of bridge building. Improvements in metallurgical processes, some of which were stimulated by military requirements, the development of steam power and the adoption of coked coal furnaces, began to reduce the cost of iron and make the casting of large-scale iron members possible for use in construction. As we have seen, Strutt was aware of Thomas Paine's iron bridge design, and the destruction of a series of manufactories by fire such as the Evanses' mill at Darley Abbey in 1788 and Richard Arkwright's Nottingham mill in 1781 made the vulnerability of traditional wooden structures apparent. The cost of insurance and replacement emphasised the commercial benefits to be had from devising buildings with greater resistance to fire. In 1792 Strutt obtained detailed information from John Walker, an architect, about Victor Louis's Palais-Royal Theatre in Paris (1785–90), which had succeeded in combining Soufflot's wrought-iron roof framing with St. Fart's hollow-pot vaulting to eliminate timbering from the structure. Strutt's response, inspired by the French innovations, was a six-storey Derby calico mill (1792–93), which Sylvester described as the 'first fire-proof mill that was ever constructed', soon followed by a four-storey warehouse at Milford of similar design. The Derby mill had brick arches with hollow earthenware pots and was paved with brick; the pots formed part of the structure rather than being merely to 'block out' sections of a floor as at Paris. The girders still used wood (Baltic fir) but were cased in iron. The four-storey Milford warehouse still survives and is therefore the oldest surviving example of a purpose-built, partially iron-framed, fire-resistant building in the world.[21] Other mill building occurred at Belper and elsewhere, where the design was developed in conjunction with Strutt's friend, the Shrewsbury cotton manufacturer Charles Bage (1752–1822). While corresponding with Strutt, Bage took the design a stage further by eliminating structural timber altogether and using cast-iron beams supported by cast-iron columns in a five-storey flax mill at Shrewsbury between 1796 and 1797, which was the first true iron-framed building in the world and influenced the construction of other fire-resistant mills. At around the same time, Strutt was also corresponding with the Bentham brothers on engineering matters, particularly Samuel, while they were attempting to realise their iron and glass Panopticon design. He shared the Benthams' interests in refrigeration and glass houses, and later utilised a Panopticon-type design in the Belper Round Mill, while his son Edward later became an associate of Bentham.[22] Indeed, Sylvester retrospectively stated that the guiding principle that had inspired the Derby hospital was 'to lessen the number of evils to which we are liable, and to increase the sum of our natural and social enjoyments' – which he considered the 'end of all philosophical enquiry'.[23]

Towards the end of the eighteenth century iron also became more common in public buildings such as churches, beginning to form part of the structure rather than being used for decorative effects. Sir John Soane, for instance, used hollow-pot vaulting in the Bank Stock Office in London (1792–93), and covered the 7m (23ft) diameter oculus over his Consols Office with a twelve-sided iron and glass lantern. The value to the military of incombustible buildings on the plan of the textile mills at Derby and elsewhere was evident, as Sir John Rennie observed in 1807 when advising on the reconstruction of dockyard buildings.[24] The 'fireproof' mill design became most widely known from Strutt's still-surviving Belper north mill (1803–4), which was iron-framed throughout, 127 feet long, 31 feet wide and 63 feet high, and had five storeys and an attic. It became a model for mill construction, an illustration and description by John Farey junior appearing in Rees's *Cyclopaedia*.[25]

Hence, by the time the Derbyshire Infirmary was under construction, Strutt had already designed various fire-resistant buildings. By April 1806, Strutt and the design committee had decided to incorporate an iron and glass dome over the centre of the Infirmary with iron-framed windows and a hollow-pot vaulted ceiling over the baths.[26] Iron was chosen for the dome as this allowed six skylights to be incorporated, each consisting of three rows of glass panels, thus maximising the amount of light shining on to the central hall and staircase, extending the day, and facilitating operations and institutional control. Indeed, the dome was regarded as one of the main innovations that excited admiration because 'being ... most difficult of execution', it appeared 'nevertheless to possess the most perfect strength and solidity'.[27] Apart from Strutt's experience utilising iron in construction, it is also significant that Forester, Saxelbye the chemist and Nathaniel Edwards the Derby attorney – two of the other members of the original committee – were partners in one of the largest iron-working businesses in the county, the Riddings Ironworks at Somercotes, which was awarded the dome contract. In 1805, just when this decision was being made, they appointed David Mushet (1772–1847), one of the principal metallurgists in Britain, as manager. Although only at Riddings for about three years, Mushet and his brother transformed the complex into the most efficient in the county. They utilised an improved quality of coke, roasted the ore in kilns instead of the open air, used yellow Derbyshire fluorspar to increase the fusibility of the ore, and roasted the coke, ore and flux in a more efficient conical-shaped stone furnace.[28] A decade later Sylvester wrote that 'were the Derbyshire Infirmary now to be erected, it would probably be done without any wood being used in its construction; and without even iron pillars and beams'. As it was, the significance of the Infirmary needs to be recognised as probably the earliest general hospital in Britain to incorporate hollow-pot and iron-frame technology originally developed for textile manufactories.[29]

The Infirmary was cubical in form, with the central part being drawn into a conical form, terminating in the dome, which was surmounted by a giant statue of Aesculapius by local sculptor William Coffee (1773–c.1846). A large staircase rose through the centre of the building, surrounded on each of the three floors by small wards and other rooms. The basement contained the baths, cellars, kitchen and wash-house, and the upper storey housed most of the wards (for between four and eight beds each only), operating rooms, convalescent rooms and the fever wing. Significantly, the middle storey, which was fronted by the main public entrance, besides housing a couple of wards and the outpatients' room and chapel, contained the board room, servants' quarters and the rooms of the medical staff. This ensured that any visitors who came through the imposing millstone grit Doric-column portico, passed the board room and matron's quarters on two sides before going anywhere else, and it was almost impossible for patients to leave the building without passing by the medical staff through the illuminated central staircase and hall.[30] Initially, the medical staff was small, consisting of a matron, two nurses, a resident house apothecary and a secretary, and the three physicians and four surgeons attended periodically and according to a rota of duty days for emergencies. As was common in Georgian general hospitals, apart from emergencies, patients could only be admitted through the recommendation of subscribers or senior medical staff, with the number of recommendations being proportional to the amount subscribed. Again, as was commonplace, many types of patient were excluded, such as those who could afford their own treatment, pregnant women, children under the age of seven, prostitutes and those convicted of criminal offences ('without evidence of reformation of character'), the motive being to limit the institution to those most amenable to cure among the 'industrious' and 'deserving' poor.[31]

The greatest technological innovations were considered to be in the interior, which also followed industrial models. In 1792 Strutt had devised a type of warm-air stove and arrangement of flues and turncaps to warm and ventilate a mill at Belper. The inspiration for this probably came from the systems of heating and ventilation that had been designed by the textile manufacturers Richard Arkwright and Jedediah Strutt, themselves probably based upon the warming system of 'fire engine' and wall cavities devised for Lombe's Derby silk mill (1721) – the prototype of all British textile manufactories.[32] Strutt is also likely to have been influenced by the efforts of Lunar Society members to devise steam and air heating and ventilation systems for manufactories, public buildings and private houses.[33] John Whitehurst, for example, had worked on improvements to domestic design, and suggested separate openings for ventilation besides chimneys and devised a heating and ventilation system for St Thomas's Hospital in London using underfloor and wall ducts and hypercausts.[34]

In Strutt's system at the Infirmary, air was introduced by means of a 4-square-foot-wide, 70-yard-long duct and heated by an iron-plated cubicle 'cockle' or stove in the basement encased in brick and placed over the fire under a grate. Warm air was directed through a series of ducts throughout the three levels of the building. The smoke from the fire escaped from the cockle into the flues by passing downward through two long, narrow slits on opposite sides of the cockle. An outlet on the roof was provided with a turncap for the escape of foul air by flues connected with all the rooms for patients. A second turncap away from the main building and connected to it by an underground culvert was controlled using a vane, which turned it into the wind. The first turncap vane turned it always away from the wind. By these means, a current of air always passed through the wards.[35] Circulation and temperature were controlled by situating the stove in the basement and calculating how much coal would produce different approximate temperatures. By experimentation Sylvester and Strutt discovered the optimum economic value for the system. They considered the system of making the air a 'medium or vehicle for supplying caloric' at the Infirmary as proven to be much more safe, effective and, especially, more economical than steam heating systems; thus, according to Sylvester, the same coal burnt in warming by steam was inferior to the air-warming system by a ratio of 6 to 1.[36]

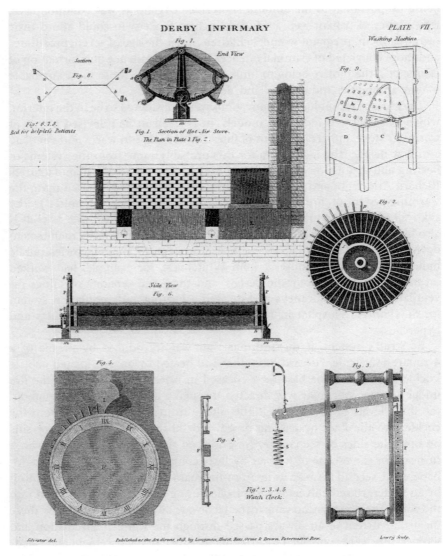

31 Plate VII from C. Sylvester, *Philosophy of Domestic Economy* (1819).

Derby Wakes:

FOR THE BENEFIT OF THE

I N F-R M-R Y.

THE Nobility, Gentry, and Public of Derby and the Neighbourhood, are respectfully informed that the Evening's Amusements in honor of the above Anniversary will take place by postponement in the L-nc-str--n Sch--l Room, D-rby, on   the   day of   next, when will be presented, a Scientific, Democratic, Literary, Dramatic, Tragic, Comic, Operatic, Musical, Farcical, and Pantomimic Entertainment by the CELEBRATED COMPANY whose public performances have lately attained so much notoriety. They have already represented, with universal eclat, "The Comedy of Errors," "Much ado about Nothing," "All in the Wrong," "Raising the Wind," "The Budget of Blunders," "The School for Arrogance," "Duplicity," &c., &c.

On this occasion the amusements of the evening will commence with a *Select Dramatic Divertisement*, in the course of which the celebrated *Derby Amateurs* of Philosophy will successively pourtray the outlines of the following characters.

Sir Francis Wronghead, Justice Shallow, Marcus Brutus, Lord Touchwood, Sir John Loverule, Bombastes Furioso, Don Testy, Shatterbrain, and Midas, by } The Monk of the Hill.

"O Tremendous Justice Midas!
Who dare oppose great Justice Midas."

Sir Giles Overreach, Old Growley, Casca, Lovegold, Crabtree, Snarl, Restive, Gripe, and Sir Jealous Traffick.................................. } Lord St. Helen's.
Sir Epicure Mammon, Jeremy Diddler, Sir Paul Pliant, Jaffier, Sir Macaroni Virtue, and Orator Mum .... } Knight of St. Peter.
Sir Andrew Mar Text, Proteus Puff, Archer Pierre, Cassius, Caleb Quotem, and Dr. Cantwell .... } Black Lettered Leer.
Romeo, Atall, Ranger, Lothario, Courtall .... Sir Thomas Bentick.
Merlin, Noodle, Sir Andrew Aguecheek, Subtle, and Jobson . ... Necromancer.
Jemmy Jumps, Runfasto, Captain Flash, and Glib, ... Professor Hoaks.
Dr. Rosey, Sir John Bull, Harmony, and Stedfast, ... Knight of St. Michaels.
Quaver, Rattle, and Rosin,.... .... Minstrel of St. Michael.
Quildriver, Crack, Fag, Endless, Mr. Smirk, and Old Ferret.. .... Temple Knight.
Solus, Dangle, and Faddle .... Murphy Delany.
Dr. Suitall, Bronze, Fainwoud, and *The Hon. Tom Shuffleton*, ... Dr. Wrong.
Sir Toby Belch, Clinker, Tony Lumpkin, and Robin Roughhead, ... Johnny Rattlepate.

At the end of the Divertisement the Rev. Mr. Fairplay will deliver a new Literary, Humorous, and Argumentative Discourse, called "*The Philosopher's Apology*," from the following text:—

Doubtless the pleasure is as great
In being cheated as to cheat.
As lookers on feel most delight
That least perceive the juggler's slight.

A new piece called "*The Philosopher's Jaunt*," will be recited by The White Helmed Knight, shewing how certain philosophers went to Nottingham to see Mr. Saddler's Balloon, how they were discovered in the SIXPENNY WHARF YARD, and the danger they were in of being blindfolded during the Ascent, and afterwards soused in the Canal, but which they escaped by being happily recognised and rescued by their Cousin *Ned Lud*.

AFTER WHICH

A Grand Miscellaneous Concert

Of Vocal and Instrumental Music.

Leader of the Band and First Fiddle .... Monk.
Double Bass,.............. .... Baron.

32 'Derby wakes for the benefit of the Infirmary', satirical broadside
(13 November 1813).

The Infirmary incorporated other innovations inspired by industrial technology and designed to fulfil the utilitarian objectives of Strutt and the committee. There were baths in the basement warmed by steam, and a wash-house. This contained a revolving-drum washing machine powered and heated by the steam engine that had been designed by Strutt. The baths were advertised in the local press and opened to the middle-class public, who were charged two shillings a time.[37] A laundry dried the linen by the action of warm air from the heating system, while wet linen was passed into the hot closet on sliding horses with grooves to run across rails. In the kitchen Strutt incorporated a more efficient roaster of his own design and a new type of steaming apparatus. Other innovations that became well known included the water closets and an adjustable sick-bed, the former being based upon that of the Yorkshire inventor Joseph Bramah (1748–1814). Strutt's toilets flushed automatically by the action of the door upon leaving the room and simulta-neously admitted fresh air into the space, though it is not fully clear how the waste was disposed of.

Medical impact

By far the most frequent justification given for the Infirmary design, especially the application of novel technology, was the supposed economic advan-tages. The basis of the institution was 'the principle of effecting the greatest good by the least expensive means'. Thus it was constantly reiterated in the annual reports that treatment was provided 'on the lowest practicable terms'. Similarly, the expense of the warm baths would 'in time … be amply repaid with interest, either by the advantages, economy, or absolute income which they will produce: as may be exemplified by the ventilator, laundry, and warm bath respectively'. The efficacy of the thermo-ventilation system was thought to be so great that it required six times less coal than a steam heating system. The fact that despite all the rhetoric of 'economy'– as contemporary critics noticed – the final cost of the Infirmary at £17,870 was 70% above the original estimate of £10,500 provided another reason to emphasise the long-term economic, medical and public health benefits. The subscriber could be sure that 'tender care and kind treatment' were available 'at an expense … of little more than that of the food and medicine with which [patients] are supplied; and that these too [are] procured on the lowest practicable terms'. The spectre of fever epidemics was raised, it being emphasised that wealthy subscribers would be, in effect, protecting themselves and their workforce by supporting the institution, as each fever case successfully treated in the infirmary 'might have been … a focus of infection, diffusing disease and perhaps destruction all around'. Treatment prevented fever stalking 'through the dwellings of the poorer classes of the community whose crowded apartments and wretched accommodations add strength to disease and give wings to infection ... whole districts are thus speedily exposed to the ravages of death – nor age, nor sex, nor condition has the means of security'.[38]

In the earliest specifications of the design committee, clinical considerations were also given some weight, supported by the physicians Forester and Bilsborrow. Spatial control was achieved through the provision of small specialised wards to 'enable the medical attendant to separate acute from chronic diseases and the former from each other as may kill the nature of their complaints'.[39] It was claimed that through the operation of the thermo-ventilation system 'many lives' would be 'preserved, which owing to a certain state of the air generally pervading hospitals, might have been inevitably lost'.[40] Clinically inspired technological innovations continued to be made after the opening, such as the adjustable bed mechanism, invented by Strutt after he had noticed nurses experiencing difficulty turning over patients without causing great pain. Strutt's bed utilised a toothed wheel, ratchet and spring mechanism, allowing patients to be moved to any sideways position and held in place for comfort and easy treatment.[41] In 1824 a new vapour bath was constructed 'by which vapour and fumes of sulphur and other substances can be applied to the skin', while in 1825 a foot warmer was installed, heated by steam and covered by carpet, which could be used by eight people. The same year a small bath was constructed to a design by Sylvester for the use of the public, whose temperature was precisely regulated by the user 'according to their own pleasure'. Also in 1825 the house surgeon and inventor Francis Fox demonstrated a steam vacuum cupping ball at a lecture to the Derby Mechanics' Institute given by his brother, the Infirmary surgeon Douglas Fox, which had been developed utilising the hospital's piped steam supply. It was asserted that it had been used 'for many months ... in our excellent Infirmary ... with great success in the operation ... and consequently less [pain]'.[42] In 1826, to these were added a 'portable [fire] extinguishing engine' and, for the use of patients or the public, a 'convenient, cheap, and simple steam or vapour bath', also designed by Francis Fox. To use the bath the patient sat upon an ordinary chair and was 'surrounded up to the neck by an atmosphere of vapour at any desired temperature'; this was said to have proved 'very efficacious' in the treatment of 'rheumatic affections', while all the innovations were considered to 'have been found fully to answer the purposes for which they were designed', and could recoup money for the institution.[43]

The clinical impact of the technology and design is difficult to assess, particularly the thermo-ventilation system, because of the quantity and indeterminacy of pertinent variables. Statistics were produced for the annual reports for the satisfaction of subscribers, providing the number of patients cured, relieved, excluded or who had died, including outpatients treated by the dispensary, which had also been opened in 1810. Some of these categories, such as that of patients discharged as 'relieved', were highly subjective, and many local factors need to be taken into consideration in assessing the significance of mortality rates, such as changing social, economic and demographic characteristics. However, analysis of the state of in-patients discharged between 1809 and 1839 from the annual reports gives an average mortality rate of 4.3%; this compares quite favourably with other hospitals founded between 1740 and 1820, which generally had rates closer to 10%. A comparison with quinquennial samples of the statistics of the state of in-patients discharged between 1839 and 1869 demonstrates that although the proportion of those cured remained constant at around half of patients discharged, the average mortality rate rose by 35%. Furthermore, a comparison between the average mortality rates of the decades between 1809 and 1839 and the quinquennial samples from 1839 to 1869 reveals that though the average mortality rate remained at around 3.5% between 1809 and 1829, it rose significantly during the 1830s to 5.4% and remained at this level or higher for the ensuing decades.[44]

The relatively low mortality rate suggests that the Infirmary was fairly successful by the standards of the period, although how much this was due to the novel technology or other aspects of the design remains an open question. It is possible that early problems with the thermo-ventilation system may have contributed to the increase in mortality rate from the 1830s. There were early complaints from patients about the dryness giving 'an unpleasant sensation'; however, means were found, probably by Sylvester, of regulating the hygrometrical qualities of the air 'without trouble or expense'.[45] In 1831 a pulley had to be fitted to one of the skylights in the operation room to improve the airflow, smoke from the stew hearths escaped into the wards, and there were difficulties ensuring a free air flow into the ventilation tower. In 1842, the master had to be instructed to encourage patients to use the convalescent rooms more frequently, and to see that windows were open more often.[46] However, by far the most important factor is likely to have been the fact that by the 1830s the Derby population was expanding at a rate of 60%, the most rapid in the history of the town.[47] A series of government health reports blamed the poor health of the towns' inhabitants – shown most dramatically in outbreaks of typhus fever and childhood mortality rates – on the proliferation of poor-quality court housing, the nature of factory employment, and pollution caused by smoke and poor drainage. It was noted that Derby's mortality rate at this time was 2.6%, some 0.6% above the national average, almost half of which was made up by child deaths under five years of age, and that the average life expectancy of labourers and artisans had fallen to twenty-one years by 1840 as compared to forty-nine years for gentry and professionals. This resulted in a greater utilisation of hospital capacity (though not at this stage overcrowding), which would have reduced the efficacy of hospital treatment and increased the chances of infection.[48] This high mortality rate was despite the foundation of a Self Supporting and Charitable Dispensary in 1830 and medical assistance offered by the Poor Law Union from 1837. The former was served by a consulting physician and eight surgeons, with the local gentry enrolled as patrons, the facilities of the dispensary, being, like the Infirmary, open to members of the working classes not supported by the parish.[49]

Wider impact

According to Edward Strutt, the Derby Infirmary had 'in many respects' served as a model for similar institutions in England and 'obtained a well-deserved celebrity even on the Continent'. Sylvester referred to 'the general admiration in which [the Infirmary] has been held, and the frequent applications' from around Britain and abroad for additional information, particularly concerning the celebrated heating and ventilation system.[50] Although it is difficult to confirm this without a search of European and American hospital archives, there is evidence to support this contention. The impact of the design is evident in three contexts: published descriptions, accounts of visits, and, of most importance, incorporations of aspects of the design into other public and private buildings, including medical institutions. Around 1820, for instance, Richard Forester seems to have guided the eminent French physician Laurent Theodore Biett (1781–1840) on a tour of the Derbyshire Infirmary, and he was presented with a specially inscribed copy of the *Philosophy of Domestic Economy*. A leading physician and tutor at the Hôpital Saint-Louis in Paris, after studying with Thomas Bateman, physician at the Carey Street Public Dispensary in London, Biett became the principal expert on dermatology in early nineteenth-century France.[51]

The Infirmary and the houses of members of the Strutt family provided practical demonstrations of the economy and efficacy of the application of industrial technology to domestic economy. Descriptions of the Infirmary appeared in local newspapers, topographical and descriptive volumes such as Davies's *New Historical and Descriptive view of Derbyshire*, Glover's *History of Derbyshire* and Simpson's *History and Antiquities of Derby*. Accounts also appeared in national works such as Cooke's *Description of Derbyshire*, Lyson's *Magna Britannia*, the *London Encyclopaedia* and Phillips's *Tour through the United Kingdom*.[52] All of these accounts were favourable, although Phillips bemoaned the dependency of the institution on acts of charity such as the music festivals, which became quite well known in the region in their own right.[53]

A succession of visitors toured the hospital, including natural philosophers, engineers, writers, medical men, royalty, government figures and architects. After William Edgeworth had visited the Strutts, his father Richard Lovell Edgeworth wrote to William Strutt referring to his house (in which technological innovations similar to those at the Infirmary had been applied to domestic economy), 'where you seem to have exhausted all the resources of ingenuity in procuring domestic comfort for yourself and your friends'. He gently condemned Strutt's 'culpable indolence' for not publishing an account of the Infirmary 'with the many admirable improvements that you have made in its economy'. He thought highly of the washing machine but suggested that 'perhaps a larger wheel (or drum) producing a greater fall in the linen and projecting pliable interrupters to turn the linen over in its fall might be advisable'.[54] On seeing St Helen's House, Sir Richard Phillips remarked that 'steam, gas, heat, hot air, philosophy and mechanics are all brought to bear on these premises, on every branch of domestic economy'.[55] Maria Edgeworth described how she had seen both the Strutt mills and the hospital:

> Seven hours of the day Mr. Strutt, and his nephew Jedediah gave up to showing us the cotton mills: and another whole morning he gave up to showing us the Derby Infirmary; he built it – a noble building; hot air from below conveyed by a cockle all over the house. The whole institution a most noble and touching sight; such a great thing planned and carried into successful execution in so few years by one man.

On the same trip she met Forester, Elizabeth Darwin, Edward Strutt and Sylvester, whom she considered, 'a man of surprising abilities, of a calm and fearless mind, an original and interesting character'.[56]

The visits of royalty provided publicity for the Infirmary, receiving notice in the local and national press. In 1816 the Grand Duke Nicholas, later Nicholas I, Tsar of Russia (1796–1855), a friend of the Duke of Devonshire, before he visited anything else, was conducted 'through the whole [Infirmary] building' with his entourage, the Duke, and Sir William Congreve (1772–1828), the natural philosopher, engineer and comptroller of the Woolwich Laboratory, which they 'inspected with minute attention'.[57] The Grand Duke, reportedly, 'highly admired the many ingenious and useful plans adopted in every department of the Institution' and was 'much gratified by the evident attention which had been paid to the convenience and comfort of the patients'. In 1819 he was followed by Prince Leopold of Saxe-Coburg, later King Leopold I of Belgium (1790–1831), who was conducted on a tour by Forester.[58] With Leopold were, once again, the Duke of Devonshire, Sir William Gardiner, Prince Karl August von Hardenburg (1750–1822) the Prussian Chancellor, and Christian Friedrich, Baron Stockmar (1787–1863), the German diplomat and physician later to become a close friend of Queen Victoria and Prince Albert.

In both cases the royal parties and their entourages proceeded to tour local manufacturing and industrial premises after viewing the hospital, especially the Strutt cotton mills at Derby and Belper. Perhaps one of the most interesting results of Leopold's visit was the impact of Strutt's fireproof and iron-frame industrial architecture upon Hardenburg and Stockmar, and through these upon the Prussian architect Karl Friedrich Schinkel (1781–1841). Inspired by the accounts of his friend Stockmar, and probably Hardenburg and Peter Christian Beuth (1781–1853), Schinkel toured England in 1826 with the express intention of observing and sketching the latest examples of industrial and civil architecture, with particular emphasis on the new 'fireproof' building techniques and the use of iron.[59] After his own tours of the manufactories of England, Beuth excitedly told Schinkel in 1823 that the 'wonders of recent times … [were the] engines and buildings to contain them, named factories [where] the columns are of iron. The beams which rest on them as well. The perimeter walls are thin as paper and … not even two-and-a-half feet thick.' Schinkel was refused permission to observe the Belper mills by George Strutt, but he went on immediately to Joseph Strutt's house in Derby and was escorted around 'the famous Infirmary', from which he produced detailed sketches. In his notebook Schinkel described it as a

> fine, pleasant building in every way. Magnificent staircase. The steps faced with lead plates. The famous hot-air heating, water-closet with shutters, movement of air in and out of the rooms, the stale air is drawn off by a rotating ventilator on the roof. Very practical cooking equipment. Magnificent baths, the anteroom through a canvas curtain, warmed by air wafted in from the bath … everything thought out to the last detail.[60]

He also detailed the features of the washing and laundry system and described another Strutt-inspired stove and thermo-ventilation system at the Derby Lancasterian School. Schinkel was clearly excited by the application of industrial technology and the possibilities of iron-framed utilitarian, functionally inspired buildings and, as Hermann Lebherz has observed, drew on the inspiration of British industrial buildings to design one of his greatest buildings, the Berlin Bauakademie (Academy of Building), which in turn, influenced the Rationalist stream of German architecture down to the Modern Movement.[61] In this respect, Strutt's 'fireproof' Derby and Belper manufactories and the Infirmary with its innovative industrial-inspired technology, bathing rooms, iron-framed windows and striking iron and glass dome contributed to the important influence that British industrial buildings exerted on European architecture.

On 26 June 1817, two years before the publication of Sylvester's book, Strutt was made a fellow of the Royal Society, his five proposers being James Watt, Marc Isambard Brunel (1769–1849), James Lawson, Richard Sharp and Peter Mark Roget (1779–1869). In 1831, the President of the Royal Society, the Duke of Sussex, described him as

> the author of those great improvements in the construction of stoves, and in the economical generation and distribution of heat, which have of late years been so extensively and so usefully introduced in the warming and ventilation of

hospitals and public buildings. He possessed a very great knowledge of practical mechanics, and employed himself through the whole course of a very active life in the furtherance of objects of public ... utility.[62]

The impact of the technology at the Infirmary is also evident from the number of other hospitals that incorporated design aspects. The structure and design of the Pauper Lunatic Asylum at Wakefield, Yorkshire most closely followed that of the Derbyshire Infirmary; however, there is evidence that other hospitals, including the Leicester Infirmary, the North Staffordshire Infirmary, the Nottingham Lunatic Asylum and the Bristol Infirmary, also copied some of the ideas, though the problems of incorporating Strutt/ Sylvester thermo-ventilation systems in buildings already constructed was a limiting factor. In the United States, Charles Bulfinch (1763–1844), the architect of the Massachusetts General Hospital in Boston, read and annotated a copy of Sylvester's book while contemplating his own hospital designs.[63] The governors of the Leicester Infirmary were one of the first medical institutions to express an interest, when John Flint, the secretary of the Board, sent a letter to the Derby governors requesting information 'on several points relative to the government of the hospital', though an offer by James Fox of Derby to install a steam engine system similar to the one he had supplied the Derby hospital with in 1812 was apparently not accepted. In 1814, Flint was asked by his governing committee to write to infirmaries at Bedford, Derby, Nottingham and Northampton requesting information concerning the water closets because of sanitary problems at the Leicester Infirmary. Strutt's water closets were considered the best, and the Leicester committee's architect, Thomas Cooke, visited the Derbyshire Infirmary to see if their system could be adapted. Reporting back favourably, he installed water closets on the Strutt model with ventilators, completing the work in July 1815.[64] Later, when a new fever house was being planned for the Leicester Infirmary, the architect William Parsons made a personal inspection of the Derbyshire Infirmary before designing a three-storey building with water closets on the second and third floors and a thermo-ventilation system modelled on that at Derby, which was opened in 1820.[65] The design of the original North Staffordshire Infirmary at Etruria (1819), produced by county surveyor Joseph Potter, also has some similarities with the Derby hospital, with its symmetrical stone structure and central stair and light-well surrounded by wards. A Strutt/Sylvester thermo-ventilation system was installed with kitchen and laundry apparatus on the Derby plan; however, Sylvester complained that the system was not as effective as it should have been because of mismanagement on the part of the builder.[66]

At Nottingham, John Storer FRS, chairman of Governors for the County Asylum, corresponded with Edward Long-Fox of Bristol on matters of iron fireproof structure, the need for galleries, and ventilation systems. Yet, curiously, despite the fact that Storer had been an external member of the Derby Philosophical Society and would have known Strutt personally, the resulting Nottingham County Asylum (1810–12), designed by Richard Ingleman, did not originally contain a satisfactory heating or ventilation system. Edward Staveley, architect and surveyor, and Mr Dale, a member of the building committee, tried to install stoves, pipes, chimneys and other heating equipment. However, Sylvester was subsequently called in to advise and had furnished a stove and thermo-ventilation system for £100 (excluding materials) by 1815, which he suggested succeeded 'very well'. This was despite the fact that the original building was, in his opinion, 'of a most inconvenient form for warming and 'would have been more complete, if flues had been properly constructed in the first instance'.[67] The governors of the Bristol Infirmary also took an interest in the Derbyshire Infirmary technology, writing to the Derby governors in July 1816 requesting information concerning the thermo-ventilation system. This was passed to Sylvester for an answer, which must have had some effect, for a governor of the Bristol Infirmary travelled to Derby in 1818 to consult Strutt about proposed improvements to the Bristol institution.[68]

Sylvester also acted as consultant to the Pauper Lunatic Asylum at Wakefield (1816–18), designed by architects Watson and Pritchett of York, which featured a thermo-ventilation system, kitchen and washroom technology on the Derby pattern. Before passing through the stoves, as at Derby, fresh air gained admittance through flue entrances in two round towers, on the top of which were turncaps to direct the inlets into the wind. It was then directed through flues in the bedrooms before escaping through another turncap in the roof similar to that in the dome of the Derby hospital. The separate male and female kitchens and laundry followed the Derby plan, with a 2-horsepower steam engine pumping the water, driving the washing machine and mangle, and providing steam from its enlarged boiler for the kitchen, baths and water heating. The kitchen contained roasters, soup boilers and a steam-heated water tank, while the laundry included a drying closet with sliding horses as at Derby, though with the addition of a machine-driven washing machine and a new wringing machine which had been invented by Strutt after the Derby Infirmary kitchen had been constructed.[69]

The publication of the *Philosophy of Domestic Economy* was partly designed to promote Sylvester's business activities, which were continued by his son John (1798–1852), who took out several patents after his father's death between 1832 and 1845. By 1817, as Strutt told Richard Edgeworth in a letter, Sylvester had almost completed a plant for the manufacture of improved air stoves, utilising the work that he and Strutt had done on domestic economy, which he continued on moving to London in 1820.[70] Before this, Sylvester had supplied various private homes and non-medical public buildings, including the New Jail at Maidstone, Leek Parish Church, Manchester College, York and the homes of Strutt, Benjamin Gott, Samuel Shore, Sir Morton Disney, and the sculptor Francis Chantrey. Indeed, Sylvester stoves were installed on board ships such as the *Erebus* and the *Terror*, used by the Arctic explorer William Edward Parry. So technology originally developed for the Derbyshire Infirmary was later utilised in the search for a north-west passage.[71]

The Strutt/Sylvester stove and thermo-ventilation system was quite widely publicised, with detailed descriptions appearing in the works of Robert Meikleham (under various pseudonyms) and in Rees's *Cyclopaedia*.[72] Strutt/Sylvester-type thermo-ventilation systems were incorporated into the Hunterian Museum in London in 1810 (by Boulton and Watt), the Derby Lancasterian School, the Derbyshire County Hall and the Derbyshire County Prison, where John Sylvester's hot water heating and ventilation system could warm the cells to 34 degrees above that of the external air in a few minutes.[73] The County Gaol, in turn, with the original Derbyshire Infirmary, Gloucester Asylum, Nottingham Asylum and Wakefield Asylum, all influenced the design of Henry Duesbury's Derby Lunatic Asylum at Mickleover (1844–51). This featured iron roofs with brick-arched 'fireproof' ceilings, a thermo-ventilation system erected under the guidance of John Sylvester, baths, kitchens, and laundry rooms containing washing machines, boiling vessels and a steam-operated wringing machine. All these were driven or supplied with water or steam by a 15-horsepower engine in the basement. The air was heated by hot water pipes and then circulated through the wards before being expelled through flues in the ceiling, with the ventilation system operating through similar ducts and flues.[74]

Conclusion

The design of the Derbyshire Infirmary is most significant as an early example of the application of industrially inspired technology to general hospitals. As enlightenment philanthropists and philosophers such as Howard and Bentham recognised, the problems of internal hospital spatial organisation needed to service the sick members of an expanding and industrious workforce required efficient and economic control. Hospitals isolated the sick from the healthy labouring population, as the rhetoric of hospital annual reports frequently emphasised.[75] These requirements were satisfied in the Infirmary through the provision of a thermo-ventilation system, automatic flushing toilets and washing and drying apparatus, which allowed each wing of the institution to be self-sufficient. Small wards with separate lavatory provision and convalescent rooms positioned around an illuminated central staircase allowed control over patient movement and limited infection through isolation according to prevalent miasmic theory. Strutt also utilised his watchman's clock from the family cotton mills to regulate the labourers who constructed the infirmary.[76]

Through the application of industrial systems and experimental rationality, which allowed regulation of temperature, humidity and air flow, medical staff, acting for bourgeois subscribers, had almost complete control of the patient environment in the 'curing machine'.

Notes

1 Sections of this chapter are based upon P. Elliott, 'The Derbyshire General Infirmary and the Derby philosophers: the application of industrial architecture and technology to medical institutions in early nineteenth-century England', *Medical History*, 46 (2002), 65–92, and I am grateful to the editors for permission to use the material here.

2 C. Sylvester, *Philosophy of Domestic Economy* (Nottingham, 1819); C. L. Hacker, 'William Strutt of Derby (1756–1830)', *Derbyshire Archaeological Journal*, 80 (1960), 49–70; V. M. Leveaux, *The Derbyshire General Infirmary* (Cromford: Scarthin, 1997). Most hospital documents are at Derbyshire Record Office, Matlock. On hospitals see more generally: J. Woodward, *To Do the Sick No Harm: A Study of the British Voluntary Hospital System to 1875* (London: Routledge & Kegan Paul, 1974); E. M. Sigsworth, 'Gateways to death? Medicine, hospitals and mortality, 1700–1850', in P. Mathias (ed.), *Science and Society, 1600–1900* (London: Cambridge University Press, 1972), 97–110; J. D. Thompson and G. Goldin, *The Hospital: A Social and Architectural History* (New Haven: Yale University Press, 1975); G. Cherry, 'The role of English provincial voluntary general hospitals' and C. Webster, 'The crisis of the hospitals during the industrial revolution', in E. G. Forbes (ed.), *Human Implications of Scientific Advance* (Edinburgh: Edinburgh University Press, 1978); S. V. F. Butler and J. V. Pickstone, 'The politics of medicine in Manchester, 1788–1792: hospital reform and public health services in the early industrial city', *Medical History*, 28 (1984), 227–49; J. V. Pickstone, *Medicine and Industrial Society: A History of Hospital Development in Manchester and its Region* (Manchester: Manchester University Press, 1985); L. Prior, 'The architecture of the hospital', in H. Richardson (ed.), *English Hospitals, 1660–1948* (Swindon: Royal Commission on the Historic Monuments of England, 1998), 86–113; L. Granshaw and R. Porter (eds.), *The Hospital in History* (London: Routledge, 1989); J. Taylor, *Hospital and Asylum Architecture in England, 1840–1914* (London: Mansell, 1991); L. Granshaw, 'The Hospital', in W. F. Bynum and R. Porter (eds.), *Companion Encyclopedia of the History of Medicine*, 2 vols. (London: Routledge, 1993), I, 1180–203; A. Berry, 'Patronage, Funding and the Hospital Patient, c.1750–1815: Three English Regional Case Studies' (DPhil thesis, University of Oxford, 1995).

3 M. Foucault, *The Birth of the Clinic* (London: Routledge, 1989); M. Foucault, 'The eye of power: a conversation with Jean-Pierre Bardou and Michelle Perrot' and 'The politics of health in the eighteenth century', in C. Gordon (ed.), *Michel Foucault: Power/Knowledge, Selected Interviews and other writings, 1972–1977* (Brighton: Harvester Press, 1980), 146–65, 166–82; 'Space, power and knowledge', and interview with Paul Rabinow in P. Rabinow (ed.), *The Foucault Reader* (Harmondsworth; Penguin, 1986); L. Prior, 'The architecture of the hospital'.

4 I. S. L. Loudon, 'The origins and growth of the dispensary movement in England', *Bulletin of the History of Medicine*, 55 (1981), 322–42; B. Croxson, 'The public and private faces of eighteenth-century London dispensary charity', *Medical History*, 41 (1997), 127–49.

5 Foucault, 'The politics of health', 169. Foucault argued that hospitals were technical and material settings within which discursive formations unfolded and simultaneously constituted the objects to which the discourse was addressed. Space therefore constituted as well as represented social and cultural existence, with what Foucault described as the 'medical gaze' becoming institutionalised and inscribed in social space, particularly in post-revolutionary France but also, to a lesser extent, in Britain.

6 Foucault, 'The politics of health', 179–82.

7 C. Hannaway, 'Environment and miasmata', and M. Pelling, 'Contagion/germ theory/specificity', in Bynum and Porter (eds.), *Companion Encyclopedia*, I, 292–308, 309–35; C. Hamlin, 'Predisposing causes and public health in early nineteenth century medical thought', *Social History of Medicine*, 5 (1992), 43–70.

8 'Ventilator', in A. Rees, *The Cyclopaedia; Or, Universal Dictionary of Arts, Sciences and Literature*, 39 vols. (London, 1819), XXXVI; R. Evans, *The Fabrication of Virtue: English Prison Architecture, 1750–1840* (Cambridge: Cambridge University Press, 1982), 94–117.

9 'Fever' and 'Fever Wards', in Rees, *Cyclopaedia*, XIV; F. M. Lobo, 'John Haygarth, smallpox and religious dissent in eighteenth-century England', in A. Cunningham and R. French (eds.), *The Medical Enlightenment of the Eighteenth Century* (Cambridge: Cambridge University Press, 1990), 217–53.

10 Foucault, *Birth of the Clinic*.

11 H. Marcuse, 'Some social implications of modern technology', in A. Arato and E. Gebhardt (eds.), *The Essential Frankfurt School Reader* (New York: Urizen Books, 1978), 138–62; T. Adorno and M. Horkheimer, *Dialectic of Enlightenment* (London: Verso, 1986); D. Held, *Introduction to Critical Theory: Horkheimer to Habermas* (London: Hutchinson, 1980), 65–70, 148–74. This can be perceived as an early example of what Weber and the theorists of the Frankfurt School described as the application of instrumental rationality (or means/end rationality) to all forms of moral, political and economic behaviour: in other words, the shaping of scientific practice according to a model of the natural sciences, the mathematisation of experience and knowledge, and the extension of scientific rationality into society and culture. In this kind of industrial and manufacturing system, individuals were motivated, guided and measured by external standards, being rewarded according to efficiency and technical proficiency, forces which undermined revolutionary enlightenment rationalism, but also partly sprang from it. Actions were dissolved into a series of semi-spontaneous reactions to prescribed mechanical norms. In industrial society, the manufacturing process was beginning to impose patterns of mechanical behaviour, with expediency, convenience and efficiency being rewarded by the competitive economy. Teleological and theological dogmas interfered less in the human struggle with matter, while philosophical experimentalism served to develop a higher efficiency of hierarchical control. Hence, technological rationality was beginning to become the servant of power, spreading its values

and forms of organisation through society and culture and threatening established notions of individual identity.

12   E. Halévy, *The Growth of Philosophic Radicalism* (London: Faber & Faber, 1934); M. Foucault, *Discipline and Punish: The Birth of the Prison* (Harmondsworth: Penguin, 1991), 195–228.

13   'Panopticon; or, the inspection house' and postscripts, in J. Bowring (ed.), *The Works of Jeremy Bentham*, 11 vols. (London, 1838–43), IV, 37–172; Halévy, *Philosophic Radicalism*, 81–5; Evans, *Fabrication of Virtue*, 225–6; Foucault, 'The eye of power'.

14   Abstract of the returns ... for the expense and maintenance of the poor in England, *Parliamentary Papers*, 13, 175 (1804), 94–5. This had increased to fifteen by 1830, including a Female Friendly Society founded in 1816, S. Glover, *The History, and Gazetteer, and Directory of the County of Derby*, 2 vols. (Derby, 1829, 1833), II, 539.

15   All Saints' parish, for instance, appointed an apothecary for the sick poor in 1727 chosen by the overseers and churchwardens for ten pounds per year, All Saints' Parish, Book of Order, 11 October 1727, quoted in M. Hodgkinson, 'Poor Relief in All Saints' Parish, Derby, 1722–1836, with special reference to the workhouse' (BA dissertation, University of Oxford, c.1960), 26.

16   E. Krause, *Erasmus Darwin, with a Preliminary Notice by Charles Darwin* (London, 1879), 54–5; Darwin was also surgeon-extraordinary to the Staffordshire General Infirmary from 1783 to 1801, see 'Origin and early history of the Staffordshire General Infirmary', in J. L. Cherry (ed.), *Stafford in Olden Times* (Stafford, 1890), 27–8.

17   Meeting of general committee of £50 subscribers, 9 April 1804; Meeting of general committee, 2 May 1805, Derbyshire General Infirmary papers, Derbyshire Record Office, D1190/1/1–4. Educated at Cambridge, Bilsborrow was another protégé of Darwin, composing poems in his praise and an obituary letter, *Monthly Magazine*, 13 (1802), 548–9. He apparently left an unfinished life of Darwin, D. King-Hele, *Erasmus Darwin: A Life of Unequalled Achievement* (London: Giles de la Mare, 1999), 288.

18   Papers on the establishment of the Derbyshire Infirmary, Derbyshire Record Office, D1190/1/1–4; Drawings and outline plans of the Infirmary by John Rawstorne of York and Moneypenny of London, 1804, Derbyshire Record Office, D1190/89/1–3, D1190/90, D1190/91, D1190/92/1–4.

19   Papers on the establishment of the Derbyshire Infirmary: report of sub-committee to consider plans, 4 August 1804; Derbyshire Infirmary: 'Report of the committee', 23 March 1805, Derbyshire Record Office, D1190/1/1–4; *Derby Mercury* (30 January 1805).

20   'Memoir of William Strutt', manuscript, Derbyshire Record Office, D2943; *Derby Mercury* (12 January 1831); T. Bannister, 'The first iron-framed buildings', *Architectural Review*, 107 (1950), 231–46; H. R. Johnson and A. W. Skempton, 'William Strutt's cotton mills, 1793–1812', *Transactions of the Newcomen Society*, 30 (1955–57), 179–205; A. W. Skempton and H. R. Johnson, 'The first iron frames', *Architectural Review*, 119 (1962), 175–86; M. C. Egerton, 'The Scientific and Technological Achievements of William Strutt FRS' (MSc dissertation, Manchester Institute of

Science and Technology, 1967); R. Fitzgerald, 'The development of the cast iron frame in textile mills to 1850', *Industrial Archaeology Review*, 10 (1988), 127–45; A. Menuge, 'The cotton mills of the Derbyshire Derwent and its tributaries', *Industrial Archaeology Review*, 12 (1993), 38–61.

21 *Derby Mercury* (20 July 1853); R. S. Fitton and A. P. Wadsworth, *The Strutts and the Arkwrights, 1758–1830: A Study of the Early Factory System* (Manchester: Manchester University Press, 1958), 201, 205.

22 In 1794 Jeremy Bentham wrote to Strutt concerning methods of heating using steam and Samuel Bentham provided him with accounts of steam heating in private houses and discussed the requirements of copper tubing for experiments. He visited Strutt in Derby on various occasions and attended a meeting of the Derby Philosophical Society (Strutt correspondence, Derby Local Studies Library, D125).

23 Sylvester, *Domestic Economy*; Fitton and Wadsworth, *The Strutts and the Arkwrights*, 181.

24 Bannister, 'The first iron framed buildings', 244.

25 Fitton and Wadsworth, *The Strutts and the Arkwrights*, 211–12; J. Farey jun., 'Manufacture of cotton', in Rees, *Cyclopaedia*, XXII and II of plates. The Belper mill with its teagle or powered lift were also illustrated in A. Ure's *Philosophy of Manufactures* (London, 1835).

26 Meeting of the sub-committee, 3 April and 4 September, 1806, Derbyshire General Infirmary papers, Derbyshire Record Office, D1190/1/1–4.

27 D. P. Davies, *A New Historical and Descriptive View of Derbyshire* (Belper, 1811), 240.

28 Saxelbye and Edwards sold their two-thirds shares to Mushet, who subsequently sold them to James Oakes (1750–1828) in 1808, who continued the improvements, while Forester remained in partnership with Oakes, R. Johnson, *A History of Alfreton* (Alfreton, 1969), 127–130; 'David Mushet', *ODNB*; J. Farey, *A General View of the Agriculture and Minerals of Derbyshire*, 3 vols. (London, 1811–17), I, 399–400; R. M. Healey (ed.), *The Diary of George Mushet, 1805–1813* (Chesterfield: Derbyshire Archaeological Society, 1982).

29 Dr. Edward Fox, however, incorporated fireproof construction, including iron window frames, staircases, joists and doors, at Brislington House private asylum near Bristol *c*.1804 (Richardson, *English Hospitals*, 157–8).

30 Sylvester, *Domestic Economy*, 2–3.

31 Glover, *History and Gazetteer*, II, 514–17.

32 *Gentleman's Magazine*, 21 (1732), 940, 985–6.

33 R. E. Schofield, *The Lunar Society of Birmingham* (Oxford: Clarendon Press, 1963), 338–9.

34 J. Whitehurst, *Observations on the Ventilation of Rooms, on the Construction of Chimneys; and on Garden Stoves* (London, 1794); Schofield, *Lunar Society*, 338–9; M. Craven, *John Whitehurst of Derby: Clockmaker and Scientist 1713–88* (Ashbourne: Mayfield, 1996).

35 Sylvester, *Domestic Economy*, 4–5; R. Simpson, *A Collection of Fragments Illustrative of the History and Antiquities of Derby*, 2 vols. (1826), I, 451; Thompson and Goldin, *The Hospital*, 146–9.

36 Sylvester, *Domestic Economy*, iv, 10, 11, 12, 13, 16, 20; A. F. Dufton, 'Early application of engineering to the warming of buildings', *Transactions of the Newcomen Society*, 21 (1940–41), 99–117; M. C. Egerton, 'William Strutt and the application of convection to the heating of buildings', *Annals of Science*, 24 (1968), 73–87.

37 Sylvester, *Domestic Economy*, 6–8; *Derby Mercury* (5 August 1813); notices for the public baths, 1833, Derbyshire Record Office, D1190/210/1.

38 Third annual report, 29 September 1812, Derbyshire General Infirmary papers, Derbyshire Record Office, D1190.

39 Report of the sub-committee, 4 August 1804, Derbyshire General Infirmary papers, Derbyshire Record Office, D1190/1/1–4.

40 Davies, *Derbyshire*, 242–3.

41 Sylvester, *Domestic Economy*, 48–9, plate v, 47–8, plate vii, figures 6 and 7.

42 Sylvester, *Domestic Economy*, 31–41; D. Fox, *Notes of the Lectures on Anatomy and Chemistry, delivered by Mr. Douglas Fox* (Derby, 1826), 45–6.

43 Derbyshire Infirmary: fifteenth, sixteenth and seventeenth annual reports, 1824, 1825 and 1826, Derbyshire General Infirmary papers, Derbyshire Record Office, D1190.

44 At Shrewsbury and Liverpool, for instance, the mortality rate rarely exceeded 7% between the 1740s and 1820; Woodward, *To Do the Sick No Harm*, 153–8; A. Borsay, 'An example of political arithmetic: the evaluation of spa therapy at the Georgian Bath Infirmary, 1742–1830', *Medical History*, 45 (2000), 149–72.

45 Second annual report of the Derbyshire Infirmary, 29 September, 1811, Derbyshire General Infirmary papers, Derbyshire Record Office, D1190.

46 Leveaux, *Derbyshire General Infirmary*, 30, 33.

47 Davison, *Derby*, 323. Between 1811 and 1821 the percentage increase was 33.5%, from 1821 to 1831 it was 35.6%, and between 1841 and 1851 it was just 8.5%.

48 W. Baker, 'Report on the sanitary condition of the Town of Derby', in *Local Reports on the Sanitary Condition of the Labouring Population of England* (London, 1842), 162–82; J. R. Martin, 'Report on the state of … Derby', in *Second Report of the Commissioners for Inquiry into the State of Large Towns and Populous Districts*, Appendix, part II; E. Cresy, *Report to the General Board of Health on a Preliminary Enquiry into the Sanitary Condition of the Inhabitants of the Borough of Derby* (London, 1849).

49 *Derby Mercury* (4 August 1830); Glover, *History and Gazetteer*, II, 517.

50 Strutt, 'Memoir of William Strutt', *Derby Mercury* (12 January 1831); Sylvester, *Domestic Economy*, preface, iii.

51 P. L. A. Cazenave and H. E. Schedel, *Abrégé pratique des maladies de la peau d'après les auteurs les plus estimés, et surtout d'après les documents uisés dans les Leçons cliniques de M. Biett* (Paris, 1828); N. J. Levell, 'Thomas Bateman MD FLS', *British Journal of Dermatology*, 143 (2000), 9–15. The inscribed copy of the *Philosophy of Domestic Economy* presented to Biett was offered for sale by Bernard Quaritch Ltd, a London bookseller, in 2002.

52 Davies, *Derbyshire*, 239–44; D. and S. Lysons, *Magna Britannia*, V: *Derbyshire* (London, 1817), 126; Simpson, *History and Antiquities*, I, 446–53; R. Phillips, *A Personal Tour through the United Kingdom* (London, 1828), no. 2, Derbyshire and Nottinghamshire, 112–13; G. A. Cooke, *Topographical and Statistical Description*

of the County of Derby (London, 1820), 71–3; Glover, *History and Gazetteer*, II, 506–17; *The London Encyclopaedia* (London, 1839), II, 173–4; S. Glover, *History and Directory of Derby* (Derby, 1843), 34–6.

53 W. Gardiner, *Music and Friends*, 3 vols. (Leicester, 1838, 1853), I, 490.

54 *Derby Mercury* (12 January 1831); R. L. Edgeworth, Letter to William Strutt, 1811, DLSL, Strutt Correspondence, D 125.

55 Phillips, *Personal Tour*, 111–17.

56 M. Edgeworth, letter to H. Edgeworth, 26 April 1813, in A. Hare (ed.), *The Life and Letters of Maria Edgeworth* (London, 1894), 52.

57 *Derby Mercury* (12 December 1816).

58 *Derby Mercury* (14 October 1819); *Leicester Journal* (8 October 1819); *Nottingham Journal* (9 October 1819); *Nottingham Review* (15 October 1819).

59 B. Bergdoll, *Karl Friedrich Schinkel: An Architecture for Prussia* (1994), 172–6.

60 K. F. Schinkel, *'The English Journey': Journal of a Visit to France and Britain in 1826*, ed. D. Bindman and G. Riemann (New Haven: Paul Mellon, 1993), 134–6.

61 H. Lebherz, 'Schinkel and industrial architecture', *Architectural Review*, 184 (August 1988), 41–6.

62 Letter from S. Lee to W. Strutt, 1 December 1817, Strutt Correspondence, DLSL, D 125; Strutt obituary notice, *Abstracts of the Papers Printed in the Philosophical Transactions of the Royal Society of London, 1830–1837*, III (London, 1860), 84.

63 L. K. Eaton, 'Charles Bulfinch and the Massachusetts General Hospital', *Isis*, 41 (1950), 8–11.

64 Minutes and order books of the Derbyshire General Infirmary, 11 March 1811, Derbyshire General Infirmary papers, Derbyshire Record Office, D1190; E. R. Frizelle and J. D. Martin, *The Leicester Royal Infirmary, 1771–1971* (Leicester, 1971), 73.

65 Frizelle and Martin, *Leicester Royal Infirmary*, 362–3.

66 Sylvester, *Domestic Economy*, preface, viii–ix; R. Hordley, *A Concise History of the Rise and Progress of the North Staffordshire Infirmary* (Stafford, 1902), 9–16.

67 Sylvester, *Domestic Economy*, preface, ix; I. Inkster, 'Notes on the finance, building and technology in the early history of the lunatic asylum near Nottingham', Nottinghamshire Record Office, Nottingham Archives, SO/40/1/50/4/1; I. Inkster, 'Early history of the Nottingham Lunatic Asylum', *Trent Vale Review*, 9 (1972), 20–3.

68 Minutes and order books of the Derbyshire General Infirmary, 1 July 1816, Derbyshire General Infirmary papers, Derbyshire Record Office, D1190; W. Strutt, letter to the Revd. W. Burslem, 20 April 1818, Strutt correspondence, DLSL, D125.

69 Sylvester, *Domestic Economy*, appendix, 57–62; C. Watson and J. P. Pritchett, *Plans, etc., and Description of the Pauper Lunatic Asylum* (York, 1819).

70 Egerton, 'William Strutt and the application of convection to the heating of buildings', 84; Sylvester, *Domestic Economy*, vii. Taking advice from Strutt, Richard Edgeworth employed iron-frame construction on buildings in Ireland, and his son William brought over the Strutt/Sylvester stove.

71 Sylvester, *Domestic Economy*, viii–ix; Egerton, 'William Strutt and the application of convection to the heating of buildings', 84; I. M. Leslie, *Rosser and Russell Limited: The First Two Hundred Years* (London, 1974).

72 'Stove', in Rees, *Cyclopaedia*, 34; W. H. Dickinson and A. A. Gomme, 'Robert Stuart Meickleham', *Transactions of the Newcomen Society*, 22 (1941), 161–9.

73 Glover, *History and Directory* (1843), 49–51; Schinkel, *'The English Journey'*, 136. Similar systems were also installed at Kent County Asylum, Bath Gaol and the Model Prison at Pentonville.

74 Thompson and Goldin, *The Hospital*, 74–6; H. Duesbury (attrib.), *Description of Design for the proposed Derby Pauper Lunatic Asylum* (Derby, 1844); Derby proposed Lunatic Asylum: copies of the objections of the Commissioners in Lunacy (1846), Derbyshire Record Office; G. H. Gordon, *History of the Pastures Hospital, Derby* (undated, *c*.1970).

75 Outdoor relief remained common in England, at least until the 1830s, and the Derbyshire Infirmary generally treated more outpatients than in-patients.

76 An example of what E. P. Thompson called 'time sense in its technological conditioning', time-measurement 'as a means of labour exploitation': 'Time, work discipline and industrial capitalism', in E. P. Thompson, *Customs in Common* (Harmondsworth: Penguin, 1993), 352–403.

9

Evolution: Erasmus Darwin, William George Spencer and Herbert Spencer[1]

Introduction

Considerable work has now, of course, been undertaken into the origins of Charles Darwin's theory of evolution, 'Darwinian' evolution of natural selection by random mutation now being recognised as just one possible form of developmentalism prevalent during the nineteenth century rather than as a 'Whiggish' yardstick with which to judge rival theories. This chapter provides further evidence for the importance of the influence of enlightenment ideas in the provincial context in the acceptance of evolutionary theories. Although the indirect role of Erasmus Darwin's developmentalism in stimulating his grandson has long been acknowledged, it is here argued that the elder Darwin arguably exerted a greater influence on Herbert Spencer through the activities of the Derby scientific community.[2] This stimulus is evident, for instance, in Spencer's highly original and influential associationist evolutionary psychology. It provides a much more satisfactory explanation for the origins of some aspects of *The Principles of Psychology* (1855) than Robert Perrin's assertion that Spencer 'literally evolved it from the inner reaches of his own mind after having read only a smattering of formal psychology'.[3]

Some of Spencer's Derby background was explored by Peel, who was interested in his status as a founding father of sociology and recognised the possible relationship between Darwin and Spencer, but only advanced the argument tentatively:

> [I]t is hard not to think that in his rapid adoption of Lamarck's ideas Spencer was not only seizing on a theory of biological evolution which lends itself well to sociological use, but was reverting to an older source of evolutionary influence – that of Erasmus Darwin, mediated through his father and the other 'Darwinians' of the Derby Philosophical Society.[4]

Peel considered that there were convincing internal and biographical reasons to 'suppose that Spencer was more subject to Erasmus Darwin's influence than Charles Darwin was', but although his analysis of the Derby background was the first study of Spencer to give due consideration to his provincial origins, he did not explore the nature of this relationship. The work that has been done on rival developmental theories prior to the publication of *The Origin of Species* and on provincial urban society and scientific culture provides an opportunity to reassess this valuable relationship.

Complementing Desmond's analysis of the relationship between Lamarckism, political radicalism and medical education in the anatomical schools of London and Edinburgh and Secord's work on the reception of the *Vestiges of the Natural History of Creation*, this chapter confirms that there was an important provincial dimension to the emergence and acceptance of biological developmental theories. It argues that to be fully appreciated, these cannot be divorced from emerging broader developmental worldviews.[5] Consideration of the origins of the evolutionary aspects of Darwinian and Spencerian theories in the context of English provincial scientific culture will indicate the kind of social, economic and intellectual conditions under which developmental theories arose, revealing the degree to which Spencer's early scientific and political activities were stimulated by the concerns of the Derby philosophers. British provincial urban cultural renewal and industrialisation were important factors in the emergence of a distinctive developmental worldview, helping to explain the reception accorded to the *Vestiges*, particularly among reformers and dissenters.

Erasmus Darwin's developmentalism

With his medical practice and reputation fully confirmed, the Derby Philosophical Society established, and using the experience he gained working upon the translation of Linnaeus under the auspices of the Lichfield Botanical Society, Darwin felt able to embark upon the publication of prose and poetical works. The two parts of *The Botanic Garden*, the *Loves of the Plants* and the *Economy of Vegetation* made him famous as a writer, while the *Zoonomia* outlined his medical philosophy, the *Phytologia* his horticultural and agricultural ideas, and the posthumously published *Temple of Nature* his grand progressive evolutionary philosophy. Five interconnected aspects of Darwin's enlightenment evolutionary worldview may be defined: geological developmentalism, biological evolutionism, developmental psychophysiology, cosmological developmentalism, and scientific and political progressivism. Even if not undertaken explicitly by evolutionists, geological studies, of course, had an important impact on the development of biological evolutionary theories, as is demonstrated by the inspiration that Lyell's *Principles of Geology* gave to both Charles Darwin and Herbert Spencer. However, the stimulation that geology gave to early forms of developmentalism is evident in the earlier work of enlightenment philosophers such as Buffon and Lamarck, and in the relationship between the emergence of a developmental geology of Derbyshire, and Darwin's evolutionary speculations.

By the 1820s, with its varied geological and topographical character, Derby-shire was one of the most comprehensively surveyed and studied counties in Britain, and this helped to stimulate the geological studies of the Derby philosophers. As we have seen, this stimulation was manifest in a number of ways, including agricultural, industrial and manufacturing exploitation of county resources, especially lead mining, the thriving fossil and mineral trades, the perceived medical efficacy of Derbyshire springs, the Wonders of the Peak tradition, and the growing tourist industry, stimulating by changes in landscape aesthetics and values as much as new forms of leisure.[6] Darwin made extensive use of the local miners and mineralogical traders, assisted by friends including Whitehurst, the Tissington brothers and White Watson.

Darwin's developmental geology held that material was produced by the digestion and secretion of organised beings and had 'given pleasure in their production', after decomposition being accumulated at the bottom of oceans. On the actions of 'submarine fires' they had been thrust up and constituted 'the immense racks and unmeasured strata of limestone, chalk and marbles'. Thus the limestone gorges and the coals, sand, iron, clay and marl were 'originally the products chiefly of vegetable organization', making them 'MONUMENTS OF THE PAST FELICITY OF ORGANIZED NATURE! – AND CONSEQUENTLY OF THE BENEVOLENCE OF THE DEITY!'[7] Darwin saw evidence for the 'submarine fires' in the action of earthquakes and in the warm springs of the Peak, which he detailed in Pilkington's *History of Derbyshire*, while an earth-quake that struck Derbyshire and adjoining counties in 1795 prompted him to outline his theory in local newspapers. Here, he argued that it had resulted from the striking of a 'boiling cauldron of lava' condensing steam in the rocks and forcing in sea-water, creating an immense body of steam 'which bursts its way in all directions, raising up rocks and mountains, produces a concus-sion for 50 or 100 miles, and an undulating vibration over many degrees of the globe'.[8]

The importance of the influence of naturalists such as John Ray, Buffon and particularly Linnaeus on the biological theories of Darwin has been well established. In their taxonomies, Ray and Linnaeus had established means of differentiating between different species, and Buffon considered that varia-tion or mutation was possible within species. In the *Loves of the Plants*, refer-ence was made to the Linnaean theory that variation within species and hybridisation occurred. In the *Economy of Vegetation*, while discussing plant sexuality, Darwin asked whether 'all the productions of nature' were 'in their progress to greater perfection? An idea countenanced by the modern discov-eries and deductions concerning the progressive formation of the solid parts of the terraqueous globe.' He also noted that the uselessness of 'incomplete appendages' to plants and animals appeared to show that they had 'gradually undergone changes from their original state; such as the stems without anthers and styles without stigmas of several plants ... and the paps of male animals; thus swine have four toes, but two of them are imperfectly formed'. This made him consider whether 'all the supposed monstrous births of nature' were 'remains of their habits of production in their former less perfect state or attempts towards greater perfection'.[9] But although the emphasis on the power or tendency of species to change their form was novel, the *Loves of the Plants* exhibited merely 'a tentative and vague eighteenth-century faith in cosmic progress'.[10]

However, in the *Zoonomia*, the *Phytologia* and particularly *Temple of Nature*, evolution and progress moved from being only an occasional component to becoming the overarching theme, with the apparent wastefulness, destructiveness and prodigality of life being taken as evidence for overall progress. Comforting his audience through occasional deistic exclamations and the structural use of neo-platonic classical imagery, Darwin recounted the production and reproduction of life from its oceanic origins, with the gravitational and chemical forces of attraction being equated to the inanimate forces of nature. Contraction, unique to life, resulted from the actions of the ethereal spirit of animation as life began:

> Hence without parent by Spontaneous birth
> Rise the first specks of animated earth;
> From nature's womb the plant or insect swims,
> And buds or breathes with microscopic limbs.

Enlightenment chemistry was combined with a materialist theory of the generation of life, rendering nature self-sufficient and divine intervention unnecessary.[11]

Using the anatomical knowledge acquired as a physician, Darwin marshalled a recapitulationist argument as evidence for the oceanic origins of life. Development in the womb was analogous to the way that animals 'in their embryonic state' were 'aquatic' and thus 'may be said to resemble gnats and frogs'. In the placenta of the foetus the 'fine extremities of the vessels which permeate the arteries of the uterus, and the blood of the foetus become oxygenated from ... maternal arterial blood; exactly as is done by the gills of the fish' from water that they pass through. Sexual reproduction was crucial to the progress of the higher forms of life, producing descendants of 'superior powers'. This was the

> chef-d'oeuvre ... of nature; as appears by the wonderful transformations of leaf eating caterpillars into honey eating moths and butterflies, apparently for the sole purpose of the formation of sexual organs, as in the silk-worm, which takes no food after its transformation, but propagates its species and dies.[12]

Animals competed with each other to satisfy 'three great objects of desire' – lust, hunger and security – and competition for the females necessitated the acquisition of 'weapons to combat each other … as the very thick, shield-like horny skin on the shoulder of the boar is a defence only against animals of his own species'. Thus the 'final cause' of this masculine contest was that 'the strongest and most active' should 'propagate the species, which should thence become improved'.[13] Competition allowed weaker forms of life to be progressively removed, ensuring that

> Immortal matter braves the transient storm,
> Mounts from the wreck, unchanging but in form.

Reproduction triumphed over decay:

> The clime unkind, or noxious food instils
> To embryon nerves hereditary ills;
> The feeble births acquired diseases chase
> Till death extinguish the degenerate race.[14]

'Eternal war' was waged on diseases such as the gout, mania, consumption and scrofula, but weaker forms of life would be removed.[15]

In the *Zoonomia*, dedicated to the medical profession, Darwin detailed an associationist psychophysiology which demonstrated how geological, biological and cosmological evolution were reflected in the progress of human culture and society. The *Zoonomia* was designed to 'reduce the facts belonging to animal life into classes, orders, genera, and species, and by comparing them with each other, to unravel the theory of diseases'. It utilised Darwin's medical experience and knowledge of the British associational tradition to formulate a 'theory founded upon nature', categorising the known types of disease for the benefit of 'legitimate' professional medical practitioners, to prevent the sick falling the 'prey of some crafty empyric'.[16] Life was more than mere laws of motion, the principles of animal chemistry, or the rudiments of a hydraulic system. As Porter argued, Darwin's animals (and plants to a lesser degree) were capable of 'entering into dialectical interplay with their environment'.[17] Darwin's psychophysiology was essentially a physiological rendering of Hartleyan psychology. It involved the action of the 'spirit of animation', an ethereal force of energy acting through the nerves with some properties akin to the actions of electricity on the muscles, which ensured that the trap of environmental determinism was avoided.[18] The spirit of animation acted through four different faculties of motions of the sensorium, defined as irritability, sensibility, volition and associability. These were the cause of all contractions of the fibrous parts of the body, with irritation being excited by external bodies, sensation by pleasure or pain, volition by desire or aversion, and the highest, association, by 'antecedent or attendant fibrous contractions'.[19] The most complex forms of behaviour required the capacity of association, which was the tendency of animal movements that had once occurred in succession or in combination to become so connected by habit that they automatically tended to succeed or accompany each other. Association explained the relationship between traits of character and life habits, and the propensity for certain diseases, with the transmission of traits underpinning the broader developmental worldview.[20] Hartley had argued that 'our immortal part acquires during this life certain habits of action or of sentiment, which become forever indissoluble', which continued in heaven. Darwin secularised this by applying it to the 'generation ... of the embryon, or new animal' which partook 'so much of the form and propensities of the parent', which, as Porter emphasised was because of his theory that the mind of the father determined the character of the child, rather than the more common eighteenth-century view that the contents of the female imagination at conception were impressed upon the embryon.[21]

The nature of human psychology and the mechanism for progressive mental development and physical inheritance ensured that cosmological development continued on the human plane, and the *Temple of Nature* was originally entitled the 'Origin of Society'. As we saw in his opening address to the Derby Philosophical Society, cultural and intellectual progress was best demonstrated in Darwin's view by the heroic natural philosophers, inventors, benefactors and industrialists of enlightenment Europe. They ensured and inspired the progressive march of civilisation towards greater political freedom despite the depredations of war and tyranny.[22] New laws had been framed by Newton; new planets, stars and satellites had been discovered by Herschel and others; progress had occurred in the physical sciences, technology and industry. Savery had developed the steam engine, later much improved by Darwin's friend Watt, which had in turn been applied to spinning by his other associates, the Strutts and the Arkwrights. Darwin described Richard Arkwright as an 'uncommon genius' who by 'persevering industry, invented and perfected a system of machinery for spinning', which 'by giving perpetual employment to many thousand families has increased the population, and been productive of greater commercial advantages to this country, and contributed more to the benefit of mankind, in so short a period of time, than any other single effort of human ingenuity'.[23]

Darwin's cosmological developmentalism provides excellent support for the nebular hypothesis, which, in Moulton's view, revealed the influence that astronomy had upon evolutionary biology.[24] The plurality of life worlds vastly increased the number of possible beings in the universe, perhaps threatening the position of humanity at the summit of creation, while the nebular hypothesis and Herschelian astronomy inspired by Linnaean taxonomy suggested a temporally and spatially vast universe, where stars passed through life cycles akin to animals and vegetables. Newtonianism and the nebular hypothesis encouraged natural philosophers to accept the idea of gradual change by arguing that planets had been formed by a combination of gravitational, rotational and thermal effects. Enthused by Herschel's work, Darwin was the first thinker with a broad evolutionary philosophy to incorporate a cosmological developmental theory as a major component. As a translator of Linnaeus into English, he appreciated Herschel's analogy between the naturalist and the astronomer, between the life cycles of stellar bodies and terrestrial bodies, and his evidence for the great age of the universe, as his famous description of the Newtonian cosmos in the *Botanic Garden* proclaimed.[25] In *The Temple of Nature*, this grand cosmological vision of progress had a greater cyclical element. After the destruction of the globe by 'a general conflagration', the sun's sinking into 'one central chaos', new earths might once again emerge:

> Thus all the suns, and the planets, which circle around them may again sink into one central chaos; and may again by explosions produce a new world; Which in process of time may again undergo the same catastrophe![26]

The simultaneous adoption of developmental theories across Europe during the 1790s such as Darwin's evolutionary cosmology and geology by some philosophers, including Lamarck, Goethe and Geoffroy Saint-Hilaire, was encouraged, as Charles Darwin considered, by the progressive model of 'enlightenment' science in tandem with fundamental socio-economic changes. The clamour for social and political reforms was stimulated by a perception of human individual, social and institutional mutability.[27] Although as a manifestation of polite elite culture, science could be perceived in certain contexts as a patriotic, public-spirited and utilitarian activity – for instance, in its natural theological manifestations or through the activities of the Society of Arts – there undoubtedly were significant associations between political reformism and urban improvement. As we have seen, these are evident in Darwin's works and in the urban improvement and political activities of the Derby philosophical community.

Herbert Spencer's developmentalism and Derby scientific culture

Herbert Spencer spent much of the first three decades of his life in Derby apart from periods at Bath with his uncle, the Rev. Thomas Spencer, and working as a railway engineer.[28] In the 1840s, he worked mainly in the midlands as an engineer during the period of 'railway mania', but also spent much time engaged in abortive career moves, political campaigns and other projects. His father's teaching activities, which we have already examined, played a major role in his intellectual development. Much of the information on Spencer's life in Derby comes from the *Autobiography* he wrote as an old man, when the volumes of the *Synthetic Philosophy* already adorned many library shelves and he had been fêted on tours of the United States. It is, therefore, necessary to be cautious about accepting some of the pronouncements at face value. One of Spencer's habits, with crucial consequences for any interpretation of the origins of his philosophical ideas, was his tendency to disparage in later life the stimulus received from his provincial Derby upbringing. He was equally apt, even when it flew in the face of the obvious, as contemporaries noted, to downplay the importance of other influences, such as those of Comte and the Positivists, finding it generally difficult to 'acknowledge intellectual debts'.[29]

William George Spencer's position as one of the leading Derby scientific activists was crucial to his son's development, and Herbert claimed to have submitted every single idea that he had to his father for criticism.[30] The continuity of Derby scientific culture is evident from Thomas Mozley's observation that the Philosophical Society, of which his father was a member, was 'remarkably enduring and uniform' and that 'few societies could be called so much the same thing for half a century'. Spencer described the Society, when his father was secretary, as 'fostered by William Strutt' and consisting of 'the most cultured men in the town, chiefly medical and besides a library which

33 William George Spencer when young, self-portrait using mirrors from H. Spencer, *An Autobiography* (1904), I.

34 Herbert Spencer when nineteen, from D. Duncan (ed.), *Life and Letters of Herbert Spencer* (1908).

it accumulated, mainly of scientific books, it took in a number of scientific periodicals'. These were circulated among the membership and Herbert was to make much use of the books himself: 'Beyond occasional works of popular kinds, such as books of travel, there were works of graver kinds; and there came habitually the *Lancet*, the *British and Foreign Medical Review*, and the *Medico-Chirurgical Review*.'[31]

Derby scientific institutions provided courses of lectures on scientific and artistic subjects and a forum for the discussion, with access to libraries, museum collections and discussion groups encouraging Spencer to formulate his early philosophical theories. He read national periodicals such as *The Philosophical Magazine*, *The Lancet* and *The Mechanics' Magazine*, and was introduced to works such as J. S. Mill's *System of Logic*, which had just been bought and had not yet circulated among the members, in the Philosophical Society's library – 'a large quiet room in St. Helen's Street, to which I occasionally resorted in the Afternoon'. Likewise in the early 1840s, Spencer read Carlyle's *Sartor Resartus* and Emerson's *Essays*, which he 'greatly admired', while rejecting some of their 'mystical' qualities.[32]

The Derby Literary and Scientific Society, created around 1842, was, according to Spencer, 'a small gathering of some dozen or so, meeting once a month, reading papers and discussing them, the members being 'mostly of no considerable calibre, and the proceedings were commonly rather humdrum'.[33] However, an account of meetings by Alfred Davis, whose father, the instrument maker John Davis (1810–73), was treasurer of the Society, coupled with the evidence of the public activities of the Derby scientific activists, tend to contradict this somewhat patronising assertion, revealing the breadth of their concerns.[34] Davis reminisced that Herbert, his uncle William and his father were frequent visitors at his own father's house and active members of the Literary and Scientific Society. Members 'met informally at their respective houses to read and discuss papers on various subjects', and although they remained unpublished, the character of meetings was evident from papers written by his father on subjects including 'Iron and Steel', 'The Eye', 'William the Silent', 'Water Supply', 'Pneumatics' and 'The Great Exhibition of 1851'.[35]

Spencer's description of this Society as merely a 'small gathering' and humdrum was part of the general downplaying of his Derby background. In fact, at Derby, Spencer seriously considered beginning a periodical entitled *The Philosopher* and composed a series of politico-ethical 'Essays on Principles', which hardly suggests that he received no intellectual stimulus.[36] The Literary and Scientific Society and the Philosophical Society jointly promoted a series of lectures given by influential scholars including George Dawson (1821–76) the Birmingham Baptist minister, Ralph Waldo Emerson the American poet, John Nichol the astronomer, and Thomas Rymer Jones, Professor of Natural History at King's College, London (a former pupil of W. G. Spencer).[37] The ambitions of the Derby scientific activists at this time are indicated by a call from Alderman John Barber that the British Association for the Advancement of Science hold a Derby meeting soon.[38]

The Spencers continued to undertake scientific experiments in the spirit of the empirical tradition of the Derby philosophers and George Spencer utilised an electrical machine, an air pump and chemical apparatus for teaching science with his son's assistance.[39] Spencer also attended courses of philosophical lectures on chemistry and other subjects promoted by the Derby Mechanics' Institute and other Derby scientific institutions. He saw the chemist John Murray explode the bottom out of a chair with chloride of nitrogen with 'terrific force' and read one of his textbooks.[40] During the 1840s, the scientific lectures and experiments encouraged Spencer to think that he might support himself as an inventor and, just like some of the earlier Derby philosophers, he tried devise new scientific and engineering instruments, such as a 'velocimeter' for calculating the speed of locomotives and a scale of equivalents for engineering. Spencer also worked on ideas for a universal language which recall Darwin's speculations on vocabulary, languages and sounds in the *Plan for the Conduct of Female Education* and *Temple of Nature* and George Spencer's system of shorthand.[41] Spencer worked too on clocks, recalling the activities of Whitehurst and Strutt, created levelling appliances, and a cephelograph for making phrenological measurements. Later, a binding pin designed for holding manuscripts and music sheets was actually manufactured by a London firm, and a new type of invalid bed (designed originally for his own use) was described in the *British Medical Journal* in 1867, again echoing Strutt's adjustable medical bed for the Derbyshire Infirmary.[42]

Although Spencer's theory of evolution was not worked out until the 1850s, there is some evidence of sympathy for developmentalism in his early works, providing confirmation independent of the teleological inevitability manifest in Spencer's own account of the development of his ideas in the *Autobiography*. The members of the Philosophical Society were described as 'Darwinians' by Mozley, who remarked upon the continuity of the society and remembered that George Spencer adumbrated a deistic evolutionary theory with some similarities to his son's later ideas.[43] Educated by W. G. Spencer with his brother James and sister Anne, Mozley provides the most detailed account of his teacher's methods and evolutionary ideas, perhaps more credible for the detailed criticisms with which it is leavened. He regarded George Spencer as a highly original thinker who maintained, at that time, a deistic evolutionary theory akin, in his view, to that of Epicurus and Lucretius, but 'applied to moral existence, to opinions, habits, customs, ideas, and individual characters'. Like Lutheran Protestantism, absolute individuality and originality were its basic creeds, along with a 'conception of right and wrong' founded upon examination of the individual conscience, though it was accepted that other characters were formed by 'the almost inevitable process of custom, accretion, assimilation, and balance of forces'.[44] As Mozley presents it, his individualistic ethical, educational and political philosophy was remarkably similar to the views enunciated by Herbert Spencer in works such as *The Study of Sociology, Man Versus the State* and *Principles of Ethics*. Ultimately, however, Mozley claimed that 'the more the opinions of the original Dr. Darwin are inquired into the more will they be found to comprise all the philosophies' that emanated from the Philosophical Society that he had founded, implying that the Spencers' ideas were unoriginal.[45]

Mozley's claims annoyed Herbert Spencer, who wrote calling for a retraction, and later devoted a long appendix of his *Autobiography* to their refutation. A curious feature of the *Autobiography*, and indeed of all of Spencer's major works, is the lack of discussion of Darwin's evolutionary theory.[46] That Spencer recognised that Darwin was being suggested to be the source for his father's theories is clear from his comment that his father was aged only twelve when Darwin died and therefore he 'knew nothing of his ideas'.[47] Copies of Darwin's works were, of course, retained in the libraries of the Derby Philosophical Society and Mechanics' Institute, and Darwin's opening address, which, as we saw, sketched arguments later to appear in the *Zoonomia*, was reprinted with every edition of the library catalogue. Unless it can be believed that the secretary never read a copy of his own society's rules and library catalogue, Spencer's claim that his father knew nothing of Darwin's theories must be considered false. It is likely that Spencer's laborious attempt to rubbish the idea that his father was at one time sympathetic to evolutionary views and, by implication, that he had appropriated Darwinian ideas, was part of the general paranoia about his own intellectual independence which characterised his later years.

It is certainly true that Darwin's ideas continued to be remembered by the Derby philosophers. As Mozley said, the tone of the Philosophical Society was 'remarkably enduring, continuous and uniform; few societies could be called so much the same thing for half a century'.[48] The Derby lawyer and amateur experimenter Charles James Flack, his friend the attorney Nathaniel Edwards, and the Anglican clergyman William Ulithorne Wray, for example, kept copies of Darwin's works in their libraries.[49] Erasmus Darwin's son the physician and explorer Sir Francis Darwin, who lived near Matlock, remained a supporter of local scientific culture and Derby Philosophical Society member during the 1840s and 1850s, a subscriber to the Mechanics' Institute, and donor of specimens to the 1839 and 1843 exhibitions.[50] Busts and portraits of his father were prominent at the 1839 exhibition among those of other local scientific 'worthies', and the 1843 Museum and Natural History Society exhibition included two portraits and one bust of Darwin donated by Forester, Edward Sacheverell Chandos Pole and Francis Darwin.

According to Mozley, William Strutt was sceptical of creationism and sympathetic to evolutionism.[51] At one Philosophical Society meeting when Henry Mozley happened to discuss the creation as the only account which he could give of some natural phenomenon, Strutt 'dryly' replied that 'we know nothing about that'. On another occasion, writing to his son Edward in 1819, Strutt was pleased to hear that one of his tutors approved of Malthusian theory, remarking, 'I supposed his opponents, if the subject were in any way connected with what they call the benevolence of the Deity would controvert the elements of Euclid.' Although this hardly confirms evolutionary ideas, Strutt was here ridiculing Paley's form of creationism, which had been a direct response to his friend Darwin's brand of developmentalism. As we saw, Sylvester was also highly sceptical of religion and could never 'understand why religious persons should be unwilling to have this subject openly and fully discussed'.[52] Sympathy for evolutionary theory probably persisted among the Derby philosophers as part of the progressive enlightenment worldview manifest in Darwin's psychophysiology and geology. This would have been reinforced by Lamarckism and Geoffroyean morphology, which dominated the medical schools and the journals by the 1830s and fuelled the evolutionary debate that has been described by Desmond. The creationist *Bridgewater Treatises* were published between 1833 and 1836, and the case for the developmental worldview was made by works such as Chambers's *Vestiges of Creation* (1844). The importance of medical men in the Derby scientific community has already been indicated, and comparative anatomy, of course, featured in the medical education of surgeons and physicians.[53] The Philosophical Society library acquired Lyell's *Principles of Geology* and, although not evolutionary as such, helped to introduce Lamarck to a wider British audience.[54]

As we have seen, the importance attached to scientific education and the pedagogical methods adopted by the Spencers were inspired by dissent and the enlightenment progressivism of the Derby philosophers. Continental educational philosophies were also important, especially those of Rousseau and Pestalozzi, and Beatus Heldenmaier, the founder of one of the leading British Pestalozzian schools, at Worksop, Nottinghamshire, became a friend of W. G. Spencer and a Derby Philosophical Society member. The second chapter of Herbert Spencer's most popular work, *Education: Intellectual, Moral, and Physical* (1861) was originally intended as a review of Edward Biber's *Henry Pestalozzi and his Plan of Education* (1831), whilst Pestalozzian ideas are also drawn upon to justify children's rights in *Social Statics* (1850). Similarly, W. G. Spencer's *Inventional Geometry* (1860) used Pestalozzian methods to introduce Euclidean theory to the child.[55] Herbert Spencer adopted the Pestalozzian theory of natural education as the cornerstone of his educational scheme and in 1848 intended, with help of his father, to found a school modelled on the Worksop establishment.[56] Heldenmaier remained part of the Derby scientific circle for at least twenty years and in 1857 purchased a copy of the first edition of Spencer's *Principles of Psychology* (1855) from George Spencer before returning to Switzerland.[57]

During the 1840s when immersed in Derby scientific culture, Spencer began to adopt a developmental worldview. Despite the conservatism associated with his laissez-faire individualism by the late-Victorian period, half a century before it was regarded as politically radical. Encouraged by their nonconformity, both Spencer and his father were committed supporters of local campaigns for political reforms. W. G. Spencer supported the borough petitions against the Corn Laws and accompanied the Rev. Noah Jones, Unitarian minister at the Friargate Chapel, on a deputation arguing for the right of the Chartists to address urban meetings in 1842.[58] His son became honorary secretary of the Complete Suffrage Union, presided over by the Birmingham Quaker Joseph Sturges, in which his uncle, Thomas Spencer, was also involved.[59]

Despite Spencer's misleading claim that he had been first exposed to Lamarckian theory through reading Lyell's *Principles of Geology* in 1840, and that Lyell's arguments against Lamarck had converted him to the latter, it is highly likely that he was fully aware of evolutionary theories prior to this. In his *Autobiography*, Spencer admitted that before 1840 he had 'been cognisant of the hypothesis that the human race has been developed from some lower race'.[60] The *Letters on the Proper Sphere of Government* (1843) for Edward Miall's *The Nonconformist* argued against government interference with the 'natural' equilibrium of society, which it justified through analogy with 'nature'. Every animate creature stood 'in a specific relation to the external world in which its lives' with instincts and organs only being preserved 'as long as they are required'. If the 'animal species' was placed in a situation where 'one of their attributes is unnecessary' and activity and exercise were diminished, 'successive generations will see the faculty, or instinct or whatever it may be, become gradually weaker, and an ultimate degeneracy of the race will inevitably ensue. All this is true of man'.[61] Some acceptance of developmentalism is also implied in Spencer's article on the theory of reciprocal dependence in the animal and vegetable creations, which combined geological knowledge with natural history and chemistry.[62] This referred to 'a gradual change in the character of the animate creation' that helped to cause – and resulted from – gradual changes in composition of the atmosphere such as increases in oxygen content.[63]

As Desmond and Secord have shown, Lamarckian ideas generated much controversy during the 1830s and 1840s, and it would be extraordinary if the subject had not been discussed within Derby philosophical circles. George Spencer was a friend and correspondent of Thomas Rymer Jones (1810–80), who had been one of his pupils. Though not a supporter of evolution, it is likely that Jones would have discussed the arguments for and against evolution then occurring in medical schools and scientific institutions. Rymer Jones was a surgeon and zoologist who studied at Guy's Hospital and in Paris, qualifying in 1833. In 1836 he was appointed first professor of comparative anatomy at King's College, London and was Fullerian Professor of physiology at the Royal Institution from 1840 to 1842. Jones was a protégé of the transcendental anatomist Richard Owen, who was, in some ways, an opponent of the evolutionists, though he later became a supporter of some aspects of Darwinism. That W. G. Spencer had been introduced to Owen, probably by Jones, is clear from the fact that Spencer wrote to the anatomist from the Isle of Wight in 1841 recounting the capture of a sun-fish and suggesting that it might be useful for dissection. On his first visit to London in 1834, Herbert Spencer was taken by his father and Rymer Jones to the Zoological Society gardens in Regent's Park. In the 'Filiation of Ideas' Spencer suggested that reading Rymer Jones's *General Outline of the Animal Kingdom* (1838–41) had introduced him to German physiologist Karl Ernst von Baer's notion of progressive animal development from homogeneous to heterogeneous functions. He had then 'recognised the parallelism between it and the truth presented by low and high types of societies', the 'earliest foreshadowing of the general doctrine of evolution'.[64] Jones gave lecture courses in Derbyshire towns, and George Spencer urged the Derby Mechanics' Institute committee to invite him to give a course in 1841. At Matlock that year Jones demonstrated the structure and adaptation of fishes to aquatic life, 'the transition from aquatic to terrestrial vertebrata' and 'the metamorphosis of the tadpole', subjects highly susceptible to a Lamarckian interpretation. In 1850, again at the invitation of George Spencer, he gave two lectures 'On the Curiosities of Natural History' to the Derby Literary and Scientific Society, illustrated by specimens provided by the museum of the Derbyshire Natural History Society.

Given the 'sensational' reception of Chambers's *Vestiges* and the controversy it provoked in the mid-1840s, the Derby Literary and Scientific Society lecture programme for 1847 suggests sympathy for Lamarckism. The Derby lectures and newspaper controversy together with Secord's analysis of the reception of the *Vestiges* in Liverpool demonstrate the strength of feeling aroused by evolutionary ideas that had been re-ignited by Chambers's book during the politically charged 1840s. They also demonstrate how different religious and political groups aligned themselves for and against developmentalism, while deliberately misrepresenting each other.[65] As at Liverpool, evolutionism in Derby tended to be received more sympathetically by political reformers and some dissenters such as the Unitarians. The evolutionists (who were associated with Combe and Emerson) were accused by the Tory *Derby Mercury* of holding to the 'view that all nature, physical and moral, exhibits a law of progressive improvement'. They clung to the 'exploded' nebular hypothesis and the 'Lamarckian dogma' that 'vegetables improved into animal parts'

which became shellfish and 'from an innate principle of improvement gradually acquired fins'. These fins 'enlarged into paddles' which 'assumed the shape
of legs', and 'the fore-legs of the Chimpanzee became arms'. So in this way
'man was introduced to the world, ready to burst with principles and laws of
progressive development'.[66]

It was claimed that Lamarckism had been 'exploded' by Owen's transcendental anatomy using Karl Ernst von Baer's non-recapitulatory embryology,
which was later incorporated into Spencer's progressive evolutionary theory.[67]
The so-called 'freethinkers' were worshippers of man, yet in the view of the
editorial, the 'ridiculous dogma of progressive improvement in the organic
world' had 'long since been discarded' by Lyell, Murchison, Sedgwick and
others, 'as inconsistent with the ... tenor of geological discoveries', which exhibited 'the same perfection in animal organisation in the days of the trilobite, as
in the nineteenth century'.

Psychophysiology and phrenology

The importance of Derby scientific culture in Spencer's intellectual development is also apparent in the impetus he received from the study of phrenology.
De Giustino and Cooter have argued that phrenology facilitated the adoption
of an evolutionary psychology given its emphasis upon conceptions of race
and development, a contention amply supported by Spencer's developmental
psychology during the 1840s and 1850s, which emerged as remarkably similar
to Erasmus Darwin's evolutionary psychophysiology. Young has concluded
that Spencer's general theory of evolution and the biological, evolutionary basis
of his psychology 'grew out of the arguments for specialisation of functions
which he elaborated in the context of his phrenological interest'.[68] Acceptance
of the mechanisms of physiological organ development in one lifetime under
certain environmental conditions could facilitate an acceptance of augmented
evolutionary change through many generations.

As at Sheffield and other provincial towns, there is ample evidence that
some of the Derby philosophers, particularly medical men, took an interest in
phrenology.[69] By 1815, the Derby Philosophical Society had begun to acquire
phrenological texts and, as we shall see, the physician and scientific activist
Douglas Fox included a description of phrenology illustrated by an array of
skulls at the Derby Mechanics' Institute inaugural lectures in 1825.[70] Subsequently, phrenologists including William Bally, John Levison, William Henry
Crook, J. Q. Rumball, Thomas Beggs (1808–96), Cornelius Donovan (1820–72)
and Spencer Hall (1812–85) all visited Derby prior to 1850, with Hall eventually
choosing to settle there. A Derby Mechanics' Institute phrenology class was
established in 1832 three years before there was a class on chemistry, and many
phrenological works were bought for the library, including the junior section,

during the 1830s and 1840s.[71] The degree of support given to the subject is also apparent from the phrenological contributions made by the Strutts and others to the 1839 Mechanics' Institute and 1843 Museum exhibitions.

Spencer 'became a believer' when he attended a lecture by Johann Gaspar Spurzheim (1776–1832) at the Derby Lancasterian schoolroom in 1831, and he remembered having to overcome 'a considerable repugnance to contemplating the row of grinning skulls he had in front of him'.[72] Spurzheim visited Derbyshire in 1829, lecturing at Bakewell and at the Strutt's Belper schoolroom, when he stayed with Jedediah Strutt, nephew of William and Joseph.[73] During his political campaigns of the 1840s, Spencer published three slightly heretical articles on phrenology in John Elliotson's periodical *The Zoist*. These suggested alterations to functions of the phrenological organs of imitation, benevolence and wonder so that they became respectively organs of sympathy, sensitiveness and revivescence. The organ of wonder functioned to revive intellectual perception, and the organ of imitation recalled feelings.[74] This is not the place for detailed examination of Spencer's influential evolutionary psychology, except to note three important points.[75] His later psychology was indebted to the early phrenological studies, which coincided with his earliest active radical political campaigning and his first acceptance of developmentalism in print; and a debt to associationism was the final major ingredient. Phrenology encouraged Spencer to apply the concept of function to mental faculties and to search for the physiological manifestations of environmentally induced development.

The 'Synthetic Philosophy'

Spencer's evolutionary theory and its relationship with Charles Darwin's work is well documented and requires no lengthy analysis.[76] Inspired by the German writers such as Wolff, Goethe and von Baer, Spencer eventually defined evolution as 'an integration of matter and concomitant dissipation of motion; during which the matter passes from an indefinite, incoherent homogeneity to a definite, coherent heterogeneity; and during which the retained motion undergoes a parallel transformation'.[77] This was a universal cosmological law of development applicable to natural history and human culture alike, an audacious system utilising the terms evolution and progress interchangeably and summarised in essays during the 1850s.[78] The formation of galaxies, the solar system, the geology of the earth, natural history, human evolution, changes in government and religion, political systems, legislation, religious institutions, language, music and the evolution of painting and sculpture were all manifestations of progressive universal laws. From the 'earliest traceable cosmological changes, down to the latest results of civilisation', the 'transformation of the homogeneous into the heterogeneous' was clear.[79] Spencer traced the onset of evolution using modern industrial examples and their effects upon social development, increasing differentiation and specialisation being evident in geographical localisation and specialisation of British industry. Labour subdivisions facilitated greater and speedier production units, and the role of technology was emphasised and interpreted in terms of the universal postulate. The manifold effects of steam power were evident in the many mining and industrial applications – a celebration of the social impact of technology and industry reminiscent of Darwin, the Lunar Society and the Derby philosophers, which likewise tended to downplay the more oppressive aspects of industrialisation. The economic, technological, social and intellectual impact of the railway demonstrated this well, with its process of financing and construction, familiar to Spencer from the family investments, his experience as a railway engineer and the growth of Derby as a railway centre.[80]

Essentially, the 'Synthetic Philosophy' elaborated a kernel of ideas formulated during the 1840s and 1850s, although Spencer's *Autobiography* exaggerated the teleological inevitability of this process. In the *Principles of Psychology*, Spencer added to the associationism of Brown and Mill a conception of evolution, which followed a path already taken by Erasmus Darwin, though showing less dependency on the Lockean conception of mind, as for Spencer the human mind was no *tabula rasa* at birth but already the product of millennia of development. In the *Principles of Sociology*, Spencer 'naturalised' society, using the organic analogy to understand social development, married to a conception of social function and structural differentiation which was thought to explain the appearance and disappearance of social institutions and organisations. In 1862 (and not 1852 as commonly asserted), he coined the famous phrase 'survival of the fittest', later adopted by Charles Darwin.[81] In the *Principles of Biology*, Spencer argued that nature was red in tooth and claw, yet regarded this struggle for existence as progressive, with moral development hastened by competition, while individualism paradoxically resulted in greater social co-operation and social harmony.[82]

Education was crucial to social progress, a position inspired by Enlightenment philosophers and the Derby *savants*, and grounded in the family pedagogical tradition.[83] Learning was 'closely associated with change' and served as the mechanism for moral and intellectual improvement. Society, partly through education, was 'always fitting men for higher things, and unfitting them for things as they are'. According to Spencer's evolutionary ethics, already clear in the *Social Statics* of 1850, humanity moved progressively towards a state of moral perfection and complete individual harmony with society. Evil would only result from the 'want of congruity between the faculties and their spheres of action' – a position Spencer made clear in the *Letters on the Proper Sphere of Government* of 1843. Social evolution ensured that human faculties would be moulded into complete fitness for society, while immorality would eventually disappear, 'so surely must man become perfect'.[84]

Erasmus Darwin and Herbert Spencer: a comparison

There are a number of striking similarities and parallels between the evolutionary theories of Erasmus Darwin and Herbert Spencer which are especially interesting given their Derby philosophical connections, though, of course, there were many differences between the two, such as Spencer's desire to create a 'science' of society. Both favoured terms such as 'progress' and 'progressive' to describe developmental phenomena rather than 'evolution or 'evolutionary', Spencer only adopting the latter during the 1850s on the basis that the former was too anthropocentric.[85] Inspired by their progressive, enlightenment model of natural science, both embraced a developmental worldview, applying it to psychological development, the natural world and social change, Spencer holding that 'the processes of modification constituting adaptation of organic structures' was 'rendered quite comprehensible by reference to the analogous social processes'.[86] Darwin applied his conception of progressive evolution to nature to explain the changes in plants and animals documented by fossils and geology, and to society, in which, with changes facilitated by education, social harmony was growing. It is hard not to believe that Spencer was stimulated by Darwin's psychology, while phrenology encouraged him to develop a system of individualistic psychophysiology remarkably similar to that of Darwin.

Both philosophers married conceptions of linear development and cyclical change to explain the presence of phenomena not obviously progressive in tendency. Both used similar arguments to their evolutionary theories, such as the progress of science, technology and industry, geological change, the nebular hypothesis, and recapitulationism. Darwin regarded the solar system as ageing towards total collapse, when it would be reborn with other systems, and Spencer argued that gaseous matter and star clusters demonstrated the life cycle of stars and planets. Both saw individual organic evolution as replaying the collective evolutionary experience of other species, and Spencer argued that education should proceed from the simple to the complex because 'a child's mind repeats the unfolding of the mind in the human race'.[87]

Darwin believed in a somewhat elitist progressive model of scientific development, with advance being led by the high priests of science, the philosophers. As he put it, 'some must think and others labour'. Both Spencer and Darwin saw the progress of reason as one great overarching process and believed in the fundamental unity and interrelationship of the sciences. Hence, both were fascinated with electricity and electrical experiments, which were explicitly associated with investigations into the nature of biological change or social development. Unlike Charles Darwin, both were mechanical inventors with special interests in the exploitation of technology in manufacturing and industry, as well as experimenters in both the physical and biological sciences, which was reflected in their evolutionary theories. Therefore the problems of biological adaptations were considered to be similar to those posed by mechanical inventions, with, in each case, a number of contrivances having to operate harmoniously to carry out certain functions in the context of particular external environments. This blurring of the boundary between the 'natural' and human contrivance was emphasised by Spencer when he referred to his father's 'habit of speculating about causes' and how he learnt that 'the discovery of cause is through analysis – the pulling to pieces of phenomena for the purpose of ascertaining what are the essential connections among them'.[88] It is also evident in his influential 'organicist' sociology, where parallels are constantly drawn between organisms and social institutions.[89] With his father he had 'many times ... assisted in experiments with the air-pump and the electrical machine; so that ideas of physical causation were repeatedly impressed on me', and this 'flourished the more in the absence of the ordinary appeals to supernatural causes'. When discussing the origination of the synthetic philosophy he specifically referred to 'the putting together and the taking to pieces' of ideas, just as he had put together and taken apart clock mechanisms. Both Spencer and Darwin were fascinated by the adaptive mechanisms of living organisms, and, partly as a result of his phrenological studies, Spencer began to consider phenomena in terms of their function. He applied the general concept of function to characteristics of biological organisms and to features of human society such as social institutions.[90]

Although phrenology, enlightenment educational philosophies and the work of George Lewes were clearly important influences on Spencer's evolutionary psychology up to the 1850s, it appears likely that – probably through his father – Darwin's individualistic psychology was another significant influence.[91] Darwin's philosophy proceeded from the individual to the social, with individual development, a product of personal education and social interaction, serving as the foundation of society. This also remained Spencer's position, despite the severe opposition he later encountered, and it too was grounded upon individualistic psychology informed by the British associational psychology. For both philosophers, inherited characteristics partly ensured progressive natural development, and Spencer remained a Lamarckian until his death in the face of growing opposition. For both, the pressure of population forced competition which tended to ensure progressive development; indeed, the problem of population growth fascinated each man. Spencer argued that there was a rough equilibrium between animal populations and food supplies, with fertility decreasing as civilisation advanced. Thus, both Darwin and Spencer considered that the evils of natural and social debilitation and death were unfortunate but necessary components of universal progress, part of the 'survival of the fittest'.[92]

As we have seen, the Spencers and Erasmus Darwin were educational theorists and teachers, recalling the concerns of enlightenment philosophers such as Locke, Rousseau, Pestalozzi and Richard Lovell Edgeworth. Darwin published a work on female education, and Spencer became one of the most widely read educational theorists in the world. Both took a dynamic, experiential view of knowledge acquisition, constructed through individual activity, a dialectical process where ideas were not passively imbibed but tested by argument or in action in the schoolroom or laboratory. Both saw science as the key to social advancement, placing special emphasis on scientific education (indeed, members of the Darwin family attended the Spencer family school). The Derby philosophers were instrumental in the foundation of educational institutions such as the Lancasterian School, Mechanics' Institute, Arboretum and county museum, which promoted 'rational recreation' and scientific education for all social classes. Darwin attacked the domination of impractical classicism in education, which ought to promote the 'comparison of things with each other' and the examination of 'ideas of causes and their effects'. His son Charles had chosen to leave Oxford because 'the vigour of the mind languished in the pursuit of classical elegance ... and sighed to be removed to the robuster exercises of the medical schools of Edinburgh'.[93] Darwin argued that girls should be given an education that included vigorous exercise, study of the natural sciences and excursions to view the manufactories of the industrial midlands. Likewise, Spencer, following the methods developed by his father and the Pestalozzians, attacked at length the practice of learning classical languages merely for the purpose of polite exhibition. Science and mathematics needed be learnt by experiential means proceeding from concrete examples to the more abstract, with the subjects of education being chosen on functional and utilitarian terms, through analysis of the requirements of the individual and society.[94]

Finally, it is certainly possible that Spencer's early conception of the importance of sympathy by imitation may have been influenced by Darwinian ideas and phrenology.[95] According to Darwin's theory, sympathy was central to social and individual psychological development, which ultimately saw the progressive increase of wisdom and social happiness. Likewise, Spencer placed sympathy at the heart of his evolutionary ethics as it had been central to his 1840s psychology. The harmony of civilisation was increasing as sympathy facilitated social congruity. Experience of pain and pleasure was at the heart of Spencer's psychology and ethics, as it had been of Darwin's evolutionary psychophysiology, and these experiences moulded individuals into a state of greater social congruity as character structures were shaped by society and contrawise.[96] Imitation and sympathy brought greater altruism and harmony, with virtues being naturalised through internalisation. The 'remoulding of human nature into fitness for the requirements of social life' would eventually make 'all needful activities pleasurable', and all activities at variance with these requirements would become displeasurable. Activities therefore previously engaged in from 'dislike' or 'a sense of obligation' would instead 'be done with immediate liking, and the things desisted from as a matter of duty [would] be desisted from because they [were] repugnant'.[97] Spencer, like Darwin, was careful to emphasise that such a state would only exist in the far distant future, due to the continuing power of war and militarism. However, even the most drug-mellowed hippie of the 1960s or 1970s could scarcely have framed such an optimistic view of eventual social harmony.

Conclusion

According to Peter Bowler it is incorrect to allow, as opponents of Charles Darwin did, that Erasmus Darwin and Lamarck had much of an impact upon nineteenth-century evolution. To a large extent, he notes, 'this acclaim was generated as a by-product of the opposition to Charles Darwin's theory of natural selection.' As in the case of Buffon, it was Samuel Butler's efforts (1879) to discredit Charles Darwin (and those of the Neo-Lamarckians) which led to an exaggerated view of the earlier worker's contributions. Bowler accepts that Erasmus Darwin and Lamarck were important because they attempted to explain organic change, but 'we should not be misled by superficial similarities into assuming that they contributed directly to the Darwinian revolution.'[98] Likewise, one sympathetic account of Spencer's philosophy argues that he was pointed towards association philosophy by Mill, which he combined with a law of development from von Baer and a theory of organic transformation from Lamarck. It was from Lamarck that Spencer took up the notion of structural and functional change in organisms over successive generations and 'expanded it into a mechanism to produce progress in mind and society'.[99] It should be clear from the analysis in this chapter, from Desmond's work on the association between Lamarckism, comparative anatomy and political radicalism in the metropolitan context, and from Secord's study of reception of the *Vestiges*, that these views require revision.[100]

As we have seen, Spencer's own account of his intellectual development insufficiently acknowledges the debt that he owed to Derby scientific culture. His evolutionary philosophy was stimulated in important ways by his association with the Derby philosophers, some of whom continued to favour Erasmus Darwin's evolutionary worldview with its biological elements, developmental association psychology, evolutionary geology and cosmological developmentalism. Darwin's Derby scientific protégés, 'Darwinians' such as Forester, Brookes Johnson and the Strutt brothers, shared his confidence in scientific and industrial progress and his 'developmental worldview', while taking the lead in campaigns for political reform and encouraging other Derby philosophers, such as W. G. Spencer, in their own deistic evolutionary worldviews. It was also through Derby scientific culture that the Spencers were introduced to phrenology and through these predominantly reformist and dissenting networks that their experimental educational methods thrived. The Spencers' adoption of Pestalozzian 'naturalistic', experiential methods, for instance, was also encouraged by the progressive Enlightenment outlook of the Derby philosophers.

Analysis of Derby social, economic and scientific culture demonstrates how the analogies between cultural and biological evolution which shaped the 'Lamarckian' aspects of Darwin and Spencer's evolutionary theories were encouraged.[101] Their status as great naturalists and empirical observers notwithstanding (despite some claims to the contrary), this elision between culture and biology is reflected in Darwin's and Spencer's roles as inventors, artists, poets, political agitators, engineers, urban improvers and industrial apologists, which differentiate them from Charles Darwin. As Secord and Desmond have shown so well, it was this 'Lamarckism' that by the 1860s helped, through its progressive enlightenment character, to accustom the Victorian public to developmentalism in a way that the 'Darwinian' theory of natural selection by random mutation alone would probably not have done.

Far from being disembodied, placeless, abstract conceptions, the evolutionary theories of Erasmus Darwin and Herbert Spencer, which through the latter had a global impact, were rooted, shaped and developed in the social, landscape and industrial character of the English midland provinces and the Derby philosophical community.

Notes

1 Although some changes have been made, this chapter is based upon P. Elliott, 'Erasmus Darwin, Herbert Spencer and the origins of the evolutionary worldview in British provincial scientific culture', *Isis*, 94 (2003), 1–29, and I am grateful to the editors for permission to use the material here.

2 A. Seward, *Memoirs of the Life of Dr. Darwin* (London, 1804); D. King-Hele (ed.), *The Collected Letters of Erasmus Darwin* (Cambridge: Cambridge University Press, 2007); M. McNeil, *Under the Banner of Science: Erasmus Darwin and his Age* (Manchester: Manchester University Press, 1987); D. King-Hele, *Erasmus Darwin: A Life of Unequalled Achievement* (London: Giles de la Mare, 1999).

3 R. G. Perrin, *Herbert Spencer: A Primary and Secondary Bibliography* (New York: Garland, 1993), 29.

4 J. D. Y. Peel, *Herbert Spencer: The Evolution of a Sociologist* (New York: Basic Books, 1971), 134; D. Freeman, 'The evolutionary theories of Charles Darwin and Herbert Spencer', *Current Anthropology*, 15 (1974), 211–37; M. T. Ghiselin, 'Two Darwins:

history versus criticism', *Journal of the History of Biology*, 8 (1976), 121–32; J. Hodge and G. Radick (eds.), *The Cambridge Companion to Charles Darwin* (Cambridge: Cambridge University Press, 2002).

5 A. Desmond, *The Politics of Evolution: Morphology, Medicine and Reform in Radical London* (Chicago: Chicago University Press, 1989); J. A. Secord, *Victorian Sensation: The Extraordinary Publication, Reception and Secret Authorship of the Vestiges of the Natural History of Creation* (Chicago: Chicago University Press, 2000).

6 J. H. Rieuwerts, 'Derbyshire lead mining and early geological concepts', *Bulletin of the Peak District Mines Historical Society*, 9 (1984), 51–100; M. Craven, *John Whitehurst: Clockmaker and Scientist, 1713–1788* (Ashbourne: Mayfield, 1997); T. Ford and J. H. Rieuwerts (eds.), *Lead Mining in the Peak District*, fourth edition (Ashbourne: Landmark, 2000); King-Hele, *Erasmus Darwin*, 89; T. Brighton, *Discovery of the Peak District* (Chichester: Phillimore, 2004).

7 E. Darwin, *Phytologia; Or, the Philosophy of Agriculture and Gardening*, first edition (London, 1800), 559–60.

8 *Derby Mercury* (26 November 1795); reproduced in King-Hele (ed.), *Letters of Erasmus Darwin* , 486–8.

9 E. Darwin, *The Botanic Garden: A Poem in Two Parts*, second edition (London, 1791); *The Economy of Vegetation*, Part II, second edition (London, 1790), canto 1, lines 5–6.

10 E. Darwin, *The Loves of the Plants* (1789), note 7; J. Harrison, 'Erasmus Darwin's view of Evolution', *Journal of the History of Ideas*, 32 (1971), 247–64; McNeil, *Under the Banner of Science*, 88; J. Browne, 'Botany for gentlemen: Erasmus Darwin and the *Loves of the Plants*', *Isis*, 80 (1989), 593–621; A. B. Shtier, *Cultivating Women, Cultivating Science: Flora's Daughters and Botany in England, 1760–1860* (Baltimore: Johns Hopkins University Press, 1996), 22–7.

11 E. Darwin, *The Temple of Nature; or, the Origin of Society* (London, 1803), canto 1, lines 233–46, 247–50, 295–302.

12 Darwin, *Temple of Nature*, additional note VIII, 33–6.

13 E. Darwin, *Zoonomia; or, the Laws of Organic Life*, first edition, 2 vols. (London, 1794–96), I, 13–14, 244.

14 Darwin, *Temple of Nature*, canto 2, lines 33–6, 43–4, 163–6.

15 Darwin, *Temple of Nature*, additional note VIII, 37–8.

16 Darwin, *Zoonomia*, I, 1–2.

17 R. Porter, 'Erasmus Darwin: doctor of evolution?', in J. Moore (ed.), *History, Humanity and Evolution: Essays for John C. Greene* (Cambridge: Cambridge University Press, 1989), 39–69, p. 46.

18 E. Halévy, *Growth of Philosophic Radicalism* (London: Faber & Faber, 1934), 440–1; R. M. Young, 'The role of psychology in the nineteenth-century evolutionary debate', in C. Chant and J. Fauvel (eds.), *Darwin to Einstein: Historical Studies on Science and Belief* (Harlow: Longman, 1980), 155–78; P. Elliott, '"More subtle than the electric aura": Georgian medical electricity, the spirit of animation and the development of Erasmus Darwin's psychophysiology', *Medical History*, 52 (2008), 195–221.

19 Darwin, *Zoonomia*, first edition, I, 30–2.

20 Darwin, *Zoonomia*, I, 49–50.

21 Darwin, *Zoonomia*, third edition (1801), II, 270, quoted in Porter, 'Erasmus Darwin', 52–3.

22 Darwin, *Temple of Nature*, canto 4, lines 223–4, 233–6.

23 *Derby Mercury* (9 August 1792); King-Hele, *Erasmus Darwin*, 274.

24 P. Lawrence, 'Heaven and earth – the relation of the nebular hypothesis to geology', in W. Yourgrau and A. D. Breck (eds.), *Cosmology, History and Theology* (New York: Plenum Press, 1977), 253–81; S. Schaffer, 'Herschel in Bedlam: natural history and stellar astronomy', *British Journal for the History of Science*, 45 (1980), 211–39; M. Ogilvie, 'Robert Chambers and the nebular hypothesis', *British Journal for the History of Science*, 30 (1975), 214–32; S. Brush, 'The nebular hypothesis and the evolutionary worldview', *History of Science*, 25 (1987), 245–78; S. Schaffer, 'The nebular hypothesis and the science of progress', in Moore (ed.), *History, Humanity and Evolution*, 131–64.

25 Darwin, *Economy of Vegetation*, I, canto 1, lines 103–14.

26 Darwin, *Temple of Nature*, note, pp. 197–9.

27 C. Darwin, *On the Origin of Species* ([1859]; Ware: Wordsworth, 1998), 370–7.

28 H. Spencer, *An Autobiography*, 2 vols. (London: Williams and Norgate, 1904); A. Low-Beer (ed.), *Herbert Spencer* (London: Collier-Macmillan, 1969); Peel, *Herbert Spencer*; S. Andreski (ed.), *Herbert Spencer: Structure, Function and Evolution* (London: Joseph, 1971); Perrin, *Herbert Spencer*; M. Francis, *Herbert Spencer and the Invention of Modern Life* (Chesham: Acumen, 2007).

29 S. Eisen, 'Herbert Spencer and the spectre of Comte', *Journal of British Studies*, 7 (1967), 48–67.

30 Spencer, *Autobiography*, I, 43. Spencer's correspondence reveals that he submitted drafts of his works to his father for his opinions, D. Duncan (ed.), *The Life and Letters of Herbert Spencer* (London: Methuen, 1908), 64–6, 77, 81, 135; P. Elliott, '"Improvement always and everywhere": William George Spencer (1790–1866) and mathematical, geographical and scientific education in nineteenth-century England', *History of Education*, 33 (2004), 391–417.

31 Spencer, *Autobiography*, I, 87.

32 Spencer, *Autobiography*, I, 87–8, 241.

33 Spencer, *Autobiography*, I, 87–8.

34 M. Craven, *Derbeians of Distinction* (Derby: Breedon, 1998), 68–9; M. Craven, *Derby: An Illustrated History* (Derby: Breedon, 1988), 164.

35 S. Simpson, 'Reminiscences of Alfred Davis', *Nottinghamshire and Derbyshire Notes and Queries*, 107 (1892–98), 398.

36 Spencer, *Autobiography*, I, 239–40.

37 *Derby Mercury* (1–29 September, 24 November, 1, 8 December 1847; 11 April, 9 May 1849; 17 April and 1, 8, 15, 22, 29 May 1850); *Derby Mercury* (30 October 1850); A. Briggs, *Victorian Cities* (Harmondsworth: Penguin, 1968), 184–240.

38 *Derby Mercury* (29 May 1850). This may have been partly inspired by the knowledge that the British Archaeological Society were due to meet at Derby in August 1851.

39 Spencer, *Autobiography*, I, 86.

40 Spencer, *Autobiography*, I, 86; *Rules and Catalogue of the Library Belonging to the Derby Philosophical Society* (Derby, 1835), 60.

41 W. G. Spencer, *A System of Lucid Shorthand* (London, 1894).
42 Spencer, *Autobiography*, I, appendices C, D, G, H, I, vol. II, appendix D.
43 T. Mozley, *Reminiscences, Chiefly of Oriel College and the Oxford Movement*, 2 vols. (London, 1882), I, 146.
44 Mozley, *Reminiscences of Oriel College*, I, 151–3, 145–6.
45 T. Mozley, *Reminiscences, Chiefly of Towns, Villages and Schools*, 2 vols. (London, 1882), II, 172–3.
46 He is mentioned in passing in H. Spencer, 'The Factors of organic evolution', H. Spencer, *Essays, Scientific, Political and Speculative*, 3 vols. (London, 1891), I, 389–465, pp. 390, 417.
47 Spencer, *Autobiography*, I, appendix K, 459; Mozley, *Reminiscences of Oriel College*, I, 173.
48 Mozley, *Reminiscences, Chiefly of Towns*, II, 171–3.
49 *Derby Mercury* (26 September 1837); *Catalogue of the ... General and Law Library of Mr. Edwards, Solicitor* (Derby, 1826); *Catalogue of the ... Library of the Late Rev. William Ulithorne Wray* (Derby, 1808). Edwards, for instance, owned copies of Darwin's *Zoonomia, Botanic Garden* and *Temple of Nature*.
50 F. Darwin, *Travels in Spain and the East, 1808–1810*, ed. F. D. S. Darwin (Cambridge: Cambridge University Press, 1927).
51 Strutt's extensive fossil and mineral collection was utilised by J. Farey, *A General View of the Agriculture and Minerals of Derbyshire*, 3 vols. (London, 1811–17).
52 E. Strutt, Private memoir of William Strutt, Derbyshire Record Office, D2912, 40.
53 Desmond, *Politics of Evolution*, chs. 2 and 3; Secord, *Victorian Sensation*.
54 C. Lyell, *Principles of Geology*, ed. and introduced J. Secord (London: Penguin, 1997), xxx.
55 H. Spencer, *Social Statics; or the Conditions to Human Happiness Specified* (New York, 1873); Low-Beer, *Spencer*, 120–33; W. G. Spencer, *Inventional Geometry* (London, 1892).
56 F. A. Cavenagh (ed.), *Herbert Spencer on Education* (Cambridge: Cambridge University Press, 1932); S. Tomlinson, 'From Rousseau to evolutionism: Herbert Spencer on the science of education', *History of Education*, 25 (1996), 235–54. The *Education* had sold over 50,000 copies by 1900, become a basic textbook for teacher training in Britain, been translated into fifteen languages, gone through thirteen editions in France alone, and in the United States, laid the foundations for the progressivist educational movement, L. Cremin, *The Transformation of the School: Progressivism in American Education, 1876–1957* (New York: Random, 1961).
57 Mss. account book of the Derby Philosophical Society, DLSL, BA 106; W. G. Spencer, account book, 180; 'Der Pestalozzianer Beat: Rud. Friedrich Heldenmaier als lehrer und erzieher in Berlin, Worksop (England) und Lausanne', *Pestalozzianum Jahrgang*, 46 (1949), 907–8; P. Elliott and S. Daniels, 'Pestalozzianism, natural history and scientific education in nineteenth-century England: the Pestalozzian institution at Worksop, Nottinghamshire', *History of Education*, 34 (2005), 295–313.
58 *Derby Mercury* (30 January 1839, 14 September 1842).
59 Spencer, *Autobiography*, I, 217–21; Peel, *Herbert Spencer*, ch. 2.
60 Spencer, *Autobiography*, I, 175–7.

61 H. Spencer, *Letters on the Proper Sphere of government*, quoted in *Autobiography*, I, 210.

62 H. Spencer, 'Remarks on the theory of reciprocal dependence in the animal and vegetable creations as regards its bearing on palaeontology', *Autobiography*, I, 533–7.

63 H. Spencer, 'Progress: its law and cause' (1857), in *Essays*, I, 8–62.

64 H. Spencer, 'Filiation of ideas', in Duncan (ed.), *Life and Letters of Herbert Spencer*, 534–59.

65 Secord, *Victorian Sensation*, 191–221.

66 *Derby Mercury* (29 September 1847).

67 D. Ospovat, 'The influence of Karl Ernst von Baer's embryology, 1828–1859: a reappraisal in light of Richard Owen's and William B. Carpenter's palaeontological application of von Baer's law', *Journal of the History of Biology*, 9 (1976), 1–28; Desmond, *Politics of Evolution*, chs. 7 and 8.

68 D. de Giustino, *Conquest of Mind: Phrenology and Victorian Social Thought* (London: Croom Helm, 1975), 51–5; R. Cooter, *The Cultural Meaning of Popular Science: Phrenology and the Organisation of Consent in Nineteenth-Century Britain* (1984), 9–10; R. J. Richards, *Darwin and the Emergence of Evolutionary Theories of Mind and Behaviour* (Chicago: Chicago University Press, 1987), 250–3; R. M. Young, *Mind, Brain and Adaptation* (Cambridge: Cambridge University Press, 1971), 161.

69 I. Inkster, *Scientific Culture and Urbanisation in Industrialising Britain* (Aldershot: Ashgate, 1997), paper III, 273–9.

70 D. Fox, *Notes of the Lectures on Anatomy and Chemistry, delivered by Mr. Douglas Fox* (Derby, 1826).

71 *Derby Mechanics' Institute Catalogue of Books comprised in the Library* (Derby, 1851).

72 Spencer, *Autobiography*, I, 200.

73 *Derby Mercury* (19, 26 August, 22 September 1829).

74 H. Spencer, 'A new view of the functions of imitation and benevolence', *The Zoist*, 1 (1843); 'On the situation of the organ of amativeness' and 'A theory concerning the organ of wonder', *The Zoist*, 2 (1844); G. Denton, 'Early psychological theories of Herbert Spencer', *American Journal of Psychology*, 32 (1921), 5–15; Young, *Mind, Brain and Adaptation*, ch. 5.

75 H. Spencer, *The Principles of Psychology*, first edition (London, 1855), second edition, 2 vols. (London, 1870–72).

76 Richards, *Darwin and the Emergence of Evolutionary Theories*.

77 H. Spencer, *First Principles*, sixth edition (London, 1900), 367; E. Clodd, *Pioneers of Evolution from Thales to Huxley* (London, 1897), 161–85; Duncan (ed.), *Life and Letters of Herbert Spencer*.

78 H. Spencer, 'The development hypothesis', 'Progress; its law and cause', *Essays*, I, 1–7, 8–62; 'Letters on the proper sphere of government', *Nonconformist* (June–December 1842).

79 Spencer, 'Progress; its law and cause', 10; Perrin, *Herbert Spencer*, 34–7, 43–4.

80 Spencer, 'Progress; its law and cause', 56–8.

81 The social and progressive implications were apparent and Huxley thought that

Darwin had made 'an unlucky substitution', D. B. Paul, 'The selection of the survival of the fittest', *Journal of the History of Biology*, 21 (1988), 411–24, p. 419.

82 H. Spencer, *Principles of Biology*, 2 vols., first edition (London, 1864–67), I, 331–475.

83 B. Simon, *The Two Nations and the Education Structure, 1780–1870* (London: Lawrence & Wishart, 1981), 17–71; Low-Beer, *Spencer*, 1–28; Tomlinson, 'From Rousseau to evolutionism', 235–54.

84 Spencer, *Social Statics*, 59–60, 341; *Principles of Psychology*, I, 278–88; *Principles of Ethics*, 2 vols. (London, 1897), I, 79–82; Richards, *Darwin and the Emergence of Evolutionary Theories*, 295–313.

85 Perrin, *Herbert Spencer*, 34–5.

86 Spencer, 'Filiation of ideas', 557.

87 Spencer, 'Filiation of ideas', 544; Cavanagh (ed.), *Herbert Spencer on Education*, 61–113.

88 Darwin, *Zoonomia*, II, 416 and note 28; Spencer, 'Filiation of ideas', 534–5.

89 H. Spencer, *Principles of Sociology*, 3 vols. (London, 1896) , I.

90 Spencer, 'Filiation of ideas', 535, 548.

91 Young, 'The role of psychology', 155–78.

92 Darwin, *Temple of Nature*, canto 4, lines 365–410; H. Spencer, 'A theory of population deduced from the general law of animal fertility', *Westminster Review*, 2 (1852), 32–3; Paul, 'The selection of the survival of the fittest', 411–24.

93 Cavanagh (ed.), *Herbert Spencer on Education*, 44–60; C. Darwin, *An Experiment Establishing a Criterion between Mucaginous and Purulent Matter ... with a Life of the Author* [probably by E. Darwin] (Lichfield, 1780), 129–32.

94 E. Darwin, *A Plan for the Conduct of Female Education in Boarding Schools* (Derby, 1797), 3–60.

95 Spencer claimed that Adam Smith had not been his source, 'Filiation of ideas', 537.

96 D. Dennet, defending an 'ultra-Darwinist' position, has accused Spencer of 'Panglossianism' in seeking to drive a simple path to altruism from evolution. According to Dennet 'Spencer in our terms was an egregiously greedy reductionist, trying to derive "ought" from "is" in a single step', D. Dennet, *Darwin's Dangerous Idea* (Penguin, 1994), 463–4, 466.

97 Darwin, *Zoonomia*, II, 245–6; Spencer, *Principles of Ethics*, I, 183, 184, 242–57.

98 P. Bowler, *Evolution: The History of an Idea* (Fontana, 1989), 81.

99 R. Smith, *The History of the Human Sciences* (London: Fontana, 1997), 461.

100 Desmond, *Politics of Evolution*.

101 That is, evolution through the inheritance of acquired characteristics.

10

Women and Derby scientific culture, *c.*1780–1850

Introduction

'An Experiment on a Bird in the Air Pump' and 'The Orrery' by Joseph Wright, as we have seen, are wonderful evocations of the excitement and drama of philosophical discovery; however, they also represent domestic scenes with women and children present, and demonstrate how important the female audience was in Georgian scientific culture. Historical analysis of women's economic and cultural experiences during the eighteenth and nineteenth centuries has, of course, been dominated by the narrative of widening separate spheres. In *Family Fortunes*, their classic study of English middle-class family life, Leonore Davidoff and Catherine Hall argued that an ideology of the private domestic sphere became prominent between 1780 and 1850 and circumscribed female behaviour, while males were able to enjoy life in the public sphere of government and the arts. One of the main reasons for this is thought to be the shift from domestic industry and shared participation of the sexes in most aspects of agriculture and domestic production, which characterised the early modern economy, to the separation between workplace and home encouraged by the development of manufactories.[1]

Given that women were such prominent supporters of the balls, assemblies, concerts, theatrical performances, charitable and improvement ventures and literary culture of provincial urban society, it is necessary to consider if their previous neglect by historians of urban scientific culture reflected a genuine absence from such activity, and if so why this culture became so gender-differentiated. This is underscored by the fact that dissent has been associated with female participation in literary and scientific culture and campaigns for female suffrage and the prominence of dissenting families in Derby scientific culture. There is a rich record of material for elite families which reveals much about cultural life for middling sort and middle-class women in Derby. As we shall see, there is evidence that despite institutional exclusion and educational disadvantage, some middle-class women were enthusiastic for science and gained considerable scientific knowledge. Therefore, although examining women in a separate chapter rather than throughout the book has its difficulties, it allows us to collect and assess more effectively the evidence for female participation in Derby scientific culture.[2]

Taking Derby as a case study, this chapter examines the degree to which the experience of women in scientific culture between 1780 and 1850 conforms to the separate spheres model. After considering the importance of women in enlightenment Newtonian science it examines how much they participated in public culture and politics in Derby in the period. The chapter then explores the kind of scientific education offered to women in the period and their responses to it, before concluding with an analysis of female involvement in Derby scientific institutions.

Women and enlightenment scientific culture

Some historians such as Linda Colley, Amanda Vickery and Amanda Foreman have argued that the separate spheres theory requires revision. Colley concluded that 'there seems no doubt that insofar as the position of women changed in Britain in the half-century after the American War, the effect was to increase – rather than diminish – their opportunities for participating in public life.'[3] Foreman has shown in her biography of Georgiana, Duchess of Devonshire (1757–1806) that the position of women in politics and society was more complicated than this and suggests that a 'model of interlocking

35 Georgiana, Duchess of Devonshire, engraved by J. Newton (1779).

spheres' might be more appropriate, as this recognises 'the flexibility of social conventions of the era' and how 'women's lives reflected the shifting patterns of society and were equally susceptible to the pressures caused by class, locality, economy and age.'[4] Of course, aristocratic women such as Georgiana were presented with special opportunities and could, for example, stretch the limits of the domestic sphere by initiating public charitable acts. In London, Lichfield, Bath and other centres, as John Brewer and others have demonstrated, women played an important part in polite culture as writers, artists, intellectuals and consumers of musical performances, the theatre, works of art and literature.[5]

There is no doubt that some women were enthusiastic participants in enlightenment science, although scientific ideas and practices helped to reinforce gender differences. Margaret Jacob and Betty Jo Teeter Dobbs have contended that 'throughout the western world, Newtonianism emerged in the hands of its popularisers as the first body of scientific knowledge also specifically addressed to women', as the careers of continental natural philosophers such as Laura Bassi (1711–88) and Emilie Chatelet (1706–49) appear to show.[6] Women were important consumers of scientific ideas and commodities, including texts such as Elizabeth Carter's *Newton's Philosophy Explained for the Use of Ladies* (1739), or the *Ladies' Diary* periodical which contained mathematical problems and astronomical data.[7] Women were encouraged to attend lectures on natural philosophy such as those given by John Arden at Derby during the 1740s, although their participation in science as polite entertainment or accomplishment was emphasised rather than the utilitarian improvement of their education. Patrons were reassured that the sciences were 'very entertaining, and frequently the subject of conversation (of which the fair sex constitute so agreeable a part)', and so for the ladies 'all uncommon terms will be … avoided and always explained' and lectures given 'in the most easy, and intelligible manner'.[8]

On the other hand it has generally been recognised that women were increasingly excluded from formal scientific culture from the later seventeenth century, a process exacerbated by the increasing professionalisation, specialisation and institutionalisation of science, which also facilitated the exclusion of many male amateur natural philosophers and natural historians. This was part of a general pattern which saw the exclusion of many women from certain activities in the public sphere and occupations traditionally regarded as female preserves, such as dairy work and midwifery, which became 'scientised' and professionalised. The increasing tendency of science to be regarded as the exemplar of 'objective' mathematical rationality reinforced gender preconceptions and social divisions through the emphasis upon the domination and exploitation of 'nature'. Anatomical and physiological representations of female brain size and skeletal structure, for instance, gave ideological reinforcement to the process. And so, although circumscribed female participation in scientific culture was encouraged, scientific practices and ideology helped to reinforce the separation of the spheres.[9]

Women in Derby culture and politics

Female participation in urban improvement and public culture also demonstrates how the variety of women's experiences connot be summarised too rigidly in terms of the separate spheres interpretation, underscoring why alternative models have been proffered, such as that of 'interlocking spheres'. As a group women have tended to be ignored in the history of Derby, despite the fact that by 1831 they constituted over 52% of the population, the local textile industry in particular being a significant employer of female labour.[10] Public cultural and political activity reveals that while excluded from elections and the corporation, middle-class women played an important public role in urban society through religious, educational, charitable and improvement ventures. By the 1820s some women were directors of the Derby Gas Company, subscribers to the Derbyshire Infirmary and supporters of urban improvement measures. Though the majority would have acted with husbands or male relatives, some seem to have acted independently. Hundreds of women were, as we have seen, involved in debates over urban improvement matters such as the Nun's Green dispute of the 1790s, or as subscribers to improvement ventures. They were actively invited by men to take part in the debates over town improvement, it being acknowledged that they had an equal stake in town society on the matter.

In Derby as elsewhere, women, of course, participated in religious ceremonies, musical and theatrical performances, assemblies and philosophical lectures. Though they are not listed as members of Derby book clubs, or on many book subscription lists, they clearly would have had access to borrowed volumes, as records from other places demonstrate, even if circumscribed through husbands or male relatives, who probably opened subscriptions on their behalf. Where absolute numbers mattered in public culture, such as charitable or political petitions, hundreds of women were encouraged to sign independently in addition to husbands and male relatives. Elite women were presented with opportunities unavailable to the women of the labouring classes through wealth, education and social connections. Until the construction of the Pickford and Ferrers assembly rooms in the 1760s, there were two sets of assemblies, one for the gentry and one for the other townsfolk, distinctions maintained through the norms of social convention as much as overt rules and charges. In the 1740s, gentry assemblies were presided over by Mrs Barnes, known as 'Blowzabella', and from 1752 by Lady Ferrers. Social class usually triumphed over 'separate spheres'. In the 1730s, the rules for the ladies' assembly in Derby ordered that no attorny's clerk could be admitted, nor any shopkeepers, and Mrs Barnes remarked to Lady Ferrers in 1752 that 'Trade never mix'd with us Ladies.'[11]

In political terms, while the reform campaigns of the 1790s and first few decades of the nineteenth century were led by men, they received considerable support from women, even if this was confined to encouragement in the private sphere. Middle-class women, particularly Unitarians, had greater opportunity to express themselves in the political arena, and those who had independent incomes such as wealthy widows were better able to make public contributions. Thirteen of the female subscribers to the Derby bridges during the 1790s, for instance, were gentry and eight of these were widows. Isabella Douglas, Joseph Strutt's future wife, wrote enthusiastically to tell him how she had eagerly devoured historical works, Paine's *Rights of Man* and James Mackintosh's refutation of Burke's *Reflections*, the *Vindiciae Gallicae* (1791). As we have seen, it was Elizabeth Evans, elder sister of William and Joseph, and her female friends who, with revolutionary ardour, advised Joseph to read the 'divine truths' of Godwin's *Enquiry* which had 'penetrated my heart', and practise his exhortations in society. As the 'grand desideratum in politics' was 'the diffusion of knowledge and morals amongst the poor' so, as a manufacturer, he had it in his power to promote social good.[12]

Female scientific education

Until the late nineteenth century most middle-class girls were either educated domestically or in private schools. Curricula tended to be restricted to accomplishments and subjects considered appropriate for the domestic sphere such as modern languages, history and geography, the sciences being uncommon, although it should be said that many boys received little scientific education either.[13] If the sciences were taught, they tended to be botanical and domestic science, rather than the mathematical or physical sciences such as chemistry, physics and astronomy, which were considered to enhance female domestic and social attributes. As the Taunton Commission found during the 1860s, there was no educational parity between the sexes. However, some commercial schools and private tutors had been teaching distinctively modern subjects to girls from the seventeenth and eighteenth centuries, a practice encouraged by influential nonconformist educators such as Isaac Watts.[14] Indeed, Catherine Gleadle, Ruth Watts and others have argued that dissenting groups were among the first to advocate the teaching of the sciences, geography and other modern subjects to girls and were also more attuned to progressive educational philosophies that advocated greater curriculum parity between the sexes, such as Pestalozzianism. Female scientific education was championed by some headmistresses, feminist campaigners such as Emily Davies and male professionals, but also resulted from moves towards a general corporate educational system for both sexes, for example through the activities of the Endowed Schools Commission and the admittance of girls and boys to the same public examinations.[15]

Important changes in female education towards the end of the eighteenth century parallel some of the changes in male education, including greater provision of scientific subjects. There is much evidence for a significant expansion in the number and variety of girls' schools and an increase in the number of women in the teaching profession in schools and as private tutors. The Quaker educationist and botanist and prolific author Priscilla Wakefield argued that there were no natural barriers preventing women, particularly those from the nobility, from being taught natural history, botany, experimental sciences and other scientific subjects alongside more accepted topics such as modern languages and geography. However, in her *Reflections on the Present Conditions of the Female Sex* (1798) she did contend that the upper middle class should be taught somewhat less science in favour of utilitarian domestic subjects, while those in the lower middle and labouring classes ought to receive none at all so as not to interfere with their homely duties.[16] Traditional methods were still employed to teach scientific subjects and Elizabeth Ham's education, for instance, at the Misses West's school in Tiverton during the 1790s, although including grammar, history, the Bible and geography, included reciting passages by rote, which was 'not very extensive, nor very edifying'.[17] Other evidence comes from Margaret Bryan's School for Girls at Blackheath, where scientific subjects were taught, including her *Astronomical and Geographical Class Book* (1815), and Mrs Wilson's school at Crofton Hall near Wakefield, where textbooks by the schoolmistress Richmal Mangnall were used. Similarly, the school managed by Mrs Florian, wife of Jolly B. Florian, which was advertised in his *Elementary Course of the Sciences and Philosophy* (1806), shows that girls were taught geometry and trigonometry, astronomy, geography, universal history and ancient and modern geography with maps, history, chronology and natural philosophy, with the aid of maps, globes experiments and machines.[18]

The extent to which teaching of natural philosophy for girls was encouraged can be seen in Erasmus Darwin's *Plan for the Conduct of Female Education* (1797), composed for the use of his natural daughters Susanna (1772–1856) and Mary Parker (1774–1859) at their girls' school at Ashbourne, Derbyshire. They conducted the school from 1794 until Susanna's marriage to Henry Hadley in 1809, when Mary took sole charge until her retirement in 1827. The school thrived and allowed them to occupy an important position in local society, with Mary being categorised among the nobility, gentry and clergy in Pigot's *Commercial Directory for 1818, 19 and 20*.[19] Priestley, Darwin's Lunar Society friend, made a point of encouraging women to perform experiments at home using everyday domestic objects, such as glass jars, in his electrical and chemical works. Darwin's fashionable and popular *Botanic Garden* brought botany and many other scientific subjects into the drawing room and became very popular with women, although its presentation of the Linnaean system has been interpreted as reinforcing gender identities. With Brooke Boothby Darwin also enthusiastically supported the work of the botanist Maria Jacson, discussing botany and sending hundreds of specimens to female botanists such as Miss S. Lea.[20]

Although quite conservative in some respects, Darwin's *Female Education* demonstrates that subjects such as natural philosophy and modern languages were taught. However, it provides much more than curriculum details, offering recommendations for the whole of school life, including health and recreational activities, recommended clothing and school trips, most of which were equally applicable to both sexes. Darwin emphasised that older young ladies should learn aspects of botany, chemistry and mineralogy, which also contributed towards geographical education, the latter explaining the differences of soils, for example, necessary for the 'theory and practice of agriculture'. He recommended that girls should attend courses of lectures by itinerant lecturers in experimental philosophy to gain an understanding of the sciences of mathematics, astronomy, mechanics, hydrostatics, optics, electricity and magnetism. The *Botanic Garden* provided introductions to botany, meteorology and geology, especially in the notes, despite the fact that some ladies had intimated to him that the *Loves of the Plants* were described in 'too glowing colours'. Likewise, the 'various arts and manufactories' of the midland region, such as those along the Derwent or in Birmingham and Nottingham, should be 'shewn and explain'd to young persons, as so many ingenious parts of experimental philosophy'. These contributed immediately 'to the convenience of life, and to the wealth of nations' that had invented or established them.[21]

Responding to criticism that he included more branches of the arts and sciences 'than is necessary for female erudition', Darwin replied in two ways. He commented that 'as in male education' the 'tedious acquirement of ancient languages for the purpose of studying poetry and oratory' was 'gradually giving way to the more useful cultivation of modern sciences'. This meant that it would be 'of advantage to ladies of the rising generation to acquire' knowledge of the 'modern sciences', so that, as companions of males, they could be 'reciprocally well understood'. Furthermore, he criticised the notion that while men were 'generally train'd from their early years to the business or profession, in which they are afterwards to engage', it 'most frequently happened' that ladies, 'tho' destined to the superintendance of a future family', received 'scarcely any instruction'. They remained ignorant of economic matters and of the 'proper application of the things which surround them'.[22] Some of Darwin's views were shared by his friend the evangelical clergyman the Rev. Thomas Gisborne, who had been, as we saw, a founder member of the Derby Philosophical Society. Gisborne called for more women to learn some aspects of natural philosophy, noting with approval improvements in schools and private families where women of the higher classes now gained knowledge of geography, natural history, history and 'popular and amusing facts in astronomy and in other sciences'. He argued that this should be extended to more women 'of the middle ranks' and condemned the fatuousness of contemporary drawing-room conversation, where 'many women who are endowed with strong mental powers, are little inclined to the trouble of exerting them'.[23]

Unlike the Derby Philosophical Society, the Derby Literary and Philosophical Society made an effort to involve women in public scientific culture as part of its attempt to forge a public platform for science. As we have seen, the main objectives were the formation of a laboratory to conduct original scientific research, and programmes of lectures were held to help pay for this. Although the members were all male, they did appeal to women and children to attend, claiming that it was 'scarcely necessary' to observe that knowledge of subjects including chemistry, the arts and mechanics had 'now become essential to the ordinary arrangements of domestic life'. Indeed, Whitehurst, Darwin, William Strutt, Sylvester and friends such as Loudon were keenly interested in the application of natural philosophy and industrial technology within domestic economy, as the Derbyshire Infirmary and the devices incorporated into their own houses demonstrate.

The justification for encouraging women to attend lectures on natural philosophy was fully articulated in the plan for public lectures of 1812, although it never extended to encouraging women to join the society. This referred to 'the female sex', and claimed that 'notwithstanding the mental inferiority with which lordly man has been pleased to charge them', women had 'ever shewn an interest in intellectual pursuits', which had been 'too often repressed instead of encouraged'. The subjects of the lectures were therefore with 'peculiar propriety' recommended to women for, 'as connected with many of the elegant arts, and with the conveniences of domestic economy, they have to them a high value'. It is also striking that the public lectures of 1813 included two by Higginson on the 'importance of ... philosophical pursuits, as the means of intellectual improvement and of elegant amuse-ment', and the 'advantages of liberal acquirements as applicable to the ... female character'.[24] Significantly, the Derby philosophers quoted in support the comments of the Unitarian minister the Rev. William Turner, president of the Newcastle Literary and Philosophical Society and one of the leading figures in provincial scientific culture. Turner had argued in his introductory address to the Newcastle Society that the 'chief excellence' of women 'appears when we consider that it is from our parents of this sex that we obtain our earliest and most valuable knowledge'. Whatever might 'enable the affectionate and attentive mother, to discharge this important branch of their duty with greater satisfaction and success' should be allowed 'of peculiar consequence to society at large'. The mother acquainted with 'the principles of natural science' had greater opportunities to 'convert the amusement and diversions of her children into occasions of important information' and so 'prepare them for entering with greater advantage' into future courses of 'regular instruction'. Besides his credentials as leading Unitarian minister and natural philosopher, Turner was also an important promoter of educational reform and took an early and sympathetic interest in new pedagogical philosophies that emphasised the importance of female education such as Pestalozzianism.[25]

Some of the lectures on scientific subjects at the Derby Mechanics' Institute (1825), discussed in more detail in the final chapter, were also aimed at women, although female attendance seems to have been circumscribed by norms and expectations revealing the subtlety and negotiated contingency of gender boundaries between notional public and private spheres. Some of the subjects, despite reassurances, were thought to conflict with contemporary notions of femininity. Women were encouraged to attend the lectures on chemistry and anatomy by Douglas Fox in 1825, but the *Derby Mercury* regretted that those who had attended the chemistry lectures had not come to those on anatomy, as they 'would have derived valuable information from the subjects illustrated' and would have 'been no less pleased with the correctness of language observed by the modest and unassuming lecturer'.[26] Fox explained that examination of the eye and ear had been deferred to last 'in order that those exquisite parts of the body might obtain the attention' of women in the audience and promised that 'nothing either disgusting or unpleasant would be introduced'. These were 'decidedly proper for female investigation' and 'universal custom invited them to the exhibition of these delicate works of nature'.[27] Similar reassurances had to be given to encourage women to view some Florentine anatomical models exhibited in the Athenaeum in 1842, where a 'distinct inspection for ladies' was provided with a female attendant, so that they could be visited 'without offence by the most sensitive', suggesting that female attendance at these too was pushing against the bounds of polite acceptance.[28] There were female donors to the 1839 Derby Mechanics' Institute exhibition and to the 1843 Derby Town and County Museum exhibition, and in June 1847 it was agreed that women should be admitted to lectures, classes and the library at the Mechanics' Institute for half the price of members. A notice placed in the institution then stated that when a sufficient number had enrolled, 'classes will be formed for their exclusive advantage and instruction, viz. in botany, calculation, dancing, etc.'. However, the policy does not appear to have met with much success. A committee report for 1848 gave the total number of members to be 347, and the number of female members admitted under the new regulation had fallen from 49 to 35.[29] Although the total number of members had risen to 588 a year later, the figures suggest that women formed only around 10% of the membership in a borough population which had, as we saw, more women than men.

The enthusiasm of women from Unitarian families towards natural philosophy is strongly evident among the Higginson family, where, as in the Strutt family, the women took a keen interest in the sciences. The Rev. Edward Higginson's daughter Emily kept her brother Edward junior fully informed of developments in local science while he was away at Manchester College in York. She was 'in great hopes' of attending Douglas Fox's opening lectures at the Derby Mechanics' Institute in 1825, 'as chemistry is a subject on which I am at present lamentably ignorant, but very desirous of gaining all the information I possible can'. She was also 'in high hopes and expectation' of hearing William Nicol (1771–1851) 'by next spring which will be very delightful, not only for the Mechanics' Institution but for all who are desirous of improvement and who can attend'. When John Walker, author of *The Philosophy of Arithmetic* (1812) and Fellow of Trinity College, Dublin, visited her father and mentioned his wish to lecture at Derby on elementary mathematics, Emily and her sister Helen 'regretted' that due to their engagement in teaching, they were unable to see 'this clever little man'. The extent of Emily's philosophical education is also evident in her reading. With her brother Edward she read through David Hartley's *Observations on Man* (1749), probably in the abridged version produced by Joseph Priestley, considering the sections on sympathy, theosophy and the moral sense to be 'very beautiful and particularly interesting'. She also read the chapter on philosophical necessity, which was adjudged 'very intelligible in some parts but not altogether'. At the same time Emily was voraciously devouring the moral and philosophical works of Lant Carpenter, the Unitarian minister and leading intellectual.[30]

More evidence concerning the value of science in girls' private education comes from the career of William George Spencer, who was, as we have seen, father of Herbert Spencer and a schoolmaster. Adopting the practices of commercial academies and within some dissenting families, and in accordance with his religious philosophy and local demand, Spencer emphasised the importance of mathematical and scientific subjects to both sexes. Probably more than half of his pupils were women and they were taught a variety of subjects by the 1830s, including physical and human geography, astronomy and chemistry, using his extensive collection of apparatus. William George Spencer appears to have made little distinction between the sexes in educational terms among his private pupils, and he also taught many children from Tory as well as Liberal and dissenting families. The intellectual stimulation that he gave to girls is evident from the recollections of Lady White Cooper, who writing to his son Herbert, remembered his lessons with feelings of 'reverence, love and gratitude'. She retained copies of her 'wonderfully neat copy books, full of algebra, questions on Euclid, astronomy and physics', which at the time she 'well understood'. She thought Spencer to have been 'remarkable for his calmness, patience, and punctuality' with his pupils, and 'that he had power over circumstances' as 'nothing ever ruffled him'. Having been 'brought up in a strictly evangelical school', the novel ideas that Spencer suggested on religious subjects were 'most interesting' to her, while Spencer's 'facility in quoting scripture' she took to be 'evidence of his knowledge of the Bible'. His attitude towards his pupils was 'truly sympathetic' and he 'never thought it a trouble to listen to complaints or grudge time, to help one's little difficulties, suggesting ideas which seemed to expand as one's own'. She even admitted that 'as a girl I quite worshipped Mr. Spencer, and shall ever be grateful to have known and had the friendship of so truly great and good a man.'[31]

The interest and excitement that women could find in scientific subjects encouraged by Spencer is also evident from the work he undertook at a school kept in Nottingham and later Lenton Fields House, Lenton by Catherine Turner (1800–94), widow of the Rev. Henry Turner, a minister at the High Pavement Chapel. According to Constance Martineau, who attended the school during the 1840s, botany was an important focus and girls went on walks to collect flowers for dissection. The thirty or so young ladies, aged between twelve and seventeen, most of whom were from leading Unitarian families in the north and east midlands, were not only taught the usual accomplishments such as playing the piano, drawing, singing, dancing and speaking French, but 'to think, and to have an earnest purpose in life', studying texts including Burke on the sublime and beautiful, Paley's moral philosophy and Alison on taste. Jane Marcet's *Conversations on Natural Philosophy* (1819) and *Conversations on Intellectual Philosophy* (1829) were used as textbooks. With Eliza Swanwick, the Irish assistant, the girls read William Buckland's *Geology and Mineralogy Considered with Reference to Natural Theology* (1836), the Bridgewater treatise which 'first impressed' upon Constance Martineau 'that we must not look to the Bible for scientific truth, that the world was not created in six days, and that the account of the creation in Genesis is not to be relied upon'. This proved to be 'an astounding enlightenment' and inspired her no longer to view other biblical episodes as literal truths, and Moses, 'not as a miracle worker, but as inspired' and as a great and good man, which was a 'gradual' and 'painful … process of enlargement'. Spencer came to the school once a week to give 'valuable lessons in natural philosophy', an arrangement that may have dated back to the 1820s, when he lived for a short time in Nottingham. According to Martineau, the 'aim of his teaching was not so much to impart knowledge as to teach us to think, to find things out for ourselves and explain in clear language the reason of them'. The girls were not shown how to make an equilateral triangle or a circle until they could do it for themselves using compasses and without measuring and it was 'not till some time later that we could see why two equal circles, described from the two ends of a straight line, enabled us to construct an equilateral triangle'. With the 'more advanced pupils', Spencer provided lessons in astronomy and electricity, and the girls 'learnt to illustrate by diagrams the moon's motion round the earth' and the 'relative distances of the planets, etc.'. When he brought in an electrical machine and showed some experiments, Martineau thought it 'delightful', and his lessons became 'the great event of the week', requiring 'a great deal of thoughtful work to prepare for them', with Mrs Turner wondering 'how she had kept school without him'.[32]

Scientific institutions

Women were not generally allowed to become members of literary and philosophical societies in the Georgian period and often remained excluded throughout the nineteenth century, although many were admitted as members to mechanics' institutes by the 1840s.[33] This is reflected in Derby scientific institutions between 1780 and 1850. However, there are tantalising pieces of evidence that women may have been more active in scientific work than this might suggest. Whitehurst's wife Elizabeth (Gretton), whom he married on 9 January 1745, appears to have helped him with his philosophical works. In 1763, Whitehurst sent a summary of what was later to be the *Inquiry into the Original State and Formation of the Earth* (1778) with a covering letter to Benjamin Franklin. The editors of the Franklin papers puzzled over the features of the manuscript, particularly the 'short sketch' of the 'General Theory of the Earth', whose 'peculiarities', they remarked, 'remain unexplained'. Though the sketch, on a four-page sheet, began with the salutation 'Sir' it carried no dateline. The first two pages and part of the third were in Whitehurst's hand, 'then, after the first word of a new paragraph, for some unknown reason a distinctly different handwriting appears and continues to the end, about the middle of the fourth page.' In the next to last line Whitehurst inserted four words omitted by the copyist, showing how closely they were working together.[34] In his obituary of Whitehurst, Charles Hutton wrote of Elizabeth that she was 'among the first of female characters. Her talents and education were very respectable; which enabled her to be useful in correcting some parts of his writings.'[35] This suggests that it was Elizabeth who worked with Whitehurst on the sketch for Franklin, and probably that her role was more than that of mere copyist. The intermixing of the two handwritings suggests more of a collaborative than passive role.

There were no female members of the Derby Philosophical Society and there is no evidence that the question of admitting them was ever considered, although some women presumably borrowed books from the library through husbands or male relatives. Many early meetings took place in the domestic context of members' houses, and Darwin provided some experiments for the entertainment of female members of the Strutt family. According to Susannah Wedgwood,

> The last meeting was at Mr. Strutts, the Miss Strutts keeps (*sic*) their brothers house, & consequently were obliged to make tea & preside at the supper table, they did not like this at all, but Doctor Darwin with his usual politeness made it very agreeable to them by shewing several entertaining experiments adapted to the capacities of young women; one was roasting a tube which turned round itself.[36]

On at least one other occasion, in 1792, Darwin invited a woman, Marianne Sykes (*c*.1770–1815), to a meeting of the Society. She remembered his 'wonderful sallies of imagination and wit, which kept us in perpetual laughter and astonishment'.[37] Some provincial philosophical societies did allow women to attend. Although the Leicester Literary and Philosophical Society, for example, initially forbade women to come to meetings, after some poor attendances it was decided in 1838 to allow them to take part as guests of members, despite continued opposition from some male members of the governing committee. The enthusiastic participation of women at meetings was credited with reviving the fortunes of the Society, though it was not until 1870 that women were admitted as associate members, and 1886 before the first women were allowed to be elected as full members.[38]

The ways that some women did participate in Derby scientific institutions between 1800 and 1850 are evident when we examine lists of the honorary members of the Derbyshire Horticultural Society, and supporters of the Derby Mechanics' Insititute and of the Derby Town and County Museum (1835). These provide us with a list of around thirty women who made significant contributions, although, it must be admitted, the number is small compared to the hundreds of men who subscribed to these institutions, especially given the fact that, as we have seen, by 1831, 52% of the borough population was female. The list represents only those women who subscribed in their own name and excludes, of course, those contributing through family subscriptions in their husband's name, the number of whom must have been much larger. On the other hand, the Horticultural Society and the Museum were the least 'scientific' of the scientific-related societies, and no women ever joined any of the formally constituted philosophical societies active in Derby between 1783 and 1850, with the possible exception of the Derby Arboretum.[39]

Only one of these women, the botanist Lucy Hardcastle (1771–*c*.1835), appears to have published a scientific work in her own right, the *Introduction to the Elements of the Linnaean System of Botany* (1830). As we have noted, Hardcastle is thought to have been an illegitimate daughter of Erasmus Darwin. She taught elements of natural philosophy to the pupils in her school for ladies on Ashbourne Road, and assisted Stephen Glover with the botanical sections of his *History of Derbyshire* (1829). About a third of the women were referred to as gentry and would, therefore, have been more likely to have had the education and means to act independently in urban public culture, for example as widows or spinsters. Perhaps not surprisingly, the families whose male members were the most prominent in Derby scientific culture also produced influential female supporters. Female members of the Evans family, for instance, were all honorary members of the Derbyshire Horticultural society. Of the women in the Strutt family, Anne, Frances and Elizabeth, of Derby, daughters of William and Barbara and sisters of Edward Strutt, were all honorary members of the Derbyshire Horticultural Society and of the Mechanics' Institute. Similarly, Susannah Strutt of Green Hall, Belper, wife of Jedediah son of George Benson Strutt, supplied fossils to the Town and County Museum exhibition in 1843.

Others of the women classed as gentry or significant property owners include Mrs Abney, wife of the Rev. Edward Abney, a Derby Philosophical Society member, who also supported the Mechanics' Institute, donating a large number of natural history items to the 1839 exhibition. Mrs Barber, wife of Alderman John Barber, the attorney, agent to the Duke of Devonshire and mayor of Derby in 1843, was another supporter of scientific culture, being an honorary member of the Mechanics' Institute, and supplied items for the 1839 exhibition. Similarly, Miss Maria Cox, supporter of the Town and County Museum, donated mineral exhibits to the museum in 1836. Mrs Drewry, an honorary member of the Horticultural Society, was the wife of John Drewry III of Spondon and Borrowash and daughter of the wealthy Unitarian landowner Alderman Samuel Rowland. Miss Catherine Richardson of Friargate, an honorary member of the Horticultural Society, owned buildings on St Mary's Gate and rented them out as shops and lodgings. Others, such as Mrs Carol Worsley, subscribed to scientific works, in this case White Watson's *Delineation of the Strata of Derbyshire* (1811), and some, such as Miss Frances Gibson of St Mary's Gate, loaned natural history exhibits to the Town and County Museum. Finally, Mrs Poyser of Kedleston Road, widow of Thomas Poyser who had been a member of the Philosophical Society, donated a microscope and case to the 1839 exhibition, and Mrs Oakley, wife of the artist Octavius Oakley, donated items to the 1843 exhibition.

Conclusion

The lack of female scientific writers, small number of women who were donors, patrons and subscribers, and the fact that women were formally excluded from most scientific institutions suggests that the so-called 'separate spheres' continued to constrain and shape female participation in scientific culture between 1770 and 1850. On the other hand, if we include evidence concerning attendance at scientific lectures, exhibitions and visits to institutions such as the Derby Arboretum, then it is clear that women did make up a significant proportion of the local audience for science, although their participation was contingent upon the negotiation of gender norms and conventions surrounding different scientific subjects. Once we enter the realm of private education and look at the number of girls who were taught scientific subjects, a different picture emerges of exciting and enjoyable science lessons using botanical and geological specimens, scientific equipment and experimental demonstrations. The enthusiasm of women such as Lady White Cooper, Constance Martineau and Emily Higginson for scientific subjects, intellectual advancement (and scientific lectures for that matter) generally leaps from the pages of their letters. That scientific ideas could have a tremendous intellectual impact upon women is suggested by Constance Martineau's observation that it was through science lessons using Buckland's Bridgewater Treatise that she came to appreciate the necessity of grounding religious faith in studies of the natural world as well as through the Bible. Nevertheless, as analysis of the Strutt family correspondence reveals, despite the major emphasis upon female intellectual self-improvement, scientific subjects were more likely to be discussed between males and tended to be compartmentalised from domestic and emotional matters in letters between the sexes. Even in the middle-class Liberal families who were W. G. Spencer's strongest patrons, where enjoyment in the subject seems demonstrable, the suspicion lingers that science remained an extension of accomplishments and polite conversation. Women were still regarded primarily as audience for science rather than doers of science.

The evidence of middle-class female involvement in scientific culture confirms that some were able to circumvent the limitations of the domestic sphere in the context of a flourishing provincial urban culture with its emphasis upon progressive 'improvement'. Those from Liberal dissenting families received greater encouragement and had the best opportunities to participate, although it is clear that through the opportunities presented by patronage, some women of all religious and political affiliations were able to partake at a restricted level.[40]

Notes

1 L. Davidoff and C. Hall, *Family Fortunes: Men and Women of the English Middle Classes, 1780–1850* (London: Hutchinson, 1987); A. Vickery, 'Golden age to separate spheres? A review of the categories and chronology of English women's history', *The Historical Journal*, 36 (1993), 383–414.

2 Although there is not room to consider it here, working-class culture should not be underestimated, and there are, of course, important class dimensions to analyses of gender relations.

3 A. Vickery, 'Golden age to separate spheres?'; L. Colley, *Britons: Forging the Nation, 1707–1837* (London: Pimlico, 1994), 280.

4 A. Foreman, *Georgiana, Duchess of Devonshire* (London: Ted Smart, 1999), 404.

5 J. Brewer, *The Pleasures of the Imagination: English Culture in the Eighteenth Century* (London: HarperCollins, 1997).

6 M. C. Jacob and B. T. Dobbs, *Newton and the Culture of Newtonianism* (Atlantic Highlands NJ: Humanities Press, 1995), 92; M. Hunt, M. Jacob, P. Mack and R. Perry, *Women and the Enlightenment* (New York: Institute for Research in History, 1984).

7 G. S. Rousseau, 'Science books and their readers in the eighteenth century', in I. Rivers (ed.), *Books and their Readers in Eighteenth-Century England* (Leicester: Leicester University Press, 1982), 187–255, p. 215.

8 *Derby Mercury* (6 January 1744, 30 June 1749).

9 M. Alic, *Hypatia's Heritage: A History of Women in Science from Antiquity to the Late Nineteenth Century* (London: Women's Press, 1986); M. B. Ogilvie, *Women in Science: Antiquity through the Nineteenth Century. A Biographical Dictionary* (Cambridge MA: Massachusetts Institute of Technology, 1986); P. Phillips, *The Scientific Lady: A Social History of Women's Scientific Interests, 1520–1918* (Weidenfeld and Nicolson, 1990); C. Merchant, *The Death of Nature: Women, Ecology and the Scientific Revolution* (San Franciso: Harper & Row, 1979); E. Fox Keller, *Reflections on Gender and Science* (New Haven: Yale University Press, 1985); L. Schiebinger, *The Mind Has No Sex: Women and the Origins of Modern Science* (Cambridge MA: Harvard University Press, 1991); Davidoff and Hall, *Family Fortunes*; D. Valenze, 'The art of women and the business of men: women's work and the dairy industry, c.1740–1840', *Past and Present*, 130 (1991), 142–69; L. Schiebinger, *Nature's Body: Sexual Politics and the Making of Modern Science* (London: Pandora, 1994); R. Watts, *Women in Science: A Social and Cultural History* (London: Routledge, 2007).

10 Census returns (1831), quoted in 'Report of the Municipal Reform Commissioners on the Borough of Derby', *Parliamentary Papers*, 25 (1835), 1858.

11 A. W. Davison, *Derby: Its Rise and Progress* (London: Bemrose, 1906), 75–6; E. Saunders, *Joseph Pickford of Derby: A Georgian Architect* (Stroud: Sutton, 1993), 58.

12 Isabella Douglas, letter to Joseph Strutt (19 December 1791), Galton papers, Birmingham Central Library, Archives and Heritage, M3101/C/E/4/8; Elizabeth Evans, letter to Joseph Strutt (24 October 1793), copied in E. Strutt, typescript 'Memoir of William Strutt', Derbyshire Record Office.

13 J. Kamm, *Hope Deferred: Girls' Education in English History* (London: Methuen, 1965); M. Bryant, *The Unexpected Revolution: A Study in the History of the Education of Women and Girls in the Nineteenth Century* (London: University of London, 1979); A. L. Roach, *A History of Secondary Education in England, 1800–1870* (London: Longman, 1987), 148–59; M. Bryant, *The London Experience of Secondary Education* (London: Athlone, 1986), 311–58; J. S. Pedersen, *The Reform of Girls' Secondary and Higher Education in Victorian England* (New York: Garland, 1987); W. B. Stephens, *Education in Britain, 1750–1914* (London: Macmillan, 1999), 109–14; the essays in *History of Education*, 32 (2003), especially J. Goodman, 'Troubling histories and theories: gender and the history of education', R. Watts, 'Science and women in the history of education: expanding the archive', and J. Martin, 'The hope of biography: the historical recovery of women educator activists', 157–74,

189–200 and 219–32, respectively.

14 I. Watts, *The Improvement of the Mind*, ed. J. Emerson (Boston, 1833), xv.

15 Pedersen, *Girls' Secondary and Higher Education*, 84–8, 318–21; Stephens, *Education in Britain*, 109–14.

16 Kamm, *Hope Deferred*, 108–9.

17 Quoted in S. Skedd, 'Women teachers and the experience of girls' schooling in England', in H. Barker and E. Chalus (eds.), *Gender in Eighteenth-Century England* (Harlow: Longman, 1997), 101–25, p. 122.

18 N. Hans, *New Trends in Education in the Eighteenth Century* (London: Routledge & Kegan Paul, 1966), 194–208; Kamm, *Hope Deferred*, 141, 144.

19 C. Darwin, *The Life of Erasmus Darwin*, ed. D. King-Hele (Cambridge: Cambridge University Press, 2003), 64, 125–6; J. Uglow, 'But what about the women? The Lunar Society's attitude to women and science, and to the education of girls', in C. U. M. Smith and R. Arnott (eds.), *The Genius of Erasmus Darwin* (Aldershot: Ashgate, 2005), 179–94.

20 E. Darwin, letters to M. E. Jackson (24 August 1795) and S. Lea (6 June 1796) in King-Hele (ed.), *Letters of Erasmus Darwin*, 482–3, 499–500.

21 E. Darwin, *A Plan for the Conduct of Female Education in Boarding Schools* (Derby, 1797), 43.

22 Darwin, *Female Education*, 44–5, 114.

23 T. Gisborne, *An Enquiry into the Duties of the Female Sex*, first edition (London, 1797), 58–9, 106–7.

24 *Syllabus of course of Lectures … to be delivered under the direction of the Derby Literary and Philosophical Society* (Derby, 1813); *Derby Mercury* (1 April 1813).

25 Printed address, 15 February 1812, bound in Journal of the Derby Literary and Philosophical Society, 1808–16, Derbyshire Record Office, D5047; D. Orange, 'Rational dissent and provincial science: William Turner and the Newcastle Literary and Philosophical Society', in I. Inkster and J. Morrell (eds.), *Metropolis and Province: Science in British Culture, 1780–1850* (London: Hutchinson, 1983), 205–30.

26 *Derby Mercury* (26 October 1825, 30 November 1825).

27 D. Fox, *Notes of the Lectures on Anatomy and Chemistry, delivered by Mr. Douglas Fox* (Derby, 1826), 41.

28 *Derby Mercury* (9 February, 1842).

29 S. Laughton, 'Derby Mechanics' Institution … A Hundred Years of Valuable Life', in *Derby Daily Telegraph* (12, 13 February 1925).

30 E. Higginson, letter to E. Higginson junior, 23 October 1825, James Martineau papers, Harris Manchester College, Oxford, NRA 827, fos. 1–29; J. Walker, *The Philosophy of Arithmetic and the Elements of Algebra* (Dublin, 1812); D. Hartley, *Observations on Man, his Frame, his Duty and his Expectations* (London, 1749); J. Priestley, *Hartley's Theory of the Human Mind, on the Principle of the Association of Ideas* (London, 1775).

31 Lady White Cooper, letter to H. Spencer, in H. Spencer, *An Autobiography*, 2 vols. (London: Williams and Norgate, 1904), I, 50–1. Spencer was impatient with women who showed little interest in the sciences such as his wife Harriet (1794–1867), to whom, according to Herbert, he was sometimes exacting, inconsiderate and unsympathetic, Spencer, *Autobiography*, I, 56–60.

32 'M. C. M.' (M. Constance Martineau), *Memoirs and Reminiscences* (Letchworth, 1910), 9–12, 64–75; F. Barnes, *Priory Demesne to University Campus: A Topographic History of Nottingham University* (Nottingham: Nottingham University, 1993), 241–4.

33 M. Benjamin (ed.), *Science and Sensibility* (Oxford: Blackwell, 1994); D. F. Noble, *A World Without Women: The Christian Clerical Culture of Western Science* (Oxford: Oxford University Press, 1992); L. and S. Sheets-Pyenson, *Servants of Nature: A History of Scientific Institutions, Enterprises and Sensibilities* (London: Fontana, 1999), 335–49; Watts, 'Science and women'.

34 J. Whitehurst, letter to B. Franklin, 18 March 1763, in B. Willcox (ed.), *The Papers of Benjamin Franklin*, 38 vols. (1958–2007), X, 226–30.

35 C. Hutton, 'Authentic Memoirs', *Universal Magazine* (1788), 225.

36 Quoted in A. E. Musson and E. Robinson, *Science and Technology in the Industrial Revolution* (Manchester: Manchester University Press, 1969), 361.

37 R. P. Sturges, 'The membership of the Derby Philosophical Society, 1783–1802', *Midland History*, 4 (1978), 224.

38 F. B. Lott, *The Centenary Book of the Leicester Literary and Philosophical Society* (Leicester: Thornley and Son, 1935), 3, 13, 131–2; W. Gardiner, *Music and Friends*, 3 vols. (Leicester, 1853), III, 305–6.

39 Evidence compiled primarily from the *Derby Mercury*, *Derby Reporter* and town directories from the period. For the establishment of the Derby Town and County Museum in 1835, see the letter by George Watson, *Derby Mercury* (23 December 1835).

40 K. Gleadle, *Early Feminists: Radical Unitarians and the Emergence of the Women's Rights Movement, 1831–1860* (Basingstoke: Macmillan, 1995); R. Watts, *Gender, Power and the Unitarians in England, 1760–1860* (London: Longman, 1998).

11

Civic science and rational recreation: the Derby Mechanics' Institute (1825) and the Derby Arboretum (1840)

Introduction

This chapter examines two institutions associated with the Derby philoso-phers that were intended to promote scientific education and civic culture, the Derby Mechanics' Institute (1825) and the Derby Arboretum (1840). Given strong backing by the Strutts and their philosophical friends, especially the Rev. Edward Higginson, the Mechanics' Institute enjoyed considerable success in promoting scientific education, sustaining a platform for public science, as the celebrated exhibition of 1839 demonstrates. However, like the Derby Arboretum, it was managed and dominated by the middle class and ultimately seems to have failed to generate sufficient working-class support to be viable. Designed by John Claudius Loudon (1783–1843), the foremost British landscape gardener and horticulturist, the Arboretum was intended to promote his vision of urban parks and serve as a physical embodiment of his 'virtual' arboretum, the authoritative *Arboretum et Fruticetum Britannicum* (1838). However, the Arboretum was also a Liberal municipal project inspired by Joseph Strutt and the Derby philosophers, a continuation of Strutt's work as the first mayor in the reformed corporation, and a kindred institution to botanical gardens, museums and mechanics' institutes. The Arboretum and the Mechanics' Institute were originally intended to be rational recreational institutions; however, in both cases, the emphasis upon science declined.[1] This chapter examines how the Derby philosophers facilitated the creation of these institutions and how they were supposed to encourage social, moral and intel-lectual progress, yet came to be experienced and appropriated in different ways from those originally anticipated.

Industry, social squalor and class tensions

As well as being institutions intended to promote middle-class civic identity and scientific education, the Arboretum and the Derby Mechanics' Institute were both supposed to ameliorate social problems through working-class intellectual and moral improvement. Social and political tensions had become particularly acute in Derby as a result of industrialisation and rapid expansion, which strained the abilities of existing governmental and parish structures to maintain urban health and sanitation. These class tensions were manifest in the Pentrich Rebellion of 1817, the Derby Silk Mill Strike of November 1833 to April 1834 and the reform bill agitation in 1831, which caused riots in Derby and the burning of Nottingham Castle.[2] In the 1820s, middle-class reformers campaigned on a series of political issues, including slavery, church reform, Catholic emancipation, relief for dissenters and, of course, parliamentary reform, and many remained unsatisfied by the 1832 Reform Act. Some of these causes had considerable Tory and working-class support; however, from the 1830s, divisions widened between middle- and working-class political campaigns, the latter being directed towards the Chartist movement. Only from the 1850s, with the decline of Chartism, did middle- and working-class campaigners tend to unite – or at least to co-operate – under the Liberal banner.[3]

Although the election of a reformed Whig corporation in Derby in 1836 marked the start of a period of fundamental improvement, it made little headway initially and the exclusion of most Tories in the new system actually, to some extent, increased political divisions within the local elite. Guide books and town histories between 1800 and 1850 conveyed an image of social and economic progress, focusing upon the appearance of new buildings and urban institutions, and it is notable that public structures and law and order were a greater priority for the reformed corporation under Joseph Strutt than the immediate problems of health and sanitation.[4] By 1843, these included new bridges, chapels and churches, a new town hall and a recently erected ensemble of buildings, the Athenaeum, Royal Hotel, Post Office and Derby and Derbyshire Bank, which formed 'a very striking improvement to the town'. In the Athenaeum, which had been built by a public company partly funded and encouraged by the Strutts and the corporation, were a newsroom, a library, a lobby, a reading room and the Town and County Museum.[5]

Having reached 10,828 by 1801, the urban population rose rapidly over the ensuing decades to 13,043 in 1811 and 32,741 by 1841, an increase of almost 40% during the latter decade, the most rapid period of growth in Derby history. Although the rate slowed a little, the population reached 40,609 by 1851, and was considerable larger if the new suburbs are included. Much growth was caused by immigration. In 1851, 37% of the inhabitants were born outside of Derbyshire, while only 28%, 35.5% and 18%, of the populations of Leicester, Nottingham and Coventry, respectively, were immigrant. Of these, the Irish were a notable contingent, being 1,314 or 8.7% of the immigrants, and almost half the borough population was under the age of twenty.[6] This was accelerated by burgeoning railway connections, three companies having chosen to form a junction at Derby: the Midland Counties Railway, the Birmingham

and Derby Junction Railway, and the North Midland Railway. Facing corporation pressure they combined to build Francis Thompson's 1,050-foot-long,
trijunct station, claimed to be 'the most spacious, convenient, and extensive structure of the kind yet erected'. An engine shed was also included, 'so
spacious as to cover nine lines of rails', along with workshops and separate
engine houses for each company. According to Glover, Derby was now 'a great
central point in the railway communication of the country', and 'no stronger
or more decisive proof could be adduced of the energy and enterprise of the
county, and ... of its continued advancement, than the station and works ...
here briefly described'. Subsequently, the Midland Railway Company, formed
by a union of the smaller regional companies, developed Derby as centre for
the construction and repair of steam engines.[7]

The railway accelerated the development of heavy industry in Derby, which
eventually took over from the older textile and porcelain manufactories,
although for long the town retained a wide manufacturing base, including
net-lacemaking, stocking making, cotton spinning, silk weaving, china,
foundry casting, lead smelting and paper making. There were extensive lead
works, including those of Messrs Cox and Company, which had a 50-yard-
high shot tower. Slitting and rolling mills had been erected at the Morledge
in the eighteenth century, for the working of iron and copper. By the 1840s
iron works included the Derwent Foundry Company in Exeter Street and the
Britannia Foundry in Duke Street. The former made wheels, rails, bridges,
cast and wrought-iron roofs, beams, castings for steam engines, water, gas
and steam pipes, windows, gates, palisades, boilers, grates, and stoves. The
latter made castings for ornamental, architectural and domestic purposes,
including side lathes, carriage wheels, castings for steam engines, steam pipes,
cats, pans, stoves, spouts, church windows, vases and much else. The Union
Foundry, established in 1822 by Falconer and Peach, made similar items, as
did the Phoenix Foundry on Nottingham Road. Much of the ironwork for
the bridge over Nottingham Road on the North Midland Railway and for
Derby Railway Station was made at the Phoenix Foundry, along with Charles
Sylvester's patent stoves for heating churches and other public buildings.[8] The
manufactories of Joseph and James Fox of City Road came to produce lathes
which were exported all over Europe, including Germany and Poland. They
developed one of the first planing machines in 1814, a screw-cutting machine
'at a very early period' and a self-acting lathe, and the Strutt and Arkwright
firms gave them 'considerable employment'.[9]

By the mid-nineteenth century, Derby had therefore experienced a period
of unprecedented commercial and industrial expansion and population
growth, helping to explain why there were such severe problems of health
and sanitation among the labouring poor. Beneath the triumphal march of
industrial progress, behind the façade of fresh public buildings such as the

elegant Athenaeum or trijunct station was another world of poverty, poor health and sanitation, and exploitation, revealed in a series of damning health reports. One of these, by physicians William Baker and J. R. Martin, presented a picture of the situation of Derby's poor and social conditions beyond the reach, control and concern of the improvement commissioners. It was thought that occupational variety ensured that poverty and disease were not even worse 'than in most manufacturing towns' because 'it is less liable ... to have large numbers of hands thrown out of employ at any one time by the fluctuations of trade.'[10] Rapid population expansion exacerbated the difficulties, especially in the courts and alleys in the Brookside area of the town, which were characterised by dampness, disease and overcrowding.[11]

Another result of industrialisation and population growth was, of course, the loss of common lands, public walks, gardens and pleasure gardens, cemeteries sometimes being among the few preserved green spaces. With some parliamentary encouragement, especially through the report of the Select Committee on Public Walks (1833) and the 1836 Enclosure Act, and local initiatives, provision was made for new public walks and gardens.[12] As we have seen, the Nun's Green, common grounds and estate gardens disappeared, including part of the Siddals, used in 1839 and 1840 for the new complex of station and railway works. In 1825, Emily Higginson remarked to her brother Edward junior that it was scarcely possible 'to know some of our walks which were once so beautiful', almost every road around Derby having been 'spoiled for pleasant walks'. It was necessary to go twice as far to avoid the red houses and brick kilns.[13] In 1833, Glover noted that 'there are now no public walks, such as Darley Grove and the Holmes afforded the inhabitants ... formerly; and the bowling greens, the constant evening amusement of respectable tradesmen during the greater portion of the last century, are now but little encouraged.'[14]

At the same time there was also pressure on forms of leisure activity that traditionally involved all social classes, such as horse racing and Shrovetide football, with evangelical Christians and local clergy leading the attack upon the latter. The Derby football was abolished in 1845, despite considerable protest from players and many town citizens, who tried to restart the game every year despite a ban from magistrates and the presence of troops.[15] These disagreements concerning forms of leisure did not simply pit classes against each other but split them vertically. Although the football was primarily a working-class game, it was opposed by some working-class radicals, who joined with evangelical Christians and temperance crusaders against elite supporters such as Joseph Strutt, who patronised the football by symbolically throwing the ball into touch each year to begin the riotous progress. Horse racing was reinstated after a ban in 1836, with a large number of middle-class leaders, including William Eaton Mousley, the Mayor of Derby, fighting off opposition from evangelicals.[16]

The Derby Mechanics' Institute (1825)

Civic scientific culture was promoted by the Derby philosophers in response to middle-class divisions and the social and political tensions exacerbated by rapid immigration, poor health and poor sanitation. In addition to the familiar utilitarian arguments extolled in contemporary literature, public scientific culture was thought to offer many benefits for all social classes. Although, like many mechanics' institutes around the country, the middle classes came to dominate at Derby, there is some evidence that between the 1820s and 1840s the Derby Mechanics' Institute did manage to promote scientific education among the labouring classes, particularly tradesmen, skilled mechanics and craftsmen, including some women. The Philosophical Society continued to exist and the members remained leaders of local scientific culture; however, it was still a small and elitist organisation, a gentlemen's club dominated by professionals. After the demise of the Derby Literary and Philosophical Society, a short-lived Literary and Philosophical Society was also formed around 1834, and a separate Literary and Scientific Society founded in 1847 enjoyed, as we have seen, an active existence, sponsoring lectures and holding bi-monthly meetings. The period also saw the creation of the Town and County Museum and Natural History Society (1835) and smaller literary and scientific societies in Derbyshire.[17] Scientific lectures by visitors and resident philosophers and tutors continued, and Thomas Mozley remembered attending courses on chemistry, architecture, astronomy and meteorology by the Unitarian itinerant Thomas Longstaff, using an orrery, in 1818. The frequency and popularity of these and the proliferation of scientific societies are evidence for a vibrant and popular scientific culture.[18]

A hundred or so British mechanics' institutes were founded by 1826, largely confined to London, Scotland and Northern England, and a second phase between 1832 and 1842 saw the creation of smaller institutions in rural areas and smaller towns.[19] Traditional histories of mechanics' institutes have provided important local case studies, often interpreted in terms of the progressive history of 'adult education'. The importance has been emphasised of mechanics' institutes in urban scientific culture, through the dissemination of scientific and rational recreational ideas, and their role in class formation, especially middle-class identity, and class control through the promotion of rational recreational ideologies.[20] Shapin and Barnes considered mechanics' institutes to have been 'a series of failed experiments in the construction of ideologies'. After analysing the curriculum of such institutions at Edinburgh, Manchester and London in terms of their degree of success in reinforcing elite ideology, they contended that not even a small number of the 'working classes' were 'distracted ... from their own spontaneous political expressions.'[21] Recent work has cast doubt on the traditional historiographical assumptions underpinning studies of mechanics' institutes. While confirming middle-class domination in their foundation and development, Jonathan Rose, for example, has emphasised the extent to which they were preceded by clubs and societies inspired by the labouring classes, and despite their short life spans, contemporary mutual improvement societies and worker's libraries probably had a greater impact upon working-class intellectual life than official mechanics' institutes. There has also been renewed recognition of the importance of the 'self-help' movement in both working- and middle-class scientific education and practice.[22]

Nevertheless, as we shall see, there is some evidence that the Derby Mechanics' Institute and 1839 exhibition did successfully promote scientific education to the working and middle classes, although the middle class and higher working class or aristocracy of labour tended to dominate proceedings with elite patronage. Although middle-class leaders of the venture tended to be Whigs (Liberals), the Mechanics' Institute attracted some support from local Tories. Similarly, promoters of the Leeds Mechanics' Institute obtained backing from Liberals and some Tories, with at least one Tory holding the office of either president or vice-president. In other towns, mechanics' institutes came to be dominated by middle-class Liberals or stymied by political divisions. At Leicester, 'party feelings' prevented the establishment of a mechanics' institute until 1833, and when founded, the Leicester Institute was regarded as being a radical affair with the conservatives dubious or hostile from the beginning.[23]

The role of the Derby philosophers in the foundation of the Derby Mechanics' Institute is clear.[24] Leonard Horner, a founder of the Edinburgh School of Arts, one of the many contenders for a pioneering mechanics' institution, wrote to William Strutt in July 1821 for advice on the scheme of instruction which had been proposed.[25] As we have seen, with the help of friends such as the Irish poet Thomas Moore, the Strutt brothers founded the Lancasterian School in Derby, and with Edward Strutt, were firm supporters of University College, London. Richard Forester, Joseph Strutt and Edward Higginson tried to forge a public platform for science using the lectures promoted by the Derby Literary and Philosophical Society. At a dinner held in Derby for the Lancasterian School, a toast was proposed to the success of the London University College because it would allow more people to receive a university education than ever before. This followed a meeting in May 1826 chaired by William Strutt which supported the new university.[26] The Strutts reinforced the campaign with their financial muscle and William, Joseph and Edward were original subscribers. Edward was elected to the Council of London University in February 1831 and later became Vice-President of University College and President from 1871 to 1879. As MP for Derby (1830–47) and later Nottingham (1852–56) he was closely associated with Brougham, the Mills, Bentham, Macaulay and the philosophical radicals and was elected FRS in 1860 just as William had been in 1817.[27]

Although the Derby philosophers were predominantly Liberals/Whigs, dissenters and reformers, the involvement of local Tories in scientific culture between 1780 and 1850 confirms that associations between dissent, reform and natural philosophy should not be exaggerated. These included John Chatterton senior and John Chatterton junior (1771–1857) and Henry Browne (1760–1831), all of whom were skilled mechanics with interests in natural philosophy. The Chattertons were lead merchants, plumbers and glaziers who lived in Amen Alley and became central figures in Derby's urban elite.[28] John senior installed plumbing and heating systems designed by John Whitehurst for the Duke of Newcastle at Clumber Park in Nottinghamshire.[29] A mechanic and friend of Darwin and Abraham Bennet, he conducted chemical experiments and was presented with a silver coffee pot for an essay on utilising waste ground for growing crops during the Napoleonic wars. Either he or his father developed a kind of manure known as 'Chatterton's Compost'.[30]

Although he may not have joined the Derby Philosophical Society, the chemist Henry Browne was another prominent Tory inventor with interests in natural philosophy. Having inherited an apothecary business from his father, Browne expanded the concern into a chemical manufactory. This was evident at the auction of the business in 1831, which included 'many large coppers, cast-iron boilers, stills, drying frames, starch machine, crane, pulleys, blocks, two large rolling mills with shafts, etc.; complete, marble and cast-iron weights', and a large variety of chemicals, tubs, barrels, carboys, bottles and other equipment.[31] He obtained patents for 'making and preparing extract of zinc for medicinal purposes' in 1799, and for the 'construction of boilers, to effect a saving of fire and the consumption of smoke' in 1821, the latter being utilised by many manufacturers after the Act of Parliament for the abatement of nuisances from steam engine furnaces came into force that year.[32] Browne was apparently one of the first in Derby to light his house with coal gas produced in his own manufactory. Other innovations included a method for preserving seeds from destruction by vermin, a technique for making manure, and a boiler which he described as an 'evaporator' for removing moisture from substances such as malt.[33] Possibly with encouragement from Darwin, a model of the latter was presented to the Royal Society of Arts, who awarded him a gold medal after he had travelled down to London to explain the operation in person, commenting that, 'evaporators on this construction promise to be of very considerable utility in many great works in this kingdom'. The chief advantage of Browne's evaporator was claimed to be the rapidity with which it carried off moisture using a current of hot air, saving fuel, time and expense. He had two of them operating at his manufactory by 1792 and found that with them one man could 'do more than three can in the usual method'.

The Chattertons and Browne were certainly not marginal figures and made important contributions to public campaigns, charitable institutions and scientific culture. However, they stand out politically as Tories and opponents of political reform. They were prominent freemasons and members of the Tyrian Lodge, which tended to be predominantly Tory, Chatterton senior being a founder member and Browne a master. Chatterton junior was adjutant of the Derby militia, alderman and mayor in 1832, and Browne served as alderman and mayor in 1799 and 1808, chairman of the Improvement Commission, lieutenant in the provincial cavalry and a director of the Palladium Fire Insurance Company. Browne also supported the enclosure of Nun's Green, the abolition of slavery and establishment of Sunday schools, and Chatterton junior subscribed to the Derbyshire Infirmary and served as president of the Derby Philosophical Society from 1824.[34]

The London Mechanics' Institute, which served as a model for the Derby foundation, had aimed at the instruction of the members 'in the principles of the arts they practise or in the various branches of science and useful knowledge connected therewith', and had provided a library, a museum, public lectures and an experimental laboratory. In provincial urban centres such as Liverpool, Manchester and Sheffield there was often cross-party support at the outset of the establishment of mechanics' institutions.[35]

In 1824 Joseph Strutt enquired of James Shuttleworth, the Strutts' agent in Manchester, about the rules of the mechanics' institute there and the rules of the institutions at Glasgow and Edinburgh.[36] A letter in the Tory *Derby Mercury* referred to the establishment of an 'Artisans and apprentices library in this town' to enlarge the 'understanding of the mechanic'. The writer hoped to 'see the delightful picture of the Industrious workman reading for recreation, to his wife and family, by his own fireside, books of useful information'. The gentry were urged to support it, and it was emphasised that no books of 'political or religious controversy' should be admissible.[37] The Whig *Derby Reporter* also supported the 'proposed plan for a mechanics' library in the town of Derby'. A correspondent thought that 'industrious mechanics' only wanted the opportunity to 'gain knowledge' so that they could learn 'the true value of liberty', loyalty and contentment, and claimed that the institutions at Sheffield, Edinburgh and London would serve as examples.[38] Donations of books were promised, though the emphasis was on artisan self-motivation and the *Reporter* urged mechanics to get into contact so that 'those gentlemen who can and will give the necessary support, will feel that their public spirit is not thrown away on objects unworthy of their patronage'.[39] While the Liberals took the lead, their rhetoric was intended to promote conservative values and win support from all religious and political groups, although the emphasis was different in the Whig *Reporter* from the Tory *Mercury*. In the *Reporter*, the 'progress of arts' and the 'value of liberty' were cited as benefits of scientific education for the 'industrious mechanic', whereas in the *Mercury* the need for exclusion of 'political and religious controversy' was emphasised and the proposed project was more limited, being restricted to a library. The appeal was to public consensus and the 'extension of enquiry and knowledge' and not to control of the workers.[40]

In a pamphlet of 1825, Higginson treated the establishment of the Derby Mechanics' Institute as the next logical step in extension of educational provision after Bell, Lancasterian and Sunday schools. He contended that in no portion of society was the 'dissemination of knowledge more valuable' than among the 'working mechanics', whose mental and moral condition would be improved.[41] Supported by the speeches and other activities of the Strutts, like Brougham and Edward Baines in Leeds, Higginson emphasised the need for self-management by the mechanics, although in practice this usually failed to be realised.[42] Though opposing the concept of class distinctions in other contexts, in print Higginson expressed the ideals of middle-class promoters of the Mechanics' Institute in language designed to transcend religious and political divisions, while underscoring the importance of free trade exemplified by the Strutts' opposition to factory legislation.[43] With a scientific education, the mechanic would improve his mind with principles applicable in daily life and be more likely to succeed in industry without wasting himself in search of 'visionary schemes for golden patents' to the ruin of himself and his family.[44] He would thus gain in independence, have greater respect for property, laws, and the 'due subordination of rank', be less dependent upon charity and less prone to losing himself in licentious pleasures. Family and home would be revitalised and the happiness of his children, wife and family immeasurably increased. Edward Strutt suggested that the Derby Mechanics' Institute might help to reduce the likelihood of industrial protests by the workforce. His father, William, suggested that there was nothing special about science and anyone could potentially make important discoveries, while his uncle Joseph offered books from his library.[45] Such language was very similar to that employed by Liberal supporters of mechanics' institutes elsewhere. At Leeds, Edward Baines, the proprietor of the Whig *Leeds Mercury* and principal supporter of the Leeds Mechanics' Institute, wrote in 1826 that their function was one of 'practical utility' and that the 'highest of all the benefits' of the mechanics' institutes was 'their effects on the morals and habits' of the 'middle and working' or 'middle and operative classes'.[46]

Membership and control

Support from skilled labourers and different religious and political groups was evident in the first few years, and middle-class leaders such as Higginson and William Strutt strove to encourage a form of organisation in which master mechanics took control but where membership was accessible to all.[47] One master carpenter, the Congregationalist, Joseph Cooper, speaking for the mechanics, argued that the Derby Mechanics' Institute would continue the work of a Journeyman Carpenter's Society which had been created in Derby around 1800 and lasted for eight or nine years. The original committee included the builder Thomas Cooper, Anglican clergyman the Rev. Charles Birch, Unitarian lace manufacturer and machine maker William Wigston, draftsman R. V. Knight, Tory attorney William Eaton Mousley, the dissenting ministers Higginson and Gawthorne, the physician Forester, Thomas Wright, a Unitarian silk throwster, and the Strutt brothers. Rules were drawn up based upon the London Mechanics' Institution and resolutions followed the lines suggested by Higginson. A library was formed with a reading room, and lectures were to be held on natural and experimental philosophy, practical mechanics, astronomy, chemistry, literature and the arts. It was agreed that quarterly subscriptions to admit a member to the library and reading room were not to exceed 2s. 6d.[48] Subscribers included aristocrats such as the Duke of Devonshire, professionals such as Forester, and members of wealthy manufacturing and industrial families such as the Strutts and the Evanses.[49] Trustees and a management committee were created headed by William Strutt, with Joseph Strutt and J. H. Smith as vice-presidents, and Higginson as secretary. The original general committee was made up of traders and skilled craftsmen, including engineers, printers and joiners, but, as we have said, the Institute remained largely under middle-class control and patronage, and in the first few years Joseph and William Strutt and the Duke of Devonshire were the presidents.[50]

The Derby Mechanics' Institute seems to have been relatively popular and had some success in providing scientific education to some of the working class and middle class between 1825 and 1855. By April 1825, 274 members had enrolled, and attendance at scientific lectures usually numbered between 200 and 600, but could reach 800. There were membership fluctuations, particularly during recessions, manifest in non-payment of fees, resignations and the need for means testing. It became increasingly clear that the Lancasterian schoolroom was inadequate and courses of lectures were also too expensive.[51] In its first five years, the Institute offered a detailed programme of lectures on subjects including astronomy, natural history, chemistry, phrenology and electricity. As we have seen, some women also attended, and there was a junior section and classes for reading, arithmetic, French, chemistry and other scientific subjects that met on a regular basis. In 1826, for example, William Nicol lectured on natural and experimental philosophy for which he charged the large sum of £60. Non-members were admitted for £1 1s. for the course, or 2s. 6d. per lecture, with each ticket admitting two persons under the age of fifteen.[52] Douglas Fox, in contrast, was offered £20 for his services in 1825–26, only accepting half this amount. William Wilson, a reformer and Unitarian, gave gratuitous instruction on several evenings a week in mathematical subjects.[53] In March 1829, Edward Strutt gave a course of lectures on 'The Elementary Principles of Political Economy', intended to illustrate the power of laissez-faire economics and impotence of interference in industry.[54] During the 1830s, there was a membership revival and lectures were offered on a range of subjects attracting large audiences, including astronomy, slavery, mental education, comparative anatomy, phrenology, and chemistry by John Murray. Classes were organised in the 1830s on subjects such as general debating (1831), mutual instruction (1834), phrenology (1832), and chemistry (1835).[55]

The first set of lectures on anatomy and chemistry given by the physician Douglas Fox in 1825 and early 1826 received extensive local newspaper coverage and provide much more information than is usually available about the content, teaching methods, equipment and apparatus employed for these events. Thomas Noble, the Unitarian editor of the *Reporter*, published an account of the lectures based upon his own notes.[56] Although no list of subscribers appears to survive, some information concerning audience reaction is given, with some 200–400 individuals attending, the numbers rising towards the end in response to newspaper publicity.[57] The involvement of the Derby philosophers was crucial to the venture and the lectures were prepared with the assistance of Charles and Francis Fox, the Strutt brothers, Higginson and Flack.

Fox proceeded from elementary to the abstract and hypothetical using plenty of concrete examples and experimental demonstrations in accordance with contemporary progressive pedagogical theories, such as the Pestalozzian emphasis upon demonstrations and object lessons. He constantly emphasised the utility of scientific knowledge for mechanics, though it was recognised that there would be difficulties in 'adapting theoretical conclusions' to 'practical materials'. Addressing the concerns of Tory opponents and doubts concerning the relevance of mechanics' institutes, he argued that it was 'strange' to ask what mechanics had to do with chemistry and anatomy when all mechanical arts including the 'wonderful steam engine' were so 'dependent upon the chemical agencies of the materials' they employed, and when the 'strength, health, and permanent labour' of artisans was related to his anatomical construction. He also provided reassuring emphasis upon his authority as guide and director, arguing that the lectures would allow mechanics to gain 'a correct outline of the truth', from which they could proceed to other sources including the valuable library.[58]

The impact of methods employed by popular scientific lecturers is evident from the range of apparatus employed by Fox and the deliberate use of striking and original demonstrations. The lectures were illustrated by active experiments and a wealth of apparatus, models, drawings, dried specimens, new inventions and interesting objects. Each lecture had a popular element to entice the audience towards more abstract scientific analysis, such as the exhibition of frozen mercury, electrical explosions, a model of the heart, chemical explosions, meteorites and an Egyptian mummy. These were designed to be original and stimulating – even startling. Thus, with a touch of theatrical gothic frisson, the mummy had been opened the day before, and the coloured life-size plaster model of the human being had been especially constructed by Francis Fox. In the same way, when exploding the chloride of nitrogen, the danger was emphasised beforehand and Douglas had been dramatically protected by the apparatus made by his brother Francis.[59] The dramatic opening of the mummy in particular drew the crowds, and helps to explain why audience numbers rose over the weeks. The quality of the lectures judged in terms of content, teaching methods and audience size indicate that Fox's lectures had a considerable impact upon those who attended. Through the circulation of reports in the newspapers and Noble's published volume, the audience would have numbered thousands, although the precise degree of prior knowledge, social character and status of the audience is difficult to assess.[60]

Although some scientific apparatus was procured, such as the mathematical instruments purchased by William Strutt in 1827, as the members of the Derby Literary and Philosophical Society had discovered, it was difficult and expensive to maintain the apparatus for a working laboratory.[61] The accommodation problem was eased in 1833 when the Mechanics' Institute moved into new premises; however, only with the construction of a new lecture hall in 1837 in the Wardwick did it have special purpose-built premises to house the library of 6,000 volumes and instruments and provide a reading room and lecture rooms. The success of the Derby Mechanics' Institute was underscored by an ambitious exhibition in 1839, which raised some £2,119 and featured an array of artistic, literary, natural history, industrial and experimental exhibits, including electrical motors and powerful galvanic troughs. The exhibition received extensive coverage in the local newspapers and also the regional, and to some extent, national press, and was visited by almost 100,000 people from Derby and beyond, who were able to take advantage of the new railway connections.[62]

The extent to which the Derby Mechanics' Institute was a civic scientific institution is evident from the celebration of local and national scientific and mechanical talent and achievement in lectures and at the 1839 exhibition. Douglas Fox frequently alluded to Davy and Watt and emphasised, just as other 'self-help' authors such as Samuel Smiles did, that these men were now considered to be 'immortal', the saints of the industrial age, who had come from relatively humble origins. Continual reference to the work of the Fox brothers and the Derby philosophical community contributed to a local scientific 'mythology' also evident at the 1839 exhibition, where inventions and portraits relating to the Strutts, the Foxes, Darwin, Sylvester, Whitehurst, Bennet, Brooke Boothby, Arkwright and Forester were prominent.[63] In his opening lectures, Douglas Fox referred to Abraham Bennet and Francis Fox's inventions such as the capillary thermometer and steam vacuum cupping ball, the latest in what was portrayed as a local tradition of scientific and technological innovation. The mechanics were encouraged to enter into the process of invention by savouring some of the excitement of the Foxes as they actively created their own inventions. This was reinforced, at the instigation of the Strutt brothers and the committee, by cash prizes of £15 for individual improvements. One prize in June 1829 was given for an essay on 'The advantages which may be anticipated from the diffusion of knowledge among the operative classes of the community'. Others in 1830 went to a steam engine and a lock or door fastener, which had competed with models of a chain bridge and a weighing machine.[64]

The role of the Derby philosophers

As we have seen, members of the Derby Philosophical Society were instrumental in promoting civic science and rational recreational institutions. Individuals from the Strutt family, especially William, Joseph and Edward, were, of course, prominent in civic science throughout the period, as were ministers at the Friargate Unitarian Chapel, especially Edward Higginson and his successor, Noah Jones. They urged the foundation of new public scientific institutions as a moral and religious imperative. In a sermon of 1820, for instance, Edward Higginson exhorted his listeners to 'extend the most unfeigned and comprehensive charity to all who differ from you', which must 'consist in something more than candid expressions: it must shew itself in kind thoughts and benevolent actions'.[65] Similarly, Noah Jones noted of Joseph Strutt in 1844 after his death, that 'as a protestant dissenter, he was the consistent supporter of civil and religious liberty' and of 'all public institutions, which have for their object the promotion of human virtue and happiness, especially such as aim at improving the social state of the great mass of the community'.[66] Medical men and Philosophical Society members such as Richard Forester, Thomas Bent and Douglas Fox also played a leading role in the foundation of the Mechanics' Institute and, as we shall see, the Derby Arboretum, and after Strutt's death, Alderman Francis Jessop, William Mousley and Andrew Handyside continued the tradition of support from the philosophical community.[67] In 1851, Bent, then president of the Philosophical Society for instance, contributed towards the building of a new lodge at the Arboretum and towards the erection of a commemorative statue to Joseph Strutt in the park. Subsequently he donated money towards an imitation crystal palace and bequeathed £200 towards the enlargement or improvement of the Arboretum in his will, while Forester left a large legacy for the Arboretum extension fund.[68]

Town and County Museum (1835)

The success enjoyed by the Derby Mechanics' Institute and the prominence of Whig supporters helped to encourage the foundation of other civic scientific institutions more avowedly intended for the middle classes. In 1831, 'A Freeholder' called for the foundation of another scientific and literary society to reduce 'factious feeling' in Derby with a public library, reading room and newsroom where busts of worthies could be deposited.[69] Dr. W. H. Robertson and the Rev. R. Wallace continued the agitation for another scientific society, resulting in the foundation of a Chesterfield Literary and Philosophical Society with the assistance of the Fox brothers.[70] In 1834, 'Amicus', probably Dr. George C. Watson the physician, called for another scientific society to share the rooms of the Derby Mechanics' Institute and for greater co-operation between mechanics and the members of the new society towards the 'promotion of useful knowledge' in the spirit of the desires of 'Messrs Strutt', and he condemned Tories who opposed the extension of scientific education. Some Tories such as Thomas Mozley ridiculed the notion of scientific mechanics during the 1830s and 1840s, claiming that Derby was full of 'unfortunate inventors – that is, poor creatures who had lost themselves in the scent of some discovery which they could never make, or which others had made before them'. However, 'no warning would stop the madness.'[71]

The most significant result of these implicit criticisms of the Mechanics' Institute was the foundation of the Derby Town and County Museum and Natural History Society in 1835 by Dr. George C. Watson and others; it moved to premises in the newly erected Athenaeum in 1840.[72] Although the museum was able to attract support from Whigs and Tories alike, statements of aims and objectives, subscription lists, the cost of subscriptions, and the fact that non-subscribers were excluded, reveal that it was a more avowedly elite venture which enjoyed greater patronage from Anglicans and Tories than the Mechanics' Institute. The initial location of the Museum within the Athenaeum also places this institution as a more emphatically elite and civic scientific venture than the Mechanics' Institute. This also reflected the nature of the county constituency for natural history and the fact that the scientific element of this venture was, of course, less significant than for the Mechanics' Institute. In response to the success of the 1839 exhibition and the opening of the Derby Arboretum, the Museum and Natural History Society also held its own exhibition in 1843.[73]

The wide range of exhibits at the Museum collection and 1843 exhibition included geological specimens, fossils, 'skeletons in comparative anatomy', stuffed birds and animals, and an Indian chief's clothing, as well as numerous works of art and antiquities. As was the case in most other museums of the period, these seem to have been displayed quite haphazardly and not according to any unifying principles. In 1842 the Tory *Mercury* felt 'a pleasure in noticing' the Town and County Museum as a 'rising institution', considered the collection of specimens to be 'highly creditable', and argued that the time was 'not far distant when [Derby] will be possessed of a museum corresponding with the increasing magnitude and importance of the town'. In a pointed snub to the Liberal-dominated Mechanics' Institute, it remarked that only then would Derby be 'distinguished as much by its literacy and scientific institutions as it is by its commercial prosperity'. The museum provided a valuable encouragement for 'rational recreation' and natural theology, natural history being a 'source of delightful recreation and intellectual improvement' particularly for the young, leading from contemplation of nature to adoration of 'the omnipotent architect of the universe, the great first cause of all things'.[74]

Derby Arboretum (1840)

The Derby Mechanics' Institute and Town and County Museum and Natural History Society were important local manifestations of types of scientific institutions prevalent in many British towns. Although the endeavours of the Derby philosophers were important, they were also inspired by similar organisations founded elsewhere. The Derby Arboretum was much more innovative than these in national terms, although also inspired by the foundation of botanical gardens, museums and similar institutions elsewhere, particularly the work of John Claudius Loudon, the leading early-Victorian landscape gardener and horticultural writer, as well as being a response to the local social, political and economic situation. It resulted from the collaboration between Joseph Strutt, the local benefactor, and Loudon, and reflected the shared social, educational and scientific objectives of Loudon and the Derby philosophers, resulting from a relationship between the two that began long before 1839.

Loudon shared the political objectives of the Strutts and the Derby philosophers, and their progressive enlightenment belief in the value of scientific education and institutions. Inspired by Bentham and other reformers, Loudon used his many publications, especially *The Gardener's Magazine*, to campaign for the creation of urban parks, botanical gardens and educational institutions for all social classes. On the other hand, the botanical concerns of the Derby philosophers were well known and continued after Darwin's death. William Brookes Johnson and Brooke Boothby were members of the London Botanical Society, for instance, while Forester, Joseph Strutt and Thomas Bent were fellows of the Linnean Society and patrons of the Derby and Derbyshire Horticultural and Floral Society.[75] In addition, Bent and other Arboretum promoters were involved in the Midland Horticultural Society, and, not surprisingly, editions of Loudon's works were stocked in the libraries of the Derby Philosophical Society and the Mechanics' Institute.[76]

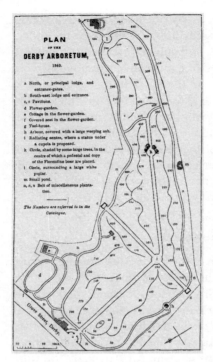

Loudon and the Derby philosophers were also eager to apply innovative technology within horticulture, particularly the development of heating systems and hot-houses for gardening and cast iron for construction. Loudon admired William Strutt's work, meeting him and other family members during his various tours, and praised the innovations undertaken by the Strutts on their estates and workers' settlements. Some of these were enthusiastically reported in John Farey's *General View of the Agriculture and Minerals of Derbyshire* (1811–17), which Loudon strongly recommended as one of the best works of its kind, including cottage-window staybars, door staybars, lodge gate fastenings, and trussed iron rod girders and rafters utilised at Derby, Milford and Belper. The systems of heating and ventilation used in the Derbyshire General Infirmary and William and Joseph Strutt's houses, kitchens and hot-houses were

36 Plan of Derby Arboretum, from J. C. Loudon, *Derby Arboretum* (1840).

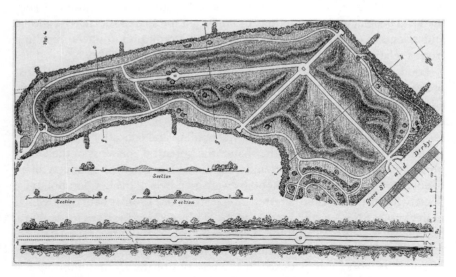

37 Relief plan of Derby Arboretum, from J. C. Loudon, *Derby Arboretum* (1840).

described and praised by Loudon, who regarded William Strutt as 'a man of most extraordinary genius' and recommended Sylvester's *Philosophy of Domestic Economy* as the basis for the heating system 'now in general use throughout Britain for large buildings'.[77]

In creating the Derby Arboretum, Strutt and Loudon wanted an institution that would facilitate access to public gardens for all social classes while providing a botanical education. Hence non-subscribers were admitted on two days each week and provision was made for visitors in lodges at the entrances, and of lawns for the erection of tents for dancing and other popular events. Strutt had originally considered a botanical garden but the idea was rejected as too costly to maintain, in favour of a special place for the display of trees and shrubs. Strutt laid down various stipulations including the retention of a plantation, flower garden, provision of lodges and public access, although the design primarily reflected the application of Loudon's 'gardenesque', intended to encapsulate the best from Humphry Repton's notion of the picturesque as applied to botanical gardens. Thus the design was supposed to have 'distinctness in the separate parts when closely examined', but, when viewed as a whole, was 'governed by the same general principles of composition as the picturesque style, the parts, though blended, being yet connected'.[78] The Arboretum was a 'living museum' dominated by a collection of one thousand foreign and native trees and shrubs and served as physical realisation of the *Arboretum et Fruticetum Britannicum*. Single specimens were carefully situated in their own space along paths with their own logic and meaning, terminating in lodges, benches or fountains and set within undulating mounds, intended to foster an illusion of size. Just as readers travelled through time and round the world surveying hundreds of trees and shrubs in the virtual arboretum of Loudon's magnum opus, so a walk through the Derby Arboretum from the North Lodge on Grove Street to Osmaston Road allowed them to survey global botanical riches. The Jussieuan system was more 'natural' and less artificial than the Linnaean, grouping plants according to the greatest number of their natural similarities of form, much of which was visible to the non-expert.[79] The specimens, some from the Horticultural Society's garden in London, were arranged to give a sense of variety, yet produce no 'violent contrasts'. They followed a logical order so that plants of the 'torrid zone' did not mix with plants of the 'frigid zone'.[80] This followed Loudon's principle of relation, an idea copied in later larger nineteenth-century parks in Manchester, Glasgow, Liverpool, London and other places.[81] Like objects in the museum, each plant was positioned, labelled, catalogued and artificially isolated for study, revealing order and meaning in the natural world, while providing botanical education.[82]

38 Statue of Joseph Strutt and Derby Arboretum anniversary festival,
Illustrated London News, 8 July 1854.

The impact of scientific institutions

Judged in terms of popularity, there is no doubt that the Derby Mechanics'
Institute, Derby Arboretum and Derby Town and County Museum were
successful. As we have seen, the audience for Mechanics' Institute activities
quickly numbered thousands, with hundreds attending lectures and classes and
using the library, many more reading accounts of lectures, and some 100,000
attending the 1839 exhibition. External observers such as Loudon praised the
exhibition as an event that could not 'fail to have an excellent effect', while
noting 'on good authority' (presumably Joseph or Edward Strutt) that after
only a fortnight, it had been visited by more than 20,000 people.[83] Likewise
the lengthy opening ceremonies for the Arboretum in September 1840, spread
over three days and including processions from the corporation, trades and
societies and children's day celebrations, had a major impact upon urban
culture and society. The creation of the Arboretum and events such as the
anniversary festivals attracted considerable attention in the local, regional and

national press, and the park was visited by many horticulturists, writers and others, and provided an important model for the creation of similar institutions in British towns and further afield.[84] The anniversary festivals, for which special trains were laid on, generated considerable income by attracting tens of thousands of visitors from northern and midland counties, boosting the local economy. The Derby Arboretum encouraged greater attention to exotics and botanical collections in public parks, and public arboretums were founded in other Victorian towns including Nottingham (1852), Ipswich (1853), Walsall (1873), Worcester (1859) and Lincoln (1872).[85] Attempts to transform the Royal Victoria Park in Bath into an arboretum, for example, including the proposed system of labelling encouraged by the surgeon Frederick Hanham, were expressly inspired by the Derby Arboretum, which was regarded as one of the most authoritative provincial British tree collections.[86] Of the other Victorian public arboretums, although the Upper Arboretum at Ipswich was designed by William Pontey who had worked on the Derby park, that at Nottingham, designed by Samuel Curtis as part of a major enclosure venture, was most closely modelled on Derby and featured a labelled collection of trees and shrubs situated on landscaped slopes.[87]

Derby observers such as the minister Noah Jones were, unsurprisingly, enthusiastic, claiming that the 'noble gift' of the Arboretum would enhance the 'rational social pleasures of mankind', particularly the 'toilworn artisan', and all should 'go and do likewise' with 'his own gifts and opportunities'. Middle-class promoters claimed that the behaviour of working-class visitors to the exhibitions and Derby Arboretum demonstrated that valuable exhibits could be viewed with respect and without damage by all social classes, encouraging civic responsibility and communal loyalty. The Arboretum commissioners concluded that the park had 'already produced a perceptible effect in improving the appearance and demeanour of the working classes and … doubtless, conferred an equal benefit upon their health'. Martin the physician argued that it had brought 'immense advantages, moral and physical, that must accrue to the inhabitants of closely-built towns by the establishment of public parks … and gardens like that presented to the town of Derby'. National experts such as Edwin Chadwick for the most part agreed, extolling the health and moral benefits and calling for munificent benefactors to come forward in other towns or suggesting that local taxation should pay for public parks. Years after his death, Joseph Strutt's example was still celebrated in the national newspapers, quotations from his inaugural speech being given to inspire similar foundations. International visitors tended to be equally enthusiastic.[88] Charles Mason Hovey, the proprietor of an American nursery in Cambridge, Massachusetts and editor of the *Magazine of Horticulture*, considered it to be easily the best he had seen in Scotland, England and France, and superior to Chatsworth.[89] The landscape gardener Andrew Jackson Downing, for instance, praised the Arboretum highly after visiting in 1850, especially Strutt's role as benefactor, and urged that his countrymen follow suit. However, he found Robert Marnock's botanical garden in Regent's Park, 'in point of tasteful arrangement and beauty of effect', could not have been more different and had 'none of the stiff and hard look of the surface of the arboretum at Derby'.[90]

Conclusion

The Derby Mechanics' Institute, Town and County Museum, Arboretum and other civic scientific institutions were utilitarian ventures, designed to be instructional and provide forms of pleasure and rational recreation to the citizens of Derby and elsewhere. As Joseph Strutt stated in his opening speech in September 1840, the thousand specimens of trees and shrubs were arranged and described for the educational benefit of public visitors, to be intellectually and morally uplifting. As the sun had shone on him in life so he wanted to give something back to those whose toil had helped to create his wealth.[91] Just as the philosophers had supported the Mechanics' Institute, schools, scientific societies, libraries and museum, so they helped to create Loudon's 'living museum'. The Arboretum and the Mechanics' Institute were responses to the problem of rational recreation that had occupied many reformers from the 1820s to the 1840s and resulted in the banning of the Shrovetide football. This posed the problem of how the strains of a rapidly industrialising society could be prevented from resulting in immorality and licentiousness among the labouring population. Thus the Arboretum can be interpreted as a paternalistic attempt to regulate the leisure activities of the middle and working classes, though such an explanation in terms of social control only partly explains the motivations of the Strutts and the Derby *savants* who were ideologically committed to promoting scientific education and social progress. Although some observers had problems with aspects of the design and the Arboretum failed, of course, to help allay problems of urban health and sanitation, the excitement of this vision meant that the institution had a national and international impact.

Notes

1 P. Elliott, 'The Derby Arboretum (1840): the first specially designed municipal public park', *Midland History*, 26 (2001), 144–76; J. C. Loudon, *The Derby Arboretum* (London, 1840); P. Elliott, C. Watkins and S. Daniels (eds.), *Cultural and Historical Geographies of the Arboretum*, special issue of *Garden History* (2007).

2 J. Wigley, 'Derby and Derbyshire during the great reform bill crisis, 1830–1832', *Derbyshire Archaeological Journal*, 101 (1981), 139–49.

3 *Derby Mercury* (14 June, 1 November 1826, 15 February, 4 March 1829).

4 S. Glover, *History and Directory of the Borough of Derby* (Derby, 1843); J. Pigot,

National Commercial Directory [for the Midlands, the North and Wales] (Manchester, 1835); L. Jewitt, *Guide to the Borough of Derby* (Derby, 1852); H. J. Dyos and M. Wolff (eds.), *Victorian City: Images and Realities* (London: Routledge & Kegan Paul, 1973); R. J. Morris (ed.), *Class, Power and Social Structure in British Nineteenth-Century Towns* (Leicester: Leicester University Press, 1986); D. Fraser, *Power and Authority in the Victorian City* (Leicester: Leicester University Press, 1979); D. Eastwood, *Government and Community in the English Provinces, 1700-1870* (Basingstoke: Macmillan, 1997).

5 Glover, *History and Directory* (1843), 56.

6 J. D. Standen, 'The Social, Economic and Political Development of Derby, 1835–1888' (MA dissertation, University of Leeds, 1959).

7 E. G. Barnes, *The Rise of the Midland Railway, 1844–1874* (London: Allen & Unwin, 1966); J. R. Kellett, *The Impact of Railways on Victorian Cities* (Routledge & Kegan Paul, 1969); G. Revill, 'Liberalism and paternalism: politics and corporate culture in railway Derby, 1865–75', *Social History*, 24 (1999), 196–214.

8 Glover, *History and Directory* (1843), 80–1.

9 S. Smiles, *Industrial Biography: Iron Workers and Tool Makers* (London, 1897), 258–60; R. S. Woodbury, 'History of the lathe to 1850', in Woodbury, *Studies in the History of Machine Tools* (Cambridge MA: Massachusetts Institute of Technology, 1972), 108–12.

10 W. Baker, MD, 'Report on the sanitary condition of the town of Derby', in *Local Reports on the Sanitary Condition of the Labouring Population of England* (London, 1842), 162; R. Rodger, *Housing in Urban Britain, 1780–1915* (Basingstoke: Macmillan, 1989); D. Reeder, 'Slums and suburbs' in Dyos and Wolff (eds.), *Victorian City*, 359–86.

11 J. R. Martin, report on Derby in *Appendix, Part II, to the Second Report of the Commissioners of Inquiry into the State of Large Towns and Populous Districts* (London, 1845), 271–8; Baker, 'Report on the sanitary condition', 164.

12 G. F. Chadwick, *The Park and the Town: Public Landscape in the 19th and 20th Centuries* (London: Architectural Press, 1966), 44 ; S. Lasdun, *English Park* (London: André Deutsch, 1991), 119–52; Loudon, *Derby Arboretum*, 83; M. L. Simo, *Loudon and the Landscape* (New Haven: Yale University Press, 1986), 236; H. Conway, *People's Parks: The Design and Development of Victorian Parks in Britain* (Cambridge: Cambridge University Press, 1991), 34–40.

13 E. Higginson, letter to E. Higginson junior, 26 April 1825, James Martineau papers, Harris Manchester College, Oxford.

14 S. Glover, *The History, and Gazetteer, and Directory of the County of Derby*, 2 vols. (Derby, 1829, 1833), II, 406.

15 A. Delves, 'Popular recreation and social conflict in Derby, 1800–1850', in E. Yeo and S. Yeo (eds.), *Explorations in the History of Labour and Leisure* (Brighton: Harvester Press, 1981), 89–127.

16 A. Delves, 'Popular recreation and social conflict', 104.

17 Glover, *History and Directory* (1843), 58; Jewitt, *Guide*, 56–7.

18 T. Mozley, *Reminiscences, Chiefly of Towns, Villages and Schools*, 2 vols. (London, 1882), I, 259; *Derby Mercury* (15 October 1818); Franklin, course of lectures in astronomy (16–18 October 1832), broadside box 14, no. 33, DLSL; I. Inkster, 'Studies

in the Social History of Science in England during the Industrial Revolution' (PhD thesis, University of Sheffield, 1977), 469.

19 M. D. Stephens and G. W. Roderick, 'The great private initiative in nineteenth century scientific and technical education: the Mechanics' Institutes', in M. D. Stephens and G. W. Roderick (ed.), *Essays on Scientific and Technical Education in Early Industrial Britain* (Nottingham: University of Nottingham, 1981), 48–50; S. Laughton, 'Derby Mechanics' Institution … a hundred years of valuable life', in *The Derby Daily Telegraph* (12, 13 February 1925); A. F. Chadwick, 'The Derby Mechanics' Institute', *The Vocational Aspect of Education*, 27 (1975), 103–5; 'The Derby Mechanics' Institute, 1825–1880' (MEd dissertation, Manchester University, 1971).

20 J. Granger, *History of the Nottingham Mechanics' Institute, 1837–1887* (Nottingham, 1887); F. B. Lott, *Story of the Leicester Mechanics' Institute, 1833–1871* (Leicester, 1935); Laughton, 'Derby Mechanics' Institution'; J. W. Adamson, *English Education, 1789–1902* (Cambridge: Cambridge University Press, 1964), 38–42, 155–70; S. J. Curtis, *History of Education in Great Britain*, seventh edition (London: University Tutorial Press, 1967), 471–8; A. D. Garner and E. W. Jenkins, 'The English Mechanics' Institutes: the case of Leeds, 1824–1842', *History of Education*, 13 (1984), 139–52; M. D. Stephens and G. W. Roderick, 'British artisan scientific and technical education in the early 19th Century', *Annals of Science*, 29 (1972), 1; G. W. Roderick and M. D. Stevens, 'Science, the working classes and Mechanics' Institutes', *Annals of Science*, 29 (1972), 4; T. Kelly, *A History of Adult Education in Great Britain*, third edition (Liverpool: University of Liverpool, 1992); B. Simon, *The Two Nations and the Educational Structure, 1780–1870* (London: Macmillan, 1976); S. Shapin and B. Barnes, 'Science, nature and control: interpreting Mechanics' Institutes', *Social Studies of Science*, 7 (1977), 31–74; C. Russell, *Science and Social Change* (London: Macmillan, 1983), 154–77; I. Inkster, *Scientific Culture and Urbanisation in Industrialising Britain* (Aldershot: Ashgate, 1997).

21 Shapin and Barnes, 'Science, nature and control'.

22 J. Rose, *The Intellectual Life of the British Working Classes* (New Haven: Yale Nota Bene, 2002), 58–91.

23 A. T. Patterson, *Radical Leicester* (Leicester: Leicester University Press, 1954), 172, 235–6.

24 Chadwick, 'Derby Mechanics' Institute', 30.

25 L. Horner, letter to W. Strutt (Nottingham, 28 July 1821), quoted in Chadwick, 'Derby Mechanics' Institute'.

26 *Derby Mercury* (6 February 1828); *Derby Reporter* (11 May 1826); Adamson, *English Education*, 89–94; Curtis, *History of Education*, 421–30.

27 Chadwick, 'Derby Mechanics' Institute', 30; E. Strutt, letter to F. Strutt (25 February 1831), Strutt Correspondence, DLSL, D 125/-.

28 Pigot, *National Commercial Directory* (1835); Glover, *History and Gazetteer*, II, 601; Derby Local Studies Library, *Queries Answered*, 153 (1966); M. Craven, *Derbeians of Distinction* (Derby: Breedon, 1998), 54.

29 *Derby Mercury* (2 October 1832).

30 *An Alphabetical List of Subscribers to the Derbyshire General Infirmary* (Derby, c.1809); Derby Philosophical Society, cash ledger (1813–47), DLSL, BA 106; Craven, *Derbeians*, 54.

31 *Derby Reporter* (1 December 1831).

32 B. Woodcraft, *List of Patents* (London, 1851), 75.

33 H. Browne, Letter to the Society of Arts on the subject of the preservation of seeds from vermin (13 March 1793), *Transactions of the Society of Arts*, 11 (1793), 143–4; H. Browne, 'Description of a furnace described as an evaporator', *Transactions of the Society of Arts*, 12 (1794), 257–62; H. Browne, Letter to the Society of Arts on the subject of manure (27 April 1798), *Transactions of the Society of Arts*, 16 (1798), 268–71; Minutes of the Committee of Chemistry, Royal Society of Arts, 3 December 1792, 15 February 1794.

34 B. Tacchella, *Henry Browne: An Old Derbeian (1760–1831)* (Derby, 1903); Craven, *Derbeians*, 54.

35 Stephens and Roderick, 'The mechanics' institutes', 48–9.

36 R. S. Fitton and A. P. Wadsworth, *The Strutts and the Arkwrights, 1758–1830: A Study of the Early Factory System* (Manchester: Manchester University Press, 1958), 186; Chadwick, 'Derby Mechanics' Institute', 42.

37 *Derby Mercury* (23 June 1824).

38 *Derby Reporter* (24 June 1824).

39 *Derby Reporter* (23 December 1824, 6 January 1825).

40 *Derby Reporter* (6 January 1825).

41 Rev. E. Higginson, *Observations ... on the Establishment of Mechanics' Institutions* (Derby, 4 March 1825), 5–6; *Derby Mercury* (9 April 1825).

42 Garner and Jenkins, 'The English Mechanics' Institutes', 145–52, though the Leeds rules were changed to allow this only in 1837.

43 E. Higginson, *The Turn Out; An Inquiry into the Present State of the Hosiery Business* (Derby, 1821), 8.

44 Higginson, *Observations on Mechanics' Institutions*, 7–8.

45 Higginson, *Observations on Mechanics' Institutions*, 9–10; E. Higginson, *Address, Delivered on the Opening of the Derby Mechanics' Institution, August 22nd, 1825* (Derby, 1825), 5–6; *Derby Mercury* (23 March 1825, 27 May 1829); Chadwick, 'Derby Mechanics' Institute', 33–4.

46 Garner and Jenkins, 'The English Mechanics' Institutes', 150–2.

47 Higginson, *Observations on Mechanics' Institutions*, 10–11; *Derby Mercury* (23 March 1825).

48 *Derby Mercury* (23 March 1825).

49 *Derby Mercury* (13 April 1825).

50 Correspondence concerning the Derby Mechanics' Institute between J. Strutt and members of the committee, Galton papers, Birmingham Central Library, Archives and Heritage, MS3101/C/E/5/32, MS3101/C/E/5/46/1 and MS3101/C/E/6; Chadwick, 'Derby Mechanics' Institute', 54.

51 Chadwick, 'Derby Mechanics' Institute', 72–6.

52 *Derby Mercury* (6, 20 May, 15, 22 August, 6 September 1826).

53 Chadwick, 'Derby Mechanics' Institute', 75, 67; Pigot, *National Commercial Directory* (1835).

54 *Derby Mercury* (25 February, 11, 25 March 1829); Higginson, *Observations on Mechanics' Institutions*, 10. The institutional records are at Derby Local Studies Library, NRA 27869, DL1–64.

55 *Derby Mercury* (24, 31 March, 7 April, 3, 10 November, 8 December 1830, 16 February 1831).

56 *Modern Mayors of Derby ... from 1835 to 1909* (Derby: Derbyshire Advertiser, 1909), I, 4–5.

57 *Derby Mercury* (7 December, 11 January 1826).

58 D. Fox, *Notes of the Lectures on Anatomy and Chemistry, delivered by Mr. Douglas Fox* (Derby, 1826), preface, vi.

59 Fox, *Notes of the Lectures*.

60 Inkster, *Scientific Culture*, paper VI, 461–65, II, 80–107, II, 100–1.

61 Derby Mechanics' Institute Minute Book, 11 June 1827, DLSL.

62 *Catalogue of Articles contained in the Exhibition of the Derby Mechanics' Institution* (Derby, 1839); *Derby Mercury* and *Derby Reporter* (May to September 1839); see also T. Kusamitsu, 'Mechanics' institutes and working class culture: exhibition movements, 1830–1840s', in I. Inkster (ed.), *The Steam Intellect Societies* (Nottingham: University of Nottingham, 1985), 33–43.

63 Derby Mechanics' Institute, *Catalogue of articles contained in the exhibition* (Derby, 1839).

64 Derby Mechanics' Institute minute book, 27 April 1829, 15 June 1829, 7 June 1830, DLSL; Chadwick, 'Derby Mechanics' Institute', 70.

65 E. Higginson, *The Doctrines and Duties of Unitarians: A Sermon Preached before the Association of Unitarian Dissenters* (Lincoln, 1820), 14–15.

66 N. Jones, *A Discourse on Occasion of the Life and Lamented Death of Joseph Strutt* (Derby, 1844), 14–16.

67 Cash book of the Derby Philosophical Society, 1813–47, DLSL, BA 106.

68 Elliott, 'Derby Arboretum'.

69 *Derby Mercury* (11 May 1831).

70 *Derby Mercury* (11 June 1834, 11 November 1835); *Derby Reporter* (10 November, 1 December 1831); Inkster, 'Studies in the Social History of Science', 483.

71 *Derby Reporter* (20 November 1834); T. Mozley, *Reminiscences, Chiefly of Towns*, I, 214, 279–80.

72 G. Crewe, *A Few Remarks upon Derby Town and County Museum addressed to the Members of the Society at Derby* (Derby, 1839); *Rules of the Derby Town and County Museum and Philosophical Society* (Derby, 1859).

73 *Derby Reporter* (25 October, 1 November 1838); *Derby Mercury* (14 December 1842, 27 September 1848, 27 October 1852); honorary members' name book of the Derby Mechanics' Institute, DLSL, DL113.36; cash book of the Derby Philosophical Society, 1813–47, DLSL.

74 Glover, *History and Directory* (1843), 56; *Derby Mercury* (14 December 1842).

75 *Derby Mercury* (22 April 1835).

76 *Catalogue of the Library belonging to the Derby Philosophical Society* (Derby, 1835); *Derby Mechanics' Institute: Catalogue of Books* (Derby, 1851).

77 J. C. Loudon, *A Short Treatise on Several Improvements Recently Made in Hot-Houses* (London, 1805); J. C. Loudon, *The Encyclopaedia of Cottage, Farm and Villa Architecture and Furniture* (1833), 1276; J. C. Loudon, *The Architectural Magazine*, 5 (1836), 34–5.

78 Chadwick, *The Park and the Town*, 57, 61–2; J. C. Loudon, 'Remarks on laying out

public gardens and promenades', *The Gardener's Magazine*, 11 (1835), 644–59; J. C. Loudon (ed.), *The Landscape Gardening and Landscape Architecture of the late H. Repton* (London: 1840).

79 Loudon, *Derby Arboretum*, 7; Simo, *Loudon and the Landscape*, 166–7.

80 Loudon, 'Remarks on laying out public gardens and promenades', 650.

81 Conway, *People's Parks*, 81.

82 Loudon, *Derby Arboretum*, 71–3.

83 Simo, *Loudon and the Landscape*, 5–6, 231–3, 247–8.

84 Loudon, *Derby Arboretum*, 90–5; *Westminster Review*, 35 (January 1841), 422–31.

85 *Derby Mercury* (1 March 1848, 16 July 1851).

86 F. Hanham, *A Manual for the Park; or, A Botanical Arrangement and Description of the Trees and Shrubs in the Royal Victoria Park, Bath* (Bath, 1857).

87 Report of the Derby Arboretum committee, minutes of the Borough of Derby, 9 November 1847, *Derby Mercury* (10 November 1847) ; P. Elliott, C. Watkins and S. Daniels, 'The Nottingham Arboretum (1852): natural history, leisure and public culture in a Victorian regional centre', *Urban History*, 35 (2008), 48–71.

88 Jones, *Discourse on the … Life and Lamented Death of Joseph Strutt*, 16; J. R. Martin, 'Report on the state of Nottingham, Coventry, Leicester, Derby, Norwich, and Portsmouth', in *Second Report of the Commissioners for Inquiry into the State of Large Towns and Populous Districts* (1845), Appendix, part II, 274; *Derby Mercury* (20 August 1851); *Westminster Review*, 35 (1841), 418–57; E. Chadwick, 'Effects of public walks and gardens on the health and morals of the lower classes of the population' in *Report on the Sanitary Condition of the Labouring Population of Great Britain* (1842), 275–76; *Illustrated London News* (1852).

89 C. M. Hovey, *Magazine of Horticulture*, 11 (1845), 122–8, quoted in Simo, *Loudon and the Landscape*, 202–3.

90 A. J. Downing, *Rural Essays* (New York, 1856), 497–557.

91 Loudon, *Derby Arboretum*; *Derby Mercury* (22 January, 9, 16, 23 September 1840).

Conclusion

This book has shown how science held a special place in Derby culture in the period between *c*.1720 and 1850, encouraged by – and closely related to – other forms of urban culture. The appearance of itinerant lecturers from around the 1730s, the careers of philosophers such as Whitehurst, Arden, Tissington and Parker, and the foundation of a philosophical society by 1770 suggest that the establishment of this provincial culture pre-dates the classic industrial revolution period. The development of turnpike road and coaching networks facilitated travel, the movement of lecturers and the importation of books and science-related objects. The expansion of scientific culture is evident from the number and popularity of scientific lectures, the formation of literary and scientific societies and the activities of scientific groups in English towns. Furthermore, while for the Derby philosophers natural philosophy was part of a progressive enlightenment philosophy that also encouraged political campaigning, it could also enjoy a broad cultural appeal, reaching beyond middle-class Whig/Liberal groups to embrace some Anglicans, Tories, women and the working class.

Although the designation of the Derby philosophers as marginal individuals using scientific culture as one means of striving for social inclusion has some validity, the picture at Derby does not entirely fit this model. As we saw, dissenters were strongly represented in the corporation and among the borough elite between 1750 and 1850. It is therefore misleading to describe such dissenters as marginal in the political context of their own towns, though they had, and were perceived to have, certain distinct characteristics as a group. The government of Derby was managed by a municipal oligarchy dominated by a few families, and prior to 1835 was fairly representative of the political and religious divisions within the bourgeoisie. Because of this there was greater continuity between the old and new corporations, including the re-election of many councillors. Partly because of the power of the Whig Dukes of Devonshire and their compromise electoral agreement with the corporation, the urban government and associational culture of Derby was fairly inclusive in politico-religious terms, in keeping with the rhetoric of public pronouncements, and quite congenial to reformers. The vibrancy of Derby literary and scientific culture between 1730 and 1850 was not simply a result of rivalry between an Anglican elite and a rising but still excluded middle class on the Manchester model but, equally, of the relative politico-religious inclusivity of urban government and society. A highly visible literary and philosophical culture was already established by the 1780s and was one manifestation of the Georgian provincial 'urban renaissance'. In the early nineteenth century, the crystallisation of working-class identity and pressure for greater reforms in national government exacerbated by rapid urban expansion and industri-alisation encouraged the foundation by the middle class of urban voluntary institutions such as the Derbyshire Infirmary, Derby Mechanics' Institute and Town and County Museum. Georgian, Regency and early-Victorian Derby, like many provincial towns of the period, was an exciting and stimulating if bewildering cultural environment, which witnessed the growth of a new kind of middle-class consciousness and identity manifest in its literary and scientific culture.

If we compare Derby with the other major east midland towns between 1750 and 1850, Lincoln, broadly speaking, with its lack of industry, scattered population and dominant agricultural industry had the least successful literary and philosophical culture. Derby with seemingly the region's earliest manifest scientific culture had probably the most varied and growing industrial struc-ture, which included – on a county level – lead mining, cotton manufacture, coal mining, engineering and iron working. Leicester and Nottingham, which had more visible literary and philosophical cultures than Lincoln, but no scientific group as prominent as the Derby philosophers, were both thriving textile towns but much more dependent upon framework knitting than the Derby economy.[1] Framework knitting employers tended to resist technological innovation because of cheap labour availability and, moreover, the industry was in decline by the end of the Napoleonic wars. Derbyshire industries such as silk, cotton, iron-working and later engineering, required and encouraged technological innovation and knowledge advancement, stimulated by the regional effects of improvements in road and canal communications, which in turn funded a prosperous middle class and those trades and professionals who supported them. The perceived and actual association between acquisition of scientific and mathematical knowledge and industrial success explained the predominance of the rhetoric of utility in the public pronouncements of the founders and supporters of local urban scientific cultural associations, with the Derbyshire lead mining industry, for instance, as we have seen, helping to stimulate geological study.

However, by the middle of the nineteenth century, there were important changes in the nature of Derby's middle-class public culture, signified by the absorption of the Derby Philosophical Society into the Town and County

Museum in 1858. Although W. G. Spencer superintended the rearrangement of the philosophical library in its new room and William Adam was appointed curator, with the task of arranging and labelling the geological and mineralogical collections according to a more 'scientific' system, the idealistic enlightenment ambition of making a contribution to international science was not promoted as central to museum activities. Aristocratic and county elements dominated and most objects presented to the collections were of antiquarian and artistic rather than scientific significance, the exception being, revealingly, botanical, geological and natural historical specimens from Derbyshire.[2]

Although it would be foolish to contend that science had disappeared from urban public culture, it played a less overt part. As the more popular institutions, the Derby Mechanics' Institute and the Derby Arboretum in particular suffered from contradictions inherent in the notion of rational recreation between the 1820s and 1850s. These reflected religious and political divisions within the middle class, most obviously manifest in the debates concerning the desirability and efficacy of working-class scientific education, and the question of financial support and access to public scientific institutions. The Town and County Museum did not face the same problems as, apart from the 1843 exhibition, it tended to discourage much of the working class from visiting by charging for admission. These disagreements concerning civic scientific institutions were widespread in British scientific culture, and there was a lengthy and rancorous dispute at Nottingham concerning the usefulness and legality of subscription charges to the Arboretum, which divided the middle class and resulted in legal action against the corporation and reductions in funding. Despite the success of the Derby Arboretum, the ambiguity and uncertainty concerning access and funding was immediately noticed, and the *Westminster Review* observed how remarkable it was that the ratepayers of Derby could not by law contribute annual funds to the park.[3] Indeed, the financial position remained uncertain, as subscriptions alone barely covered expenses. This was one reason why, despite experiments with more free days, charges were usually levied on five days a week until 1882.[4]

Access to the Arboretum and the Derby Mechanics' Institute became easier for all classes, but they became less successful as scientific cultural institutions if this is defined in terms of the scientific content of the library, lectures and classes in the latter case, and the maintenance of systematic botanically significant collections in the former. Many contemporary observers of mechanics' institutes noted that most of the working class could not afford to attend the lectures. For example, in 1826 John Marshall wrote to Henry Brougham saying that they 'are at present adapted only for the elite of the working people', and excluded the others through admission charges. Other observers such as George Barclay attributed their failure to spread scientific education among the labouring classes to middle-class domination and the lack of elementary education.[5] Analysis of the Derby Mechanics' Institute library between 1825 and 1855 reveals a decline in the scientific content by the early 1850s, paralleling similar declines in mechanics' institutions at Sheffield and elsewhere.[6] Similarly, by the 1850s the Derby Arboretum committee faced considerable problems maintaining the botanical significance of the collection, and labelling ceased altogether during the 1870s. This was partly due to the difficulties of growing delicate exotic specimens in the toxic smoky atmosphere caused by growing heavy industries, but also to the popularity of the Arboretum as an urban leisure park rather than a botanical institution, represented by the overwhelming success of the anniversary festivals and decline in numbers of subscribers.

The 1851 Derby Mechanics' Institute catalogue of books and possessions reveals that the list of scientific apparatus acquired had hardly changed since 1839. Between 1851 and 1856, there were no additions whatsoever. After the work of Davy and Faraday, the electrical apparatus still consisted of a plate discharging machine, an electric orrery, various electrometers, pith balls and insulating stools – essentially Georgian scientific apparatus.[7] The 1851 library catalogue lists works acquired during the course of printing over a few months that year. Not a single one of these twenty-one works was on the subject of modern science, and this at the time when, as we have seen, the Derby Literary and Scientific Society was still flourishing. Only *Humboldt's Cosmos*, translated by Otle, Knox *On the Races of Men* and Wilson's *Life of Cavendish* could really be considered as works of scientific interest. The remaining books were either novels or historical and philosophical works, such as Plato and Ruskin. Taking the entire class of astronomy, geology, chemistry, mathematics, architecture, mechanics and most other sciences except for natural history, there was a total number of 175 works in 1851. In 1856 this had risen to 204, and a mere twenty-nine works on these subjects were acquired over a five-year period. Of these only twenty-two were scientific works of the 1850s, eight being on the subject of chemistry including two published and donated by the local chemist Albert Bernays, *On the Application of Chemistry to Farming* (1845) and *Household Chemistry* (1852). Thus, apart from natural history, the catalogue was mostly made up of historical or fictional volumes by 1850.

By the early 1850s, lectures featured more literary, historical, political and artistic subjects. F. B. Calvert, for instance, lectured at the Mechanics' Institute on the British poets in August and September 1850, the Rev. Wilkinson lectured on the poetry of Greece, and the same year there were other lectures on 'Pindar and the lyric poetry of Greece', 'The life and labours of John Howard' and 'Ancient Egypt'. The type of scientific subjects offered also changed between 1825 and 1855. Two of the most popular lecturers at the Mechanics' Institute and smaller institutions during the 1840s and 1850s were the chemist Albert Bernays and the phrenologist Spencer Hall. Bernays was a pupil of Justus

Liebig at Giessen, where he produced a doctoral thesis on limonin, a bitter substance that he had discovered in the pips of oranges and lemons.[8] In 1845 he moved to Derby and began a career as an analyst and teacher, becoming a member of the Philosophical Society. He founded a chemical laboratory in St Peter's Street, and was a member of the Chemical Society of London and the Institute of Chemistry.[9] Spencer Hall's brand of Christian phrenology was extremely popular, judging from the number of lecture courses that he gave. First visiting Derby in 1843, he lectured on George Combe's *Constitution of Man* at the Mechanics' Institute and followed this up with lectures at a host of other institutions large and small in the Derby region, including the Temperance Society and the Railway Institute.[10] The Tory *Derby Mercury* was initially hostile in 1843, but given Hall's evident popularity and the strong theological and moral content of his phrenology, the newspaper later carried favourable reviews. After lectures on physiology, phrenology and mesmerism at Temperance Hall in 1853, which included character studies of the Duke of Wellington, Lord Brougham, Ebenezer Elliott and William Cobbett to 'illustrate the laws of physical and mental correspondence', it was said that Hall ('our townsman') clothed 'a philosophical style of thought with vigorous and often picturesque

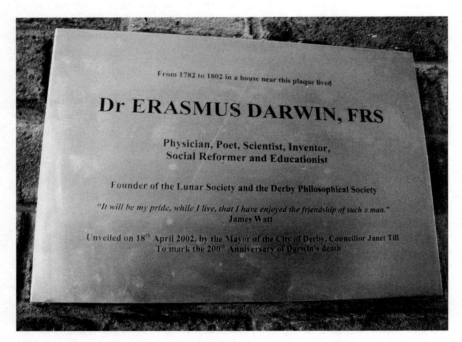

From 1782 to 1802 in a house near this plaque lived

Dr ERASMUS DARWIN, FRS

**Physician, Poet, Scientist, Inventor,
Social Reformer and Educationist**

Founder of the Lunar Society and the Derby Philosophical Society

"It will be my pride, while I live, that I have enjoyed the friendship of such a man."
James Watt

Unveiled on 18th April 2002, by the Mayor of the City of Derby, Councillor Janet Till
To mark the 200th Anniversary of Darwin's death

39 Civic celebration of the Derby Philosophers I: Plaque unveiled by the Mayor of Derby in 2002 on the wall of the Tourist Information and Box Office, Derby Market Place, commemorating the bicentennial anniversary of Erasmus Darwin's death.

and poetical language'. He 'riveted the attention of his numerous audience', though the *Mercury* remained uneasy 'with the peculiarities of Dr. Hall's isms and ologies'.[11] Although phrenology remained popular between the later 1820s and 1850s, it became increasingly diluted with phreno-mesmerism, illustrated by the difference between the physiologically grounded system of Gall and Spurzheim and the moralisms of Spencer Hall. By comparison, at Sheffield, the community of scientific activists had tended to support phrenology in the 1820s and 1830s, but by the later 1840s only one phrenologist was a member of the local scientific community.[12]

Hence, although from the 1840s larger venues such as the Derby Mechanics' Institute building, cheaper admission prices and popular literature encouraged more of the working class to attend, the scientific content of lectures and libraries had become diluted. This was due to a number of factors, including changes in the nature of science itself, especially institutionalisation, professionalisation and specialisation resulting from the formation of distinctive scientific disciplines and the increasing domination of London institutional and university-based science.[13] Beyond the world of science, it has been suggested that periods of economic depression, the formation of mature middle and working classes with their own distinctive identities, the domination of the domestic sphere in middle-class life, middle-class disillusion with the discourse of rational recreation, and the expansion of local civic control over public institutions such as libraries, museums and parks also played a role in the 'decline' of provincial public scientific culture. The fact that the working class were drawn away from elite-led institutions by Chartism, trades unionism, the co-operatives and the operatives' libraries has also been adduced as a factor. Periods of economic depression were transitory and hardly provide a long-term explanation for the decline of Derby scientific culture. Chartism declined rapidly from the early 1850s and, if anything, co-operation between working-class reformers and the middle classes increased under the banner of Liberalism in high Victorian England. Certainly, the picture of middle Victorian society among the middle class in a middle-ranking midland town painted by Derby-born economist John Atkinson Hobson (1858–1940), when 'peace, prosperity, and progress' appeared to be 'the permanent possession of most civilised nations', suggests that they felt little need to experiment with novel institutional and educational organisations. After the decline of Chartism, the game of party politics was played out at the hustings and in local newspapers but lacked the urgency occasioned by the fevered political climate of the 1820s and 1830s. According to Hobson, while there was a 'good deal of loose philanthropic talk about the amelioration of the condition of the working classes', there was 'no sincere attempt at amelioration by governmental action'.[14]

It is true that the narrative of a 'decline' of public provincial scientific culture and the polarisation between C. P. Snow's famous 'two cultures' in

British society should not be exaggerated, and the development of Victorian museums and libraries, imperial expansion and rivalry, greater provision for scientific education in the state-supported school curriculum, growth of university extension courses in towns such as Derby and Nottingham, and the requirements of local industries continued to encourage science to be taught and applied. However, the lack of a group of promoters of scientific ventures equivalent to the Derby philosophers, and also the proliferation of educational institutions either organised by, or directed towards, the working class in Derby and Derbyshire, seem to have helped to weaken the role of science in public culture. These included the temperance societies, smaller organisations such as the Belper Mechanics' Institute (1837), young men's instruction societies, the Derby Working Men's Association (1849) and the Derby Railway Literary Institute (1850). The Working Men's Association, which amalgamated with the Society of Arts in 1854, offered evening classes on subjects including geography, history and arithmetic, and members had the use of the Friargate chapel school library.[15] Although the Derby Literary and Scientific Society continued to offer a vigorous programme of scientific lectures into the 1850s, the death or removal of long-term leaders of Derby scientific culture, including Richard Forester (1843), Joseph Strutt (1844), Thomas Bent (1859) and William George Spencer (1866) and Douglas Fox (who moved to Brighton), was another major factor and helps to explain the absorption of the Derby Philosophical Society into the museum. Forester and William and Joseph Strutt had known and been inspired by Darwin and with their wealth and influence provided a continuity of leadership in Derby public culture from the late eighteenth century to the early 1840s. They were not replaced by philosophical leaders with such a broad range of scientific interests, and intellectuals who might well once have remained in a provincial town like Derby, such as Edward Strutt and Herbert Spencer, were tempted away by much more attractive metropolitan and national opportunities. With some important exceptions such as the chemist Albert Bernays, writers and thinkers who remained in Derby and played a major role in later Victorian public culture, such as Spencer Hall, Llewellyn Jewitt and Henry Howe Bemrose, were not primarily interested in science.

As we have seen, there are important similarities between Derby and scientific communities in Manchester, Liverpool, Leicester, Nottingham and other towns, including the importance of medical men and medical institutions, political reformism, dissenters and textile manufacturers. However, there are also ways in which the town was unusual, some of which were related to the special landscape, geology and topography of the county and the status of Derby as manufacturing, industrial and county town. Some of this is reflected in the usage and interpretations of the Derby philosophers in contemporary heritage, education, leisure and local government (for example the use of

40 Civic celebration of the Derby Philosophers II: Plaque placed on Exeter Bridge in 1931 commemorating the life of Herbert Spencer.

Wright, Darwin and the Strutts to promote Derby and the Derwent Valley world heritage site). It is ironic that in an age of imbalance between metropolis and provinces, where the autonomy of local government has long been eroded, the perception of the British provincial enlightenment and Victorian periods as a golden age has increased. The history of British industry, society and culture are now utilised by tourism, now one of Britain's largest industries, as a means of rekindling some vestiges of this wonderful past.

Notes

1 J. V. Beckett and J. E. Heath, 'When was the industrial revolution in the east midlands?', *Midland History*, 13 (1988), 77–93; J. V. Beckett, *The East Midlands from AD 1000* (London; Longman, 1988), 260–98.

2 *Derby Town and County Museum and Philosophical Society: Report of the Committee* (Derby, 1860); *Derby Town and County Museum and Philosophical Society: Report of the committee* ... (Derby, 1862); minute book of the Derby Town and County Museum and Philosophical Society, 1861–8, DLSL.

3 *Westminster Review*, 35 (January 1841), 430.

4 *Derby Mercury* (23 September 1840).

5 J. Marshall to H. Brougham, 13 September 1826, A. D. Garner and E. W. Jenkins, 'The English Mechanics' Institutes: the case of Leeds, 1824–1842', *History of Education*, 13 (1984), 139–52, 143–4; G. C. T. Barclay, *The Schools of the People* (1871), 306–7.

6 I. Inkster, *Scientific Culture and Urbanisation in Industrialising Britain* (Aldershot: Ashgate, 1997), paper III, 295–8.

7 Derby Mechanics' Institution, *Catalogue of Books Comprised in the Library with a List of Paintings and Engravings, and of the Philosophical Apparatus Belonging to the Institution* (Derby, 1851), 60–1 and *Catalogue* (1856).

8 *DNB*, supplement, I (1901), 183–4; I. Inkster, 'Studies in the Social History of Science during the Industrial Revolution' (PhD thesis, University of Sheffield, 1977), 484.

9 *Derby Mercury* (30 April 1845, 12 November 1845).

10 *Derby Mercury* (8, 15 November 1843, 25 May, 8, 15 June, 9, 16 November and 17 December 1853); R. Cooter, *The Cultural Meaning of Popular Science: Phrenology and the Organisation of Consent in Nineteenth Century Britain* (Cambridge: Cambridge University Press, 1984).

11 *Derby Mercury* (25 May, 15 June 1853).

12 I. Inkster, *Scientific Culture*, paper III, 297–8, II, 95–100.

13 Inkster, *Scientific Culture*, paper II, 105–6.

14 J. A. Hobson, *Confessions of an Economic Heretic* (London: George Allen & Unwin, 1938), 15, 17; R. J. Morris, 'Voluntary societies and British urban elites, 1780–1850: an analysis', *The Historical Journal*, 26 (1983), 95–118; D. Eastwood, *Government and Community in the English Provinces, 1700–1870* (Basingstoke: Macmillan, 1997), 57–90; R. J. Morris, 'Civil society and the nature of urbanism: Britain, 1750–1850', *Urban History*, 25 (1998), 289–301.

15 *Derby Mercury* (14, 28 May 1845); *The Derby Working Men's Association* (Derby, 1849), DLSL, acc. 4452, no 36; Inkster, 'Studies in the Social History of Science', 484.

Appendix
Members of the Derby Philosophical Society (1783–c.1850) and members of the Derby Literary and Philosophical Society (1808–1816)

(*DNB* denotes biographical entry in the revised edition of the *Oxford Dictionary of National Biography*, 60 vols., Oxford University Press, 2004).

Original members during Darwin's lifetime

Archdall, Richard (1746–1824): Irish parliamentarian
Beridge, Dr John (1745–88)
Crompton, Dr Peter (1762–1833): physician
Darwin, Dr Erasmus (1731–1802): physician, natural philosopher and poet, *DNB*
Darwin, Erasmus (1759–99): lawyer
Duesbury, William (1763–96), Derby China manufactory, *DNB*
Fowler, William Tancred (1764–1821): surgeon
Fox, Samuel (1765–1851): hosier
French, Richard (1739–1801)
Gisborne, Rev. Thomas (1758–1846): Yoxall Lodge, Staffordshire, St Helen's House, Derby, *DNB*
Haden, Thomas (1761–1840): surgeon
Hadley, Henry (1762/3–1830): surgeon
Hope, Rev. Charles (1733–98)
Johnson, Dr William Brookes (1763–1830): physician
Leaper, John (1754–1819): attorney
Pickering, Rev. William (1743–1802)
Pigot, Dr John Hollis (1757–94): physician
Pole, Edward Sacheverell (1769–1813)
Roe, Richard (d.1814): secretary, schoolmaster
Sneyd, John (1734–1809): of Belmont, Staffordshire
Strutt, William (1756–1830): president 1802–15, *DNB*

Non-resident members during Darwin's lifetime

Arnold, Dr Thomas (1741–1816): Leicester physician and mad doctor, *DNB*
Bage, Robert (1728–1801): of Elford, Staffordshire, paper manufacturer, novelist, *DNB*
Beaumont, Mr: of Melbourne, surgeon
Bent, James (1739–1812): surgeon, Newcastle-under-Lyme, Staffordshire

Boothby, Brooke (1744–1824): Ashbourne, scholar and political writer
Bradshawe, Francis (*c*.1760–1841): Holbrook
Bree, Dr Robert (*c*.1759–1839): Leicester, physician
Brock, Jeffrey: surgeon, Mansfield, Nottinghamshire
Buck, Dr Robert: Newark, physician
Clarke, Thomas (1758–1837): Kirkby Hardwick, Nottinghamshire
Coke, Rev. D'Ewes (1747–1811): Brookhill, Nottinghamhsire
Crofts, Mr: Tutbury, Staffordshire
Darwin, Dr Robert Waring (1766–1848): Shrewsbury, Shropshire, father of Charles
    Darwin, *DNB*
Evans, Thomas (*c*.1723–1814): Darley Abbey
Goodwin, Dr Anthony (1752–1819): Wirksworth
Haggit, Rev. George (1767–1832): Sudbury
Hunt, Mr: Loughborough, Leicestershire
Hurt, Charles (1758–1834): Alderwasley
Jackson, William (1735–98): Lichfield, Staffordshire, proctor of the cathedral court,
    member of the Lichfield botanical society
Jones, Dr Trevor (d.1832): Lichfield, Staffordshire, physician
Power, John: Market Bosworth, Leicestershire, surgeon
Riddlesden, Richard: Dove Lees, surgeon and apothecary
Riddlesden, Samuel: Ashbourne, surgeon and apothecary
Smith, Mr: Alfreton, Nottinghamshire
Stevenson, Joseph: Kegworth, Leicestershire, surgeon
Storer, Dr John (1747–1837): FRS, Nottingham, physician
Strutt, Jedediah (1726–97): Makeney
Taylor, Dr: Ashbourne and Warwick
Trowell, John (1744–1821): Long Eaton and Thornhill House, Derby, major in the militia
Walker, John junior: Ashbourne, gentleman
Wedgwood, John (1766–1844): Etruria, Staffordshire
Wedgwood, Ralph (1766–1837): Burslem, Staffordshire
White, Dr Snowden (d.1791): Nottingham, physician
Wilmot, Sir Robert (d.1834): Osmaston
Wilson, Dr: Mansfield, Nottinghamshire
Wray, Rev. William Ulithorne (1721–1808): Darley

Members of the Derby Literary and Philosophical Society, 1808–1816

Bakewell, Robert (1767–1843): geologist, *DNB*
Beasant, Hanson: assistant surgeon to the 37th Regiment of Foot
Bent, Dr Thomas (1792–1859): physician, president of the DPS
Byewater, William: of Nottingham
Clare, Peter (1781–1851): Manchester, secretary of the Manchester Literary and Philo-
    sophical Society, close friend of John Dalton, *DNB*
Cochrane, Archibald, ninth Earl of Dundonald (1748–1831): of Culross Abbey, Perth-
    shire, Scotland, chemical manufacturer, *DNB*
Dalton, Richard: Manchester and York, scientific lecturer

Davies, Rev. David Peter: of Milford/Belper

Duesbury, William III (1787–1845)

Edwards, Nathaniel (1768–1814): attorney

Edwards, William

Evans, William: surgeon of Belper

Flack, Charles James (1799–1837): of Cavendish Bridge, attorney

Forester, Dr Richard Forester (1771–1843): physician and president of the DPS

Godwin, Richard Bennet (d.1867): surgeon

Goodlad, William: of Bellevue, Bury, Lancashire, surgeon

Granger, Benjamin: of Burton-upon-Trent, Staffordshire

Haden, Charles Thomas (1786–1824): surgeon

Haden, Henry (1790–1831): surgeon

Haden, Thomas (1761–1840): surgeon

Hall, John: surgeon and house apothecary at the Derbyshire Infirmary

Hanock, William (1771–1821): surgeon

Hay, Captain: of Nottinghamshire

Henry, William Dr, FRS (1774–1836): chemist, natural philosopher, Vice-President of the Manchester Literary and Philosophical Society

Higginson, Rev. Edward (1781–1832): Unitarian minister at Friargate Unitarian Chapel

Holland, Rev. John (1766–1826): Unitarian minister, Bolton, Lancashire, *DNB*

Hope, Thomas (d.1809)

Hopper, Richard: of Papplewick, Nottingham, cotton merchant, Sheriff of Nottingham in 1815

Lowe, Charles Matthew: brewer of the Wardwick, mayor of Derby, 1831

Lowe, George: of Buxton and London

Lowe, Rev. Henry (1779–1851): mayor of Derby, 1812, 1821

Lucas, Jean-André-Henri (1780–1825): of Paris, naturalist, mineralogist, head of the galleries in the Natural History Museum and author of *Tableau méthodique des espèces minérales*, 2 vols. (Paris: d'Hautel, 1806–13)

Mousley, William Eaton (d.1853): attorney

Mushet, David (1772–1847): Scottish ironfounder and metallurgist, *DNB*

Nugent, Dr: of London

Oakes, James (1750–1828): mayor of Derby, 1820

Rawlinson, Thomas

Robinson, Dr: of London

Roe, Richard (d.1814): secretary of the Derby Philosophical Society

Sandars, Joseph (1785–1860): of Liverpool, corn merchant, industrialist and co-initiator of the Liverpool and Manchester Railway

Stonehouse, William (1793–1862): Manchester and Oxford

Strutt, George Henry (1784–1821)

Strutt, Joseph (1766–1844): mayor of Derby, 1836

Strutt, Joseph Douglas (1794–1821)

Swanwick, Thomas (1755–1814): master of the commercial academy

Sylvester, Charles (1774–1828): chemist and engineer

Traill, Thomas Stewart, Dr (1781–1862): physician at Liverpool, later Professor at Edinburgh University, founder member (1812) and first secretary of the Liverpool

Literary and Philosophical Society, *DNB*

Wallis, George (1786–1869): of Cambridge, surgeon

Ward, Rev. Richard Rowland: of Sutton Hall, Sutton-on-the-Hill, Vicar of St Peter's

Webster, John: either of the King's Arms, Derby or surgeon of Burton-upon-Trent

Wilson, William: Manchester, surgeon

Wright, John (1781–1850): surgeon

Members of the Derby Philosophical Society, *c.*1813–1859

Abney, Rev. Edward Henry (1811–92): Vicar of St Alkmund, Prebendary of Lincoln

Allsop, James: Uttoxeter, Staffordshire

Bainbrigge, Joseph Hamley (d.1825): surgeon and apothecary

Barber, John (1800–80): attorney, agent of the Duke of Devonshire, mayor of Derby, 1843

Bennett, William: surgeon

Bent, Dr Thomas (1792–1859): physician, DPS president

Bernays, Albert Jones (1823–92): chemist and teacher in Derby between 1845 and 1855, subsequently lecturer in chemistry at St Mary's Hospital and St Thomas's Hospital, London, *DNB*

Bilsborrow, Dr Dewhurst (1776–?): physician

Birch, Richard William: attorney

Boden, J. G.

Boden, W. C.

Borough, Charles (1803–78): physician

Bowditch, Mr

Brigstocke, Dr Henry: physician

Brown, Richard (1736–1816): stonemason, petrefactioner and mineralogist, *DNB* or just R. B. below

Brown, Richard junior (1765–1848): stonemason, petrefactioner and mineralogist, *DNB*

Byng, John: hosiery manufacturer

Cade, James: surgeon of Spondon

Calvert, Edward: businessman, banker

Challinor, William: of Normanton

Chappell, Graham (1768–1834): of Arnold Grove, near Nottingham, and Spondon

Chatterton, John (1771–1857): plumber and glazier, mayor of Derby

Clarke, Thomas (1758–1837): of Kirkby Hardwick

Cooper, John Douglas (d.1831) cotton-spinner of Mayfield, Staffordshire

Cox, Roger: of Spondon

Darwin, Sir Francis (1786–1859): of Buxton, physician and explorer

Davenport, Samuel: surgeon

Drewry, John: either II (d.1834) or III (d.1840)

Duesbury, William III (1787–1845): chemist and artist

Edwards, Henry (1789–1863): assistant surgeon to the 43rd Regiment of Foot and later surgeon of Tutbury, Staffordshire

Edwards, Nathaniel

Edwards, W.: may be W. E., watchmaker

Evans, David: surgeon of Belper

Evans, Mr: probably Thomas Evans (1743–1814) of Darley Abbey

Evans, Mr: probably Walter Evans (1764–1839) of Darley Abbey

Evans, Mr.: probably William Evans (1788–1856) of Allestree Hall

Every, Frederick Simon (1804–88): son of Sir Henry Every of Eggington. Emigrated to New Zealand.

Fearn, Samuel Wright (d.1870): surgeon

Flack, Charles James (d.1837): attorney

Fletcher, Mr: either Rev. Dr William or William Vicars Fletcher

Forester, Dr Richard Forester (1771–1843): physician and DPS president

Fowler, William Tancred (1764–1821): surgeon

Fox, Douglas (1797–1885): surgeon, mayor of Derby in 1834, 1838 and 1850 and MP

Fox, Dr Francis (1759–1833): surgeon

Fox, Dr Francis, junior: physician, son of above and brother of Douglas and James

Fox, J.: probably Joseph Fox, engineer

Fox, Samuel (1765–1839): hosier, of Osmaston Hall

Fox, Samuel, junior (b.1793): son of above, silk and cotton merchant

Gamble, Stephen (1780–1860): furniture dealer and mayor of Derby

Garlike, Dr William Bennet: physician

Gilbert, Jonathan (1777–1845): of The Towers, Matlock Bath

Gisborne, Henry Francis (1807–87): surgeon and mayor of Derby, 1856

Goodwin, Anthony (1752–1819): physician, Wirksworth

Goodwin, Captain Francis Green: of Wigwell Grange near Wirksworth

Granger, Benjamin (1783–1846): surgeon of Burton-upon-Trent, Staffordshire

Greaves, Dr William (1771–1848): physician of Mayfield Hall, Staffordshire, JP for Stafford

Haden, Henry (1790–1831): surgeon

Haden, Thomas (1761–1840): surgeon

Hadley, Henry (1762/3–1830): surgeon

Hall, Mr: secretary, probably Joseph Hall, spar turner and mineralogist

Hancock, William (1771–1821): surgeon

Handyside, Andrew (1805–87): Scottish ironfounder, took over the Britannia ironworks *c.*1846, *DNB*

Harwood, Thomas (1796–1867): surgeon at the Derby Dispensary

Heldenmaier, Beatus (1795–1873): Swiss master of Pestalozzian academy at Worksop, Nottinghamshire

Hewitt, Benjamin (1788–1868): of Burton-upon-Trent

Heygate, Dr James (1801–72): physician

Higginson, Rev. Edward (1781–1832): minister at Friargate Unitarian chapel

Hoare, William: surgeon

Hood, George (1788–1856): plumber and glazier of Derby and Stoke Newington

Horrocks, John: attorney

Hudson, John: schoolmaster, Full Street

Huish, Marcus: of Castle Donnington and Duffield

Hulbert, George Redmond (1774–1825): naval prize agent at Nova Scotia, of Aston Lodge; see the *Mariner's Mirror*, 87 (2001)

Hurt, Charles: Alderwasley (1758–1834)
Hurt, Charles: Wirksworth
Hurt, Edward N.: of London
Hutchinson, Mr
Jessop, Francis: attorney, mayor of Derby 1840
Johnson, W.
Jones, John (1788–1863): surgeon
Lockett, William Jeffery (1768–1839): attorney
Lowe, Charles Matthew: brewer of the Wardwick, mayor of Derby, 1831
Lowe, George: Burton and London
Lowe, Rev. Henry (1779–1851): mayor of Derby, 1812 and 1821
Mammatt, Edward (1776–1835): coalmaster of Over Seal Cottage, Measham, Derby-
    shire and Manor House, Ashby-de-la-Zouch, Staffordshire, agent to the Marquis
    of Hastings, mineral collector
Mammatt, Edward (1807–60): of Ashby-de-la-Zouch, Staffordshire, blind musician,
    manager of Burton Brewery, scientific lecturer, geologist, inventor, editor of *The
    Analyst: A Monthly Journal of Science, Literature and the Fine Arts* (1834–40), and
    author of *The History and Description of Ashby-de-la-Zouch* (1852), see *History,
    Gazetteer and Directory of the Counties of Leicester and Rutland* (1863), 435
Mason, Robert: of Burton, Staffordshire
Moore, Henry (1776–1848): artist, engraver, inventor and author of travel guides
Moss, John (1796–1860): attorney and mayor of Derby
Mousley, William Eaton (d.1853): attorney, mayor of Derby, 1845
Mozley, Henry (1773–1845): bookseller/ printer and publisher, mayor of Derby
Mugglestone, Mr
Murphy, James Brabazon: surgeon dentist
Nichol, Mr
Oakes, James (1750–1828): attorney, mayor of Derby, 1819
Oldham, George: surgeon, of Alfreton, Nottinghamshire
Orton, Richard (1765–1843): surgeon of Kegworth
Ordish, Edward (1772–1834): farmer and mechanic of Ingleby
Peach, Dr James (1785–1874): physician
Pole, Edward Sacheverell (1769–1813): 1813, Radbourne
Pole, Edward (1757–1824)
Poyser, Thomas: surgeon, Wirksworth
Severne, Francis: goldsmith and jeweller
Sharpe, Mr
Shirley, Robert Sewallis (1778–1824), Viscount/Lord Tamworth
Simpson, Mr (d.1813 or just before)
Smith, Samuel George: attorney
Spencer, William George (1790–1866): secretary from 1814, *DNB*
Stevens, Henry Isaac (1806–73): architect and surveyor
Stevenson, Joseph: surgeon of Kegworth, Leicestershire
Strutt, Anthony Radford (1791–1875): Makeney House
Strutt, Joseph Douglas (1794–1821)
Strutt, Edward (1801–80): MP and Lord Belper from 1856, *DNB*

Strutt, George Benson (1761–1843): of Belper
Strutt, George Henry (1784–1821)
Strutt, Joseph (1766–1844): mayor of Derby in 1836, *DNB*
Strutt, William (1756–1830): Derby Philosophical Society president 1802–15, *DNB*
Sylvester, Charles (1774–1828): chemist and engineer
Taylor, G. (d.1888): surgeon
Tunaley, Thomas Snape (1796–1868): dancing master
Ward, Rev. Richard Rowland (1762–1834): of Sutton Hall, Sutton-on-the-Hill, Vicar of St Peter's
Watson, Dr George Bott Churchill: physician, largely responsible for the foundation of the Derby Town and County Museum in 1835, but left Derby in 1838. Moved to Liverpool where he helped to found the Lying-In Hospital and Dispensary for the Diseases of Women and Children (1841), *Liverpool Mercury* (28 December 1847)
Weatherhead, Samuel (1784–1854): gentleman, of Weatherhead, Walters and Co., rope and twine makers, gunsmiths and ironmakers
Whitehurst, John III (1788–1855): clockmaker
Wigston, William: machine maker and lace manufacturer. Employed by the Derby Gas Light and Coke Company, declared bankrupt after legal action.
Wilmot: Sir Robert (1752–1834): of Osmaston
Wood, Mr
Wright, John (d.1850): surgeon
Wright, Samuel Job (1796–1848): silk throwster

Entries often consist of surnames only and therefore we cannot always be sure which individual is being referred to. The list of Derby Philosophical Society members is based upon R. P. Sturges, 'The membership of the Derby Philosophical Society, 1783–1802', *Midland History*, 4 (1978), 212–29; Derby Philosophical Society cash ledger (1813–45), Derby Local Studies Library, ms. 7625, BA 106, 9229–30. The list of Derby Literary and Philosophical Society members has been obtained from the Journal Book of the Derby Literary and Philosophical Society (1808–16), Derbyshire Record Office (D5047).

Select bibliography

Archival and manuscript sources: main collections

Birmingham Central Library, Archives and Heritage
Galton papers, Birmingham Central Library, Archives and Heritage (MS3101).

Derby Local Studies Library, Derby (DLSL)
Bakewell, W., untitled handbill (Derby, 14 December 1792).

Bemrose, H. H., papers.

Bemrose, W., original and other documents relating to Joseph Wright collected by W. Bemrose (1887) (12332).

Bennet, Revd. A., 'Memoranda Miscellania', commonplace book.

Bent, T., case book.

Derby Friargate Chapel minutes and accounts, 1697–1819 (BA 288, DER A.22207).

Derby Mechanics' Institute, honorary members' name book (DL113.36).

Derby Mechanics' Institute, records, 1825–1915 (NRA 27869).

Derby Philosophical Society, manuscript catalogue and charging ledger, 1785–9 and cash ledger, 1813–45 (BA 106, 9229–30).

Derby Town and County Museum, minute book of the Derby Town and County Museum and Philosophical Society, 1861–8 (NRA 20300).

Drewry's Book Society, printed list of members (from 1790s).

Glover, S., notebooks.

Gregory, W., 'Diary of William Gregory'.

Lancasterian School Committee, Annual Reports of the Royal Lancasterian School Committee (1813 onwards).

Latham, E., Latin notebook.

Nun's Green box of broadsides and pamphlets (box 27).

Paving and Lighting Commissioners, Minute Book of the Paving and Lighting Commissioners, 1789–1825 (BA 625/8, 16048).

Spencer, H., correspondence, 1892–98 (BA 921, 13105).

Spencer, W. G., account book, 1814–65 (acc. 9487).

Strutt papers and correspondence on microfilm (D125). Originals moved to Derbyshire Record Office.

Tilley, J., research notes and papers, indexes.

Wright, H. (attrib), 'Some Account of Joseph Wright of Derby', undated (11172, 12331–2).
Wright, J., correspondence.

Derby Museum and Art Gallery
Material relating to the Spencer family, including photographs, works of art and school record books.
Spencer, W. G., teaching account book, 1806–13 (acc. 528-22-35).

Derbyshire Record Office, Matlock
Bemrose Family papers (D5239).
Derby Lancasterian School, registers and other papers (D3839, D5845).
Derby Literary and Philosophical Society (originally Mutual Improvement Society), journal and papers, 1808–16 (D5047).
Derbyshire General Infirmary, printed and manuscript material including annual reports and papers on the establishment of the Infirmary (D1190/1/1–4).
Friargate Chapel, Derby, birth and baptismal registers (M690 RG4/5, M695 RG4/2033, M695 RG4/2034, M691 RG4/499), and minutes and accounts from the period (1312 D/A1, A2), c.1700–1860.
Strutt Family papers, including deeds, estate and family papers and correspondence (D1564, D2912, D2943M, D3772, D5303).

Fitzwilliam Museum, Cambridge
Edgeworth, M., letters to members of the Strutt family, 1812–43 (Gale Box).
Strutt, W., letters to M. Edgeworth, 1819–29.
Strutt papers (MS 48 – 1947).

Harris Manchester College, Oxford
Martineau papers.

Keele University Library
Wedgwood Manuscripts.

Leicestershire Record Office
Shirley family papers (26 D53).

Royal Society of London, Library and Archives
Journal Book of the Royal Society.

University of London, Senate House Library
Spencer papers (mss. 791).

Main periodicals used, 1680–1900

Derby and Chesterfield Reporter
Derby Herald, or Derby, Nottingham and Leicester Advertiser
Derby Mercury
Derby Reporter
Derbyshire Archaeological Journal
Gentleman's Magazine
Harrison's Derby and Nottingham Journal
Illustrated London News
Ladies' Diary, or the Woman's Almanack
Leicester and Nottingham Journal (continued as *Leicester Journal*)
Nicholson's Journal
Nottingham Journal
Nottingham Review
Philosophical Magazine
Philosophical Transactions
The Times

Works published before 1910

(Place of publication is London unless otherwise stated. Editions of works are listed under the name of the original author, even if the work is cited in the Notes by the later editor's name.)

Anonymous, *Poetical Attempts, consisting of an Allegorical Poem in Blank Verse entitled The Sciences: and Ode to Pleasure and some other pieces* (Derby, 1783).
—— *The Golden Age, A Poetical Epistle from Erasmus Darwin to Thomas Beddoes* (London and Oxford, 1794).
—— 'Memoirs of the Revd. Dr. Pegge', *Gentleman's Magazine*, 66 (1796), 451–4; (August 1796), 627–30; (October 1796), 803–7.
—— *On the Education of the Poor*, unsigned pamphlet (Derby, 1812).
—— *The Duster, edited by Barnaby Brush Esq., Secretary to the Political Elective, Dusting Committee, at Derby* (Derby, October–November 1832).
Alger, J. G., 'The British colony in Paris', 1792–93', *English Historical Review*, 13 (1898), 672–94.
Arden, J., *A Short Account of a Course of Philosophy and Astronomy* (1744).
Arnold, T., *Observations on the Nature, Kinds, Causes and Prevention of Insanity*, 2 vols. (Leicester, 1782, 1786).
Bage, R., *Man As He Is*, 4 vols. (1792).
—— *Hermsprong, or Man As He Is Not*, first edition, 3 vols. (1796).
Bailey, F., *An Account of the Rev. John Flamsteed: The First Astronomer Royal* (1835–7).
Baker, W., 'Report on the sanitary condition of the town of Derby', in *Local Reports on the Sanitary Condition of the Labouring Population of England* (1842).
Barclay, G. C. T., *The Schools of the People* (1871).
Beighton, H., 'A description of the water works on London Bridge', *Philosophical Transactions*, 37 (1732).

Bemrose, W., 'Wright of Derby, a biographical notice', *The Reliquary*, 4 (1864), 176–84.

—— *The Life and Works of Joseph Wright A.R.A. commonly called Wright of Derby* (Derby, 1885).

Bennet, A., 'A description of a new electrometer', *Philosophical Transactions*, 76 (1786), 26–34.

—— 'An account of a doubler of electricity', *Philosophical Transactions*, 77 (1787), 288–96.

—— *New Experiments on Electricity Wherein the Causes of Thunder and Lightning are explained ... Also, a Description of a Doubler of Electricity, etc.* (Derby, 1789).

—— 'A new suspension of the magnetic needle', *Philosophical Transactions*, 82 (1792), 81–98.

Bent, J., 'An account of a woman enjoying the use of her right arm after the head of the *os humeri* was cut away', *Philosophical Transactions*, 13 (1774), 539–41.

Bentham, J. *The Works of Jeremy Bentham*, ed. J. Bowring, 11 vols. (1838–43).

—— *Introduction to the Principles of Morals and Legislation* (Oxford, 1876).

Bernays, A. J., *On the Application of Chemistry to Farming* (1845).

—— *Household Chemistry* (1852).

Birks, J., *Memorials of the Friargate Chapel, Derby* (Derby, c.1890).

Boothby, B. B., *A Letter to the Right Honourable Edmund Burke* (1791).

—— *Observations on the Appeal from the New to the Old Whigs, and on Mr. Paine's Rights of Man* (1792).

—— *Sorrows Sacred to the Memory of Penelope* (1796).

—— *Fables and Satires, with a Preface on the Aesopean Fable*, 2 vols. (1809).

Burdett, P. P., *P. P. Burdett's Map of Derbyshire, 1791 Edition*, ed. and introduction J. B. Harley, D. V. Fowkes and J. C. Harvey (Derbyshire Archaeological Society, 1975).

Burke, E., *A Philosophical Enquiry into the Origin of our Ideas of the Sublime and Beautiful*, ed. and introd. Adam Philips ([1756]; Oxford, 1996).

—— *Reflections on the Revolution in France, and on the Proceedings in Certain Societies in London Relative to that Event* (1790).

Brown, T., *Observations on the Zoonomia of Erasmus Darwin, M.D.* (Edinburgh, 1798).

Browne, H., Letter to the Society of Arts on the subject of the preservation of seeds from vermin (13 March 1793), *Transactions of the Society of Arts*, 11 (1793), 143–4.

—— 'Description of a furnace described as an evaporator', *Transactions of the Society of Arts*, 12 (1794), 257–262.

—— Letter to the Society of Arts on the subject of manure (27 April 1798), *Transactions of the Society of Arts*, 16 (1798), 268–71.

Canning, G. *et al.*, 'The loves of the triangles', *The Anti-Jacobin Review* (13 April 1799).

Cavallo, T., 'Description of a new electrical instrument capable of collecting together a diffused or little condensed quantity of electricity', *Philosophical Transactions*, 78 (1788), 255–60.

—— 'Of the methods of manifesting the presence and ascertaining the quality, of small quantities of natural or artificial electricity', *Philosophical Transactions*, 78 (1788), 1–22.

Chadwick, E., 'Effects of public walks and gardens on the health and morals of the lower classes of the population', in *Report on the Sanitary Condition of the Labouring Population of Great Britain* (1842), 275–6.

Chambers, W., 'Biographical sketches: John Claudius Loudon', *Chambers Edinburgh Journal* (May 1844), I, 284–6.

Cherry, J. L. (ed.), *Stafford in Olden Times* (Stafford, 1890).

Clodd, E., *Pioneers of Evolution* (1897).

Clowes, W. L., *The Royal Navy: A History*, 4 vols. (1898).

Conway, M. D., *The Life of Thomas Paine*, ed. H. B. Bonner (1901).

Cooke, A., *Topographical and Statistical Description of the County of Derby* (1820).

Cox, J. C., *Three Centuries of Derbyshire Annals*, 2 vols. (Derby, 1890).

Cox, J. C. and Hope, W. St. J., *The Chronicles of the Collegiate Church ... of All Saints, Derby* (London, 1881).

Cradock, J., *Literary and Miscellaneous Memoirs*, 3 vols. (1826).

Cresy, E., *Report to the General Board of Health on a Preliminary Enquiry into the Sanitary Condition of the Inhabitants of the Borough of Derby* (London, 1849).

Crewe, G., *A Few Remarks upon Derby Town and County Museum addressed to the Members of the Society at Derby* (Derby, 1839).

Darwin, C., *On the Origin of Species by Means of Natural Selection, or the Preservation of Favoured Races in the Struggle for Life* ([1859]; reprinted Ware: Wordsworth Classics, 1998).

—— *Life of Erasmus Darwin*, ed. D. King-Hele ([1879]; Cambridge: Cambridge University Press, 2003).

Darwin, E. (ed.), *A System of Vegetables, according to their Classes, Orders, Genera, Species with their Characters and Differences*, 2 vols. (Lichfield, 1783).

—— 'Address to the Philosophical Society delivered at their first regular meeting held July 18th 1784' (Derby, 1784).

—— 'An account of an artificial spring of water', *Philosophical Transactions*, 75 (1785), 1–7.

—— *The Families of Plants, with their Natural Characters, according to the Number, Figure, Situation, and Proportion of all the Parts of Fructification*, 2 vols. (Lichfield, 1787).

—— 'Frigorific experiments on the mechanical expansion of air', *Philosophical Transactions*, 78 (1788), 43–52.

—— 'Letter on the medicinal waters of Buxton and Matlock', in J. Pilkington, *A View of the Present State of Derbyshire*, 2 vols. (Derby, 1789), I, 256–75.

—— *The Loves of the Plants* (1789).

—— *The Botanic Garden: A Poem in Two Parts*; Part I: *the Economy of Vegetation*; Part II: *The Loves of the Plants* (1791).

—— *A Plan for the Conduct of Female Education in Boarding Schools* (Derby, 1797).

—— *Phytologia; Or, the Philosophy of Agriculture and Gardening* (1800).

—— *Zoonomia; Or, the Laws of Organic Life*, first edition, 2 vols. (1794–96), third revised edition, 4 vols. (1801).

—— *The Temple of Nature; Or, the Origin of Society* (1803).

Darwin, R. W., 'New experiments on the ocular spectra of light and colours', *Philosophical Transactions*, 76 (1786), 313–48.

Davies, D. P., *A New Historical and Descriptive View of Derbyshire* (Belper, 1811).

Defoe, D., *A Tour thro' the Whole Island of Great Britain*, first edition, 3 vols. (1727), sixth edition, 4 vols. (1762).

Derby Literary and Philosophical Society, *Rules and Regulations of a Literary and Philosophical Society Founded at Derby in September 1808* (Derby, 1812).

—— *Syllabus of a Course of Lectures to be Delivered Under the Direction of the Derby Literary and Philosophical Society, at the room belonging to the Lancasterian School, in the Full Street* (Derby, 1813).

Derby Mechanics' Institute, *Catalogue of Articles Contained in the Exhibition of the Derby Mechanics' Institute, 1839* (Derby, 1839).

—— *Derby Mechanics' Institute: Catalogue of Books comprised in the Library with a List of Paintings and Engravings and of the Philosophical Apparatus belonging to the Institution* (Derby, 1851; 1856).

Derby Paving and Lighting Committee, *Report of the Committee appointed at a general meeting of the inhabitants of Derby, held on Monday November 23, 1789, to prepare a plan for more effectually paving and lighting the streets*, broadside (Derby, 1789).

Derby Philosophical Society, *An Address to Dr Priestley Agreed Upon at a Meeting of the Philosophical Society at Derby, September 3, 1791*, broadside (Derby, 1791).

—— *Laws or Regulations Agreed Upon by the Philosophical Society at Derby at their first Regular Meeting, August 7, 1784* (Derby, 1784).

—— *Laws and Regulations Agreed Upon by the Philosophical Society at Derby, Catalogue of the Library* (Derby, 1793), with supplements, 1795 and 1798.

—— *Derby Philosophical Society Rules and Regulations and Library Catalogue* (Derby, 1835).

Derby Society for Constitutional Information, *Address to the Friends of Free Enquiry and the General Good*, Talbot Inn (Derby, 16 July 1792).

—— *Rules of the Derby Society for Political Information* (Derby, 1792).

Derby Town and County Museum, *Catalogue of the articles comprised in the exhibition held for the benefit of the Derby Town and County Museum* (Derby, 1843).

Derby Town and County Museum and Philosophical Society, *Rules of the Derby Town and County Museum and Philosophical Society* (Derby, 1859).

——*Derby Town and County Museum and Philosophical Society: Report of the Committee … Meeting, held, February 14th 1860* (Derby, 1860).

—— *Derby Town and County Museum and Philosophical Society: Report of the Committee …* (Derby, 1862).

Derby Working Men's Association, *The Derby Working Men's Association* (Derby, 1849).

Derbyshire General Infirmary, *An Alphabetical List of Subscribers to the Derbyshire General Infirmary* (Derby, c.1809).

Desaguliers, J. T., *A Course of Experimental Philosophy*, second edition, 2 vols. (1745).

Dixon, F., *A Sermon against Jacobinical and Puritanical Reformations* (Chesterfield, 1793).

Downing, A. J., *Rural Essays* (New York, 1856).

Duesbury, H. (attrib.), *Description of Design for the proposed Derby Pauper Lunatic Asylum* (Derby, 1844).

Edgeworth, M., *The Life and Letters of Maria Edgeworth*, ed. Augustus Hare (1894).

Edgeworth, M. and R. L., *Essays on Practical Education*, 2 vols. (1798), third edition (1811).

——*Memoirs of Richard Lovell Edgeworth Esq., begun by himself and concluded by his daughter Maria Edgeworth*, 2 vols. (1820).

Ellicott, J., 'The description and manner of using an instrument for measuring the degree of the expansion of metals by heat', *Philosophical Transactions*, 39 (1735–6), 297–9.

Farey, J., *A General View of the Agriculture and Minerals of Derbyshire*, 3 vols. (1811–17).

——*A General View of the Agriculture and Minerals of Derbyshire*, I (1811); reprinted with an introduction by T. Ford and H. Torrens (Peak District Mines Historical Society, 1989).

——'An account of the great Derbyshire denudation', *Philosophical Transactions*, 101 (1811), 242–56.

Ferguson, J., *Astronomy Explained on Sir Isaac Newton's Principles* (1756).

——*Tables and Tracts* (1767).

——*Introduction to Electricity* (1770).

Fiennes, C., *Through England on a Side Saddle in the time of William and Mary* (1888).

Fowler, R., *Experiments and Observations Relative to the Influence Lately Discovered by M. Galvani and Commonly Called Animal Electricity* (Edinburgh, 1793).

Fox, D., 'A case of the reversed position of the thoracic and abdominal viscera', *London Medical and Physical Journal*, 51 (1824), 474–5.

——*Notes of the Lectures on Anatomy and Chemistry, delivered by Mr. Douglas Fox, Surgeon, in the Lecture Room of the Derby Mechanics' Institution, from November 4, 1825, to January 6, 1826* (Derby, 1826).

Franklin, B., *The Papers of Benjamin Franklin*, ed. B. Willcox, 38 vols. (New Haven: Yale University Press, 1959–2006).

Gandon, J., *The Life of James Gandon Esq. … from material collected and arranged by his son James Gandon Esq.* (Dublin, 1846).

Gardiner, W., *Music and Friends*, 3 vols. (Leicester, 1838, 1853).

Gisborne, T., 'On the benefits and duties resulting from the institution of societies for the advancement of literature and philosophy', *Memoirs of the Literary and Philosophical Society of Manchester*, 5 (1798), 70–88.

——*An Enquiry into the Duties of Men in the Higher Rank and Middle Classes of Society in Great Britain*, first edition (1794), fifth edition, 2 vols. (1800).

——*An Enquiry into the Duties of the Female Sex*, first edition (1797).

Glover, S., *The History, and Gazetteer, and Directory of the County of Derby*, 2 vols. (Derby, 1829, 1833).

——*History and Directory of Derbyshire for 1827–1829* (Derby, 1829).

——*History and Directory of the Borough of Derby* (Derby, 1843; reprinted Derby: Breedon Books and Derbyshire County Council, 1992).

Godwin, W., *An Enquiry Concerning Political Justice*, ed. I. Kramnick ([1793]; Penguin, 1976).

——*The Letters of William Godwin*, ed. D. Wardle (1967).

Goldsmith, O., *An History of the Earth and Animated Nature*, 8 vols. (1774).

Gordon, G. H., *History of the Pastures Hospital, Derby* (Derby, undated, *c.*1970).

Gore's Liverpool Directory (1825).

Granger, J., *The History of the Nottingham Mechanics' Institute 1837–1887* (Nottingham, 1887).

Gray, A., 'Francis Boott M.D', *American Journal of Science and Arts*, second series, 37 (1864), 288–92.

Gray, T., 'Dr. Gray's account of the earthquake felt in England, November 18th, 1795', *Philosophical Transactions* (1796).

Gregory, J., *A Manual of Modern Geography*, third edition (1748).

Griffis, W., *A Short Account of a Course of Mechanical and Experimental Philosophy and Astronomy* (1748).

Gwynn, J., *London and Westminster Improved* (1766).

Hanham, F., *A Manual for the Park; or, A Botanical Arrangement and Description of the Trees and Shrubs in the Royal Victoria Park, Bath* (Bath, 1857).

Hartley, D., *Observations on Man, his Frame, his Duty and his Expectations* (1749).

Hearne, F. S., 'An old Leicester bookseller, Sir Richard Phillips', *Transactions of the Leicester Literary and Philosophical Society*, 3 (1893), 65–73.

Higginson, E., *The Doctrines and Duties of Unitarians: A Sermon Preached before the Association of Unitarian Dissenters* (Lincoln, 1820).

—— *The Turn Out; An Inquiry into the Present State of the Hosiery Business …* (Derby, 1821).

—— *Observations Addressed to all Classes of the Community, on the Establishment of Mechanics' Institutions* (Derby, 4 March 1825).

—— *Address, Delivered on the Opening of the Derby Mechanics' Institution, August 22nd, 1825* (Derby, 1825).

Hobson, W., *The History and Topography of Ashbourne* (Ashbourne, 1839).

Hogarth, W., *The Analysis of Beauty* (1753), ed. R. Paulson (New Haven: Yale University Press, 1997).

Holyoake, G. J., 'Obituary for Herbert Spencer', *Literary Guide and Rationalist Review*, 91 (1904), 1–2.

Home, H., Lord Kames, *Elements of Criticism*, seventh edition, 2 vols. (Edinburgh, 1788), II, 432–3, 455.

Hooson, W., *The Miner's Dictionary, explaining not only the terms used by miners, but also containing the theory and practice of that most useful art of mineing* [sic], *more especially of lead-mines* (Wrexham, 1747; reprinted for Institute of Metallurgy, London: Scolar Press, 1979).

Hope, C. S., *An Address to the Derby Volunteers … at the Consecration of their Colours* (Derby, 1799).

Houghton, J., *Collection for Improvement of Husbandry and Trade*, 38 (21 April 1693).

House of Commons, *The History and Proceedings of the House of Commons from the Restoration to the Present Time* (1742).

Howell, T. B. and Howell, T. J. (eds), *A Complete Collection of State Trials*, 33 vols. (1811–26).

Hume, A. and Evans, A. I., *The Learned Societies and Printing Clubs of the United Kingdom* (1853).

Hume, David, *A Treatise of Human Nature*, ed. L. A. Selby Bigge ([1739–40]; Oxford, 1978).

—— *Enquiries Concerning Human Understanding and Concerning the Principles of Morals*, ed. L. A. Selby Bigge, third edition ([1777]; Oxford, 1975).

—— *Essays and Treatises on Several Subjects*, 2 vols. ([1770]; Edinburgh, 1825).

—— *Dialogues Concerning Natural Religion* ([1779]; reprinted Penguin, 1990).

Hunt, H., *The Green Bag Plot ... Addressed to the Reformers of England, Wales, Scotland and Ireland* (Derby, 1819).

Hutton, C., 'Some account of the life and writings of Mr. Thomas Simpson', *Annual Register* (1764), 29–38.

Hutton, J., 'Authentic memoirs of the life and writings of the late John Whitehurst FRS', *Universal Magazine* (November, 1788), 224–9.

Hutton, W., *The History of Derby: from the Remote Ages of Antiquity* (1791).

—— *Life of William Hutton and the History of the Hutton Family* (Birmingham, 1816).

Huxley, L., *Life and Letters of Thomas Henry Huxley*, 2 vols. (1900).

Jewitt, L., *Guide to the Borough of Derby* (Derby, 1852).

—— *The Ballads and Songs of Derbyshire* (Derby, 1867).

Johnson, S., *The Works of Samuel Johnson, LLD*, ed. A. Murphy, 12 vols. (1820).

Johnson, W. B., *The History and Present State of Animal Chemistry*, 3 vols. (1803).

Jones, N., *A Discourse on Occasion of the Life and Lamented Death of Joseph Strutt* (Derby, 1844).

Krause, E., *Erasmus Darwin, with a Preliminary Notice by Charles Darwin* (1879).

Lamarck, J. B., *Philosophie Zoologique* (1809), ed. H. Elliot (New York, 1963).

Lewes, G. H., *A Biographical History of Philosophy* (1892).

Locke, J., *An Essay Concerning Human Understanding*, 30th edition ([1689]; London: Tegg, 1846).

—— *Two Treatises of Government*, ed. P. Laslett (Cambridge, 1963).

Loudon, J. C., *A Short Treatise on Several Improvements Recently Made in Hot-Houses* (1805).

—— *The Gardener's Magazine* (1826–34), which became *The Gardener's Magazine and Register of Rural and Domestic Improvement* (1835–43).

—— *The Magazine of Natural History* (1828–36).

—— *The Encyclopaedia of Gardening* (1830).

—— *The Encyclopaedia of Cottage, Farm and Villa Architecture and Furniture* (1833).

—— *The Architectural Magazine* (1834–38).

—— 'Remarks on laying out public gardens and promenades', *The Gardener's Magazine*, 11 (1835), 644–59.

—— *The Suburban Gardener and Villa Companion* (1838).

—— *Arboretum et Fruticetum Britannicum*, 8 vols., first edition (1838), second edition (1844).

—— *The Derby Arboretum, containing a Catalogue of the Trees and Shrubs Included in It; A Description of the Grounds ...* (1840).

—— (ed.), *The Landscape Gardening and Landscape Architecture of the late H. Repton* (1840).

—— *Loudon's Encyclopaedia of Plants*, ed. Mrs Loudon, assisted by G. Donald and D. Wooster (1880).

Lucas, J.-A.-H., *Tableau méthodique des espèces minérales*, 2 vols. (Paris, 1806–13).

Lyell, C., 'Scientific institutions', *Quarterly Review*, 34 (1826), 153–79.

——*Principles of Geology*, ed. and introduction J. Secord ([1830–33]; Penguin, 1997).

Lysons, S. and D., *Magna Britannia*, V: *Derbyshire* (1817).

Malthus, T. R., *An Essay on the Principle of Population; or, A View of its Past and Present Effects on Human Happiness* (1803).

Marshman, J. C., *The Story of Carey, Marshman and Ward, the Serampore Missionaries* (1864).

Martin, J. R., report on Derby from *Appendix, Part II, to the Second Report of the Commissioners of Inquiry into the State of Large Towns and Populous Districts* (1845), 271–8.

Martin, W., *Outlines of an Attempt to Establish a Knowledge of Extraneous Fossils* (Macclesfield, 1809).

Mawe, J., *The Mineralogy of Derbyshire: with a Description of the Most Interesting Mines in the North of England …* (1802).

Meteyard, E., *The Life of Josiah Wedgwood* (1866).

Mitchell, W. S., 'Biographical notice of John Farey, geologist', *The Geological Magazine*, 10 (1873), 25–7.

Moritz, C. P., *Travels Chiefly on Foot through Several Parts of England* (1795).

Mozley, T., *Reminiscences, Chiefly of Oriel College and the Oxford Movement*, 2 vols. (1882).

——*Reminiscences, Chiefly of Towns, Villages and Schools*, 2 vols. (1882).

Mundy, F. N. C., 'Letters of a Derbyshire squire and poet in the early nineteenth century', ed. W. G. Clark-Maxwell, *Derbyshire Archaeological Journal*, 53 (1932).

Mushet, G., *The Diary of George Mushet, 1805–1813*, ed. R. M. Healey (Chesterfield: Derbyshire Archaeological Society, 1982).

Neville, S., *The Diary of Sylas Neville, 1767–1788*, ed. B. Cozens-Hardy (Oxford University Press, 1950).

Newton, I., *Newton's Philosophy of Nature: Selections from his Writings*, ed. H. S. Thayer (Hafner Press, 1974).

Nichols, J., *The History and Antiquities of the County of Leicester*, 4 vols. (1804).

Nicholson, W., 'A description of an instrument which by the turning of a winch, produces the two states of electricity without friction or communicating with the earth', *Philosophical Transactions*, 78 (1788), 403–7.

Nicholson, W. and Carlisle, A., 'Account of the new electrical or galvanic apparatus of Signor Alessandro Volta and experiments performed with the same', *Nicholson's Journal*, first series, 4 (1800–1), 179–87.

Nightingale, F., *Notes on Hospitals*, third edition (1863).

Oldfield, T. H. B., *The Representative History of Great Britain and Ireland*, 6 vols. (1816).

Rice Oxley, L. (ed.), *The Poetry of the Anti-Jacobin* (Oxford, 1924).

Paine T., *Common Sense: A Letter Addressed to Americans* (Philadelphia, 1776; reprinted Penguin, 1982).

——*The Rights of Man*, ed. H. Collins ([1792]; Harmondsworth: Penguin, 1969).

Parker, B., *Parker's Projection of the Longitude at Sea* (Nottingham, 1731).

——*A Journey Thro' the World: in a View of the Several Stages of Human Life*, second edition (Birmingham, 1738).

—— *Philosophical Meditations with Divine Inferences* (Birmingham, 1738).

—— *A Second Volume of Philosophical Meditations* (Birmingham, 1738).

—— *Philosophical Dissertations with Proper Reflections Concerning the Non-Eternity of Matter* (1738).

—— *A Survey of the Six Days Work of the Creation* (1745).

Payley, T., *The Principles of Moral and Political Philosophy* (1785), ninth edition (1792).

—— *Reasons for Contentment; Addressed to the Labouring Part of the British Public* (1793).

—— *Natural Theology, or, Evidences of the Existence and Attributes of the Deity, Collected from the Appearances of Nature* (1802).

—— *A View of the Evidences of Christianity*, ed. T. R. Birks ([1802]; 1838).

Payne, C. J., *Derby Churches Old and New* (Derby, 1893).

Phillips, R., *A Personal Tour through the United Kingdom* (1828), no. 2, Derbyshire and Nottinghamshire.

Pigot and Company, *Directory of Derbyshire* (Derby, 1820, 1835).

Pilkington, J., *A View of the Present State of Derbyshire*, 2 vols. (Derby, 1789).

—— *The Doctrine of Equality of Rank and Condition examined and supported on the authority of the New Testament, and on the principles of reason and benevolence* (1795).

—— 'A short account of the origin and establishment of Sunday schools in the Town of Derby', in J. Birks, *Memorials of Friargate Chapel* (Derby, c.1890), 19–20.

Priestley, J., *The History and Present State of Electricity, with Original Experiments*, first edition (1767).

—— *Hartley's Theory of the Human Mind, on the Principle of the Association of Ideas* (1775).

—— *Experiments and Observations on Different Kinds of Air*, 3 vols. (Birmingham, 1775–7).

—— *Letters to the Right Honourable Edmund Burke Occasioned by his Reflections on the Revolution in France* (Birmingham, 1791).

Rees, A., *The Cyclopaedia; or, Universal Dictionary of Arts, Sciences, and Literature*, 39 volumes (1802–19).

Rodd, T. ('Philobiblos'), *A Defence of the Veracity of Moses … Illustrated by Observations in the Caverns of the Peak of Derby* (Sheffield, 1820).

Royal Society of London, *Abstracts of the Papers Printed in the Philosophical Transactions of the Royal Society of London, 1830–1837*, III (1860).

Russell, J., *Memoirs, Journal and Correspondence of Thomas Moore*, 5 vols. (1853).

Schinkel, K. F., *'The English Journey': Journal of a Visit to France and Britain in 1826*, ed. D. Bindman and G. Riemann (New Haven: Paul Mellon, 1993).

Scott, W. (ed.), *Ballantyne's Novelist's Library* (Edinburgh, 1824).

Seward, A., *Memoirs of the Life of Dr. Darwin, Chiefly during his Residence at Lichfield* (1804).

Simpson, R., *A Collection of Fragments Illustrative of the History and Antiquities of Derby*, 2 vols. (Derby, 1826).

Simpson, T., *A New Treatise of Fluxions* (1737).

Smiles, Samuel, *Self Help* (1859).

——*Industrial Biography: Iron Workers and Tool Makers* (1897).

——*Lives of the Engineers: Boulton and Watt* (John Murray, 1904).

Smith, Adam, *The Theory of Moral Sentiments ...* (1759).

——*An Inquiry into the Nature and Causes of the Wealth of Nations*, 2 vols. (1776).

Southern, J., *A Treatise upon Aerostatic Machines containing rules for Calculating their Powers of Ascension* (Birmingham, 1785).

Spencer, Herbert, *The Principles of Psychology*, first edition, 2 vols. (1855), second edition, 2 vols. (1870–72), fourth edition (1899).

——*First Principles*, first edition (1862), sixth edition (1900).

——*The Principles of Biology*, 2 vols., first edition (1864–67), second edition (1898–9).

——*Social Statics; or the Conditions to Human Happiness Specified, and the first of them Developed* ([1850]; New York, 1873).

——*Man Versus the State* (1884).

——*Essays: Scientific, Political, and Speculative*, 3 vols. (1891).

——*The Principles of Sociology*, 3 vols. (1896).

——*The Principles of Ethics*, 2 vols. (1897).

——*Various Fragments* (1900).

——*Facts and Comments* (1902).

——*An Autobiography*, 2 vols. (1904).

——*Education: Intellectual, Moral, and Physical* ([1861]; 1905).

Spencer, W. G., *Inventional Geometry* (1860), republished in 1892 with a preface by Herbert Spencer.

——*A System of Lucid Shorthand* (1894).

Stennet, S., *Memoirs of Mr. William Ward* (1825).

Stephen, J., *Essays in Ecclesiastical Biography*, fifth edition (1867).

Strutt, J., *Catalogue of Paintings, Drawings, Marbles, Bronzes, &c in the Collection of Joseph Strutt*, privately printed (Derby, 1827, 1835).

Swanwick, T., 'Annual meteorological table for Derby, 1813', *Gentleman's Magazine* (1814), i, 241.

Sylvester, C., 'Observations and experiments to elucidate the operation of the galvanic power', *Nicholson's Journal*, 9 (1804), 179–82.

——'On galvanism', *Nicholson's Journal*, 1 (1805), 106–7.

——'Observations and experiments on galvanism, the precipitation of metals by each other, and the production of muriatic acid', *Nicholson's Journal*, 2 (1806), 94–8.

——'On the advantages of malleable zinc, and the purposes to which it may be applied', *Nicholson's Journal*, 19 (1808), 11–13.

——'Experiments on the decomposition of the fixed alkali by galvanism', *Nicholson's Journal*, 19 (1808), 156–7.

——'Further experiments and observations on potash and its base', *Nicholson's Journal*, 19 (1808), 307–9.

——'On the production of an acid and an alkali from pure water by galvanism', *Nicholson's Journal*, 13 (1809), 258–60.

——*An Elementary Treatise on Chemistry* (Manchester, 1809).

——'Observations on galvanic batteries', *Nicholson's Journal*, 26 (1810), 72–5.

—— 'On the nature and detection of the different metallic poisons', *Nicholson's Journal* (1812), 306–13.

—— *The Philosophy of Domestic Economy* (Nottingham, 1819).

—— 'On the discovery of bipersulphate of iron', Thomson's *Annals of Philosophy*, 13 (1819), 466.

—— 'On a method of expressing chemical compounds by algebraic characters', Thomson's *Annals of Philosophy*, 2 (1821), 212–16.

—— 'On the motions produced by the difference in the specific gravity of bodies', Thomson's *Annals of Philosophy*, 3 (1822), 408–15.

—— 'Observations on the presence of moisture in modifying the specific gravity of gases', Thomson's *Annals of Philosophy*, 4 (1822), 29–31.

—— 'Additional remarks on Dr. Thomson's paper on the effect of aqueous vapour on the specific gravity of gases', Thomson's *Annals of Philosophy*, 4 (1822), 360.

—— *Report on Rail-Roads and Locomotive Engines* (Liverpool, 1825).

Thompson, J., *The History of Leicester in the Eighteenth Century* (Leicester, 1871).

Traill, T. S., 'On the migration of swallows', *Nicholson's Journal*, 15 (1811), 213–14.

The Trial of the Derby Rioters, 9th October 1831 (Derby, 1831).

Ure, A., *The Philosophy of Manufactures: or, an Exposition of the Scientific, Moral, and Commercial Economy of the Factory System of Great Britain*, second edition (1835).

—— *A Dictionary of Arts, Manufactures and Mines*, second edition (1840).

Vickers, J., *An Address to the Tradesmen and Operatives of Derby* (Belper, 1832).

Volta, A., 'On the electricity excited by mere contact of conducting substances of different kinds', *Philosophical Transactions*, 90 (1800), 403–31.

Wallis, A., 'A sketch of the early history of the printing press in Derbyshire', *Derbyshire Archaeological Journal*, 3 (1881), 137–56.

Watson, R. S., *The History of the Literary and Philosophical Society of Newcastle-Upon-Tyne (1793–1896)* (Newcastle, 1897).

Watson, W., *A Delineation of the Strata of Derbyshire* (Sheffield, 1811).

Watts, I., *The Improvement of the Mind*, ed. J. Emerson (Boston, 1833).

Wedgwood, J., *The Letters of Josiah Wedgwood*, ed. K. E. Farrer, 3 vols. (Didsbury, 1973).

White's Directory of Derbyshire (Derby, 1857).

Whitehurst, J., *Inquiry into the Original State and Formation of the Earth*, first edition (1778), third edition (1792).

—— *An Attempt toward Obtaining Invariable Measures of Length, Capacity, and Weight* (1787).

—— *The Works of John Whitehurst F.R.S. with Memoirs of his Life and Writings* (1792).

—— *Observations on the Ventilation of Rooms; on the Construction of Chimneys; and on Garden Stoves* (1794).

Wilkinson, C. H., *Elements of Galvanism in Theory and Practice*, 2 vols. (1804).

Williams, F. S., *The Midland Railway, its Rise and Progress* (1875).

Wollstonecraft, M., *A Vindication of the Rights of Women* (1792; reprinted Everyman, 1995).

Wooler, W. M., *On the Philosophy of Temperance, and the Physical Causes of Moral Sadness* (1840).

Woolley, W., *History of Derbyshire*, ed. C. Glover and P. Riden (Chesterfield: Derby-shire Record Society, 1981).

Wright, T., *An Original Theory or New Hypothesis of the Universe* (1750).

Yorke, H., *Reason Urged against Precedent, in a Letter to the People of Derby* (Derby, 1793).

——*These are the Times that Try Men's Souls! A Letter Addressed to John Frost, a Prisoner in Newgate* (Derby, 1793).

——*Thoughts on Civil Government* (1794).

Index